Improving Energy Efficiency through Technology

Trends, Investment Behaviour and Policy Design

Edited by

Raymond J.G.M. Florax

Professor, Purdue University, West Lafayette, USA, VU University Amsterdam and Tinbergen Institute, The Netherlands

Henri L.F. de Groot

Professor, VU University Amsterdam, Ecorys NEI and Tinbergen Institute, The Netherlands

Peter Mulder

Assistant Professor, VU University Amsterdam and Tinbergen Institute, The Netherlands

Edward Elgar
Cheltenham, UK • Northampton, MA, USA

Published by
Edward Elgar Publishing Limited
The Lypiatts
15 Lansdown Road
Cheltenham
Glos GL50 2JA
UK

Edward Elgar Publishing, Inc.
William Pratt House
9 Dewey Court
Northampton
Massachusetts 01060
USA

A catalogue record for this book
is available from the British Library

Library of Congress Control Number: 2009940999

FSC
www.fsc.org
MIX
Paper from
responsible sources
FSC® C018575

ISBN 978 1 84542 390 2

Typeset by Servis Filmsetting Ltd, Stockport, Cheshire
Printed and bound by MPG Books Group, UK

Contents

Contributors

Kornelis Blok studied experimental physics at Utrecht University in The Netherlands and received a PhD in 1991 with a dissertation on the reduction of carbon dioxide emissions. In 1984 he was one of the founders of Ecofys, a consultancy firm specializing in energy saving and renewable energy solutions, where he is now Scientific Director. Dr Blok has extensive research and consultancy experience in the field of energy efficiency improvement and clean energy production. He has carried out studies into improving the efficiency of industrial energy use, combined generation of heat and power, and renewable energy, and he has extensive experience with quantitative energy policy analysis and climate policy development. He is also associated with Utrecht University in The Netherlands, where he is Professor of Sustainable Energy, and supervises the Master's programme 'Energy Science'. He has authored and co-authored about 80 articles in peer-reviewed scientific journals, several books, including the textbook *Introduction to Energy Analysis*, and over 200 research reports, conference contributions and other scientific publications. He was a lead author for the third and fourth assessment reports of the Intergovernmental Panel on Climate Change (IPCC) and is a guest professor at the North China Electric Power University.
Ecofys, P.O. Box 8408, 3503 RK Utrecht, The Netherlands, k.blok@ecofys.com

Raymond J.G.M. Florax holds degrees in economics and sociology, and is Professor in the Department of Agricultural Economics at Purdue University in the USA. He is also associated with the Department of Spatial Economics at VU University Amsterdam in The Netherlands, and is a fellow of the Tinbergen Institute (TI), the Spatial Econometrics Association (SEA) and the Wageningen School of Social Sciences (WASS). He has served a 10-year tenure as editor-in-chief of the journal *Papers in Regional Science*. At VU University Amsterdam, he formerly directed the MASTER-point research group engaged in research on meta-analysis in spatial, transport and environmental economics. Dr Florax coordinated the research programme on the use of energy efficient technology in small and medium-sized enterprises, which forms the basis for this volume. At

Purdue University he directs a research group focusing on 'Space, Health and Population Economics' (SHaPE). His research deals with spatial data analysis, spatial and environmental modelling, and meta-analysis, on which he has published widely.
Department of Agricultural Economics, Purdue University, 403 W. State Street, West Lafayette, IN 47907, USA, rflorax@purdue.edu

Willem J.H. van Groenendaal is Professor at the Faculty of Economics and Business Administration of Tilburg University, The Netherlands. He has worked on energy issues for almost thirty years using a wide range of models and methods. His work is not only academic. Van Groenendaal participated in the design of domestic natural gas markets and networks in Indonesia and Bangladesh, and in the design of dissemination policies for renewable energy technologies in China. He also played a leading role in research projects addressing the adoption of energy efficient technologies by Dutch energy-extensive companies, and the effects of environmental policies on the efficiency and sustainability of Dutch process technologies.
Tilburg University, Faculty of Economics and Business Administration, P.O. Box 90153, 5000 LE Tilburg, The Netherlands, w.j.h.vgroenendaal@uvt.nl

Henri L.F. de Groot is Professor in the Department of Spatial Economics at VU University Amsterdam in The Netherlands. He is also affiliated with Ecorys in Rotterdam and a fellow of the Tinbergen Institute (TI). Dr de Groot studied economics at Tilburg University in The Netherlands and graduated in 1994 (*cum laude*). In 1998, he obtained his PhD from Tilburg University with a dissertation on economic growth, unemployment and the sectoral composition of economies. Subsequently, he was appointed Post-doctoral Researcher at the Department of Spatial Economics of the VU University in Amsterdam. His main research interests are in the areas of the empirics of economic growth and trade, endogenous growth theory, agglomeration, regional economic development, environmental economics, technology adoption and meta-analysis. He has published extensively in a wide variety of scientific journals. He is programme coordinator of the BSc in Economics and Business at VU University, and strongly involved in the MSc in Spatial, Transport and Environmental Economics.
Department of Spatial Economics, VU University Amsterdam, De Boelelaan 1105, 1081 HV Amsterdam, The Netherlands, h.l.f.de.groot@vu.nl

Mirjam Harmelink (MSc) is Managing Consultant and Director at Harmelink consultancy and has over 15 years of experience in advising governments, businesses and non-governmental organizations (NGOs).

Before starting her own company she worked for the Dutch National Environmental Assessment Agency, the energy agency SenterNovem, and in several positions within the consultancy firm Ecofys. She is a policy expert and has a good overview of policies applied in the field of renewables, energy efficiency and climate change across Europe, and has advised a large number of national and international clients in this field including the European Commission, national ministries, large European energy companies and NGOs. Her work focuses on the design, monitoring and evaluation of renewable, energy efficiency and climate change policies.
Harmelink Consulting (www.harmelinkconsulting.com), Troosterlaan 36, 3571 NN Utrecht,The Netherlands, info@harmelinkconsulting.nl

Greg W. Hunter holds a doctoral degree in economics from the University of California, Santa Barbara, and is an Associate Professor in the Department of Economics at the California State Polytechnic University (Cal Poly), Pomona. He currently directs a public policy and opinion research group focused on the inland valleys of Southern California. Prior to coming to Cal Poly, Dr Hunter conducted research at Tilburg University, The Netherlands, as part of a project funded by The Netherlands Scientific Research Organization (NWO) examining energy-saving technical change. His research deals primarily with economic valuation under uncertainty and environmental policy. He has acquired extensive teaching experience at the graduate and undergraduate levels on topics related to these research interests.
Department of Economics, California State Polytechnic University, 3801 W. Temple Ave., Pomona, CA 91768, USA, gwhunter@csupomona.edu

Suzanne Joosen is Senior Consultant Sustainable Energy at Grontmij, an engineering consultancy company. She studied chemistry at Utrecht University in The Netherlands. She has more than eighteen years of experience in energy-related topics. At the beginning of her career in 1991, at Erbeko engineering consultancy, she among others analysed and advised energy contracts for large-scale industries, specifically in aluminium, chemistry and glass. In May 1996 she started working at the consultancy firm Ecofys. The first years she contributed mainly in awareness projects regarding the general aspects of renewable energy. During the period 2000–2005 she coordinated the renewable energy statistics in The Netherlands. This was an assignment by the Dutch energy agency SenterNovem, in close cooperation with Statistics Netherlands (CBS) and KEMA (a global company in energy consulting and testing and certification). Recently, she has specialized in renewable energy and energy savings in buildings, particularly in national and international projects regarding policy evaluation, and she focuses on obtaining insight in the effects of energy-saving

measures and feasible policy instruments to achieve the impact in practice. She also contributed to the building chapter of the fourth assessment report of the Intergovernmental Panel on Climate Change (IPCC).
Grontmij, Postbox 68, 3800 AB Amersfoort, The Netherlands, suzanne.joosen@grontmij.nl

Mark J. Koetse holds a PhD in economics and is Senior Researcher at the Netherlands Environmental Assessment Agency (PBL) in The Hague, The Netherlands, and Postdoctoral Researcher at the Department of Spatial Economics at VU University Amsterdam in the Netherlands. In his dissertation research he focused on the application of meta-analysis in the area of technology adoption and on methodical development of meta-analysis as a research method in economics. His current research focuses on energy economics, transport economics, environmental economics and meta-analysis.
Netherlands Environmental Assessment Agency, Oranjebuitensingel 6, 2511 VE The Hague, The Netherlands, mark.koetse@pbl.nl

Peter Mulder is Assistant Professor in the Department of Spatial Economics at VU University Amsterdam in The Netherlands, and a fellow of the Tinbergen Institute (TI). In 2003, he obtained his PhD from VU University Amsterdam with a dissertation on the economics of technology diffusion and energy efficiency, based on research carried out at the Institute for Environmental Studies (IVM) of VU University Amsterdam and the CPB Netherlands Bureau for Economic Policy Analysis. Subsequently, he has been an associate research scholar at the International Institute for Applied Systems Analysis (IIASA) in Laxenburg, Austria, and a postdoctoral researcher at the Department of Spatial Economics of VU University Amsterdam. In 2005, he interrupted academic life to work for two years in Mozambique as a consultant/energy economist for the Danish International Development Agency (Danida), based at the Ministry of Planning and Development and the Ministry of Energy of the government of Mozambique. He has published in a variety of scientific journals, mainly in the field of energy economics, and has considerable experience in undergraduate teaching.
Department of Spatial Economics, VU University Amsterdam, De Boelelaan 1105, 1081 HV Amsterdam, The Netherlands, p.mulder@vu.nl

Martin K. Patel is Assistant Professor in the Department of Science, Technology and Society (STS) at Utrecht University in The Netherlands, where he coordinates the research cluster 'Energy and Materials Demand and Efficiency' (EME). He studied chemical engineering at the University of Karlsruhe in Germany and graduated in 1992. From 1993 until mid-

2000 he was with the department of Energy Technology and Energy Policy of the Fraunhofer Institute for Systems and Innovation Research (ISI) in Karlsruhe, Germany. In 1999, he earned his PhD from Utrecht University, The Netherlands, with a dissertation on closing carbon cycles, specifically carbon use for materials in the context of resource efficiency and climate change. Martin Patel's research deals with the resource efficiency and emission reduction potentials of industrial processes from a techno-economic perspective. He has been coordinating and contributing to several national and European Union projects on energy efficiency and emission reduction, especially in the industrial sector.
Department of Science, Technology and Society and Copernicus Institute, Utrecht University, Heidelberglaan 2, 3584 CS Utrecht, The Netherlands, m.k.patel@uu.nl

Andrea Ramírez is Senior Researcher in the Department of Science, Technology and Society (STS), at Utrecht University in The Netherlands. In 1996, she obtained a bachelor's degree in chemical engineering at the National University of Colombia in Colombia, South America. From 1996 until 1998 she worked as researcher at the Solid Waste Research Program (PIRS) of the same university. In September 2000, Ms Ramirez obtained a Master of Science degree in human ecology with a *magna cum laude* distinction from the Free University of Brussels in Belgium. In 2005, she earned her PhD from Utrecht University in The Netherlands, with a dissertation focusing on monitoring energy efficiency in the food industry. Her current research deals with CO_2-capture and storage as an option for climate change mitigation. She coordinates and contributes to several national and European Union projects on this topic.
Department of Science, Technology and Society and Copernicus Institute, Utrecht University Heidelberglaan 2, 3584 CS Utrecht, The Netherlands, c.a.ramirez@uu.nl

Daan P. van Soest is Professor of Environmental Economics in the Department of Spatial Economics and the Institute for Environmental Studies (IVM) at VU University Amsterdam in The Netherlands, and a fellow of the Tinbergen Institute (TI). He holds a PhD in economics of Groningen University in The Netherlands, with a dissertation on the economics of tropical deforestation. He has worked for more than ten years on issues in environmental and energy economics as an assistant and associate professor at Tilburg University in The Netherlands. His research is both theoretical and empirical in nature, and topics that he has worked on include renewable resource use (including renewable energy), investments in energy-saving technologies, and instrument choice in environmental policy making. He has published in a wide variety of scientific journals,

and is currently associate editor of the journal *Environmental and Resource Economics*.
Department of Spatial Economics and Institute for Environmental Studies (IVM), VU University Amsterdam, De Boelelaan 1105, 1081 HV Amsterdam, The Netherlands, d.p.van.soest@vu.nl

Herman R.J. Vollebergh is Senior Research Fellow at the Netherlands Environmental Assessment Agency (PBL) in The Hague, The Netherlands, and research fellow at Tilburg Sustainability Centre at University of Tilburg, The Netherlands. Previously, he was Assistant Professor in the Erasmus School of Economics at Erasmus University Rotterdam, and in the Department of Economics and Business Administration at Tilburg University, both in The Netherlands. He has also been Visiting Research Fellow at the Smith School of Enterprise and the Environment at Oxford University in the UK. His main research work is in environmental and public economics. He is particularly interested in how the use of market-based incentive mechanisms, such as taxation, subsidies and tradable permits, affect the behaviour of economic agents such as firms or households. Much of his research has an empirical focus, including his recent research on the identification of reduced-form relationships, such as the Environmental Kuznets Curve, the role of corruption in environmental policy implementation, and the effect of environmental policy on technological change. He is strongly involved in applying his academic work to the policy community including the Organisation for Economic Co-operation and Development (OECD), the European Union (EU) and the Dutch government.
Netherlands Environmental Assessment Agency, P.O. Box 303, 3720 AH Bilthoven, The Netherlands, vollebergh@few.eur.nl

Preface

Economic development and increasing prosperity around the globe go hand in hand with a continuous surge in the demand for energy. This increase in the demand for energy has profound implications for energy security and climate change, and it also raises questions regarding the sustainability of economic development itself. The need to maximize efficiency in usage of limited and often non-renewable resources has been at the centre of a substantive collection of applied research in energy economics, and has caught the attention of policy-makers.

The mitigation of the adverse effects of increasing energy consumption through the promotion of efficient energy usage is associated with large investments in new energy-saving technologies, both in the energy sector as well as in the various end-use sectors. One would suspect that continuously rising energy prices cause firms to adopt such technologies instantaneously upon availability. However, this is often not the case. This phenomenon constitutes a rather fascinating paradox, which is generally known as the 'energy-efficiency paradox': even although energy-efficient technologies are available and cost-effective, producers are reluctant to adopt such technologies. The paradoxical phenomenon occurs in energy-intensive sectors, such as aluminium, chemicals, paper and steel industries, but also in energy-extensive sectors of the economy such as the light industry and the service sector. The latter sectors are of vital importance when it comes to meeting ambitious energy usage and associated CO_2 emission targets, for instance in relation to worldwide climate change concerns. The energy-extensive sectors are responsible for no less than 20 per cent of the economy-wide energy consumption, and are therefore a segment of economic activities that can, or should not be ignored. Not surprisingly, there is substantial knowledge about energy consumption in energy-intensive sectors of the economy. Knowledge and expertise regarding energy usage in energy-extensive sectors, in particular in view of the 'energy-efficiency paradox' is, however, largely lacking.

The research collected in this volume therefore starts from the observation that the adoption of energy-saving technologies is often a lengthy process and that, in fact, many firms do not or only slowly invest in best-practice cost-effective technologies. We have brought together a group of

13 authors to investigate trends in energy usage, investment behaviour and energy policy design in view of the goal of promoting energy efficiency. In this research we have paid particular attention to the energy-extensive sectors of the economy, specifically the light industry and the service sector, and to the role of small and medium-sized firms. The research reported in this collected volume was part of a larger research programme on stimulating the adoption of energy-efficient technologies, financed by the Netherlands Organization for Scientific Research (abbreviated NWO in Dutch) and SenterNovem, an agency of the Dutch Ministry of Economic Affairs that focuses on the promotion of sustainable development and innovation.

A key aspect of the research programme on stimulating the adoption of energy-efficient technologies was to bring together knowledge, expertise and insights from both the natural and the social sciences. Our research efforts reflect this overall aim of the programme and integrate the available knowledge in economics and the sciences with respect to investment in energy-efficient technology. Through its diverse collection of authors, this volume brings together a mix of theoretical and empirical contributions written on the basis of different disciplinary backgrounds. The volume presents new empirical material and uses innovative methodological approaches, such as spatial econometrics and meta-analysis, to empirically assess actual energy-efficiency trends and investment behaviour. In several chapters, the authors combine and compare economic and physical indicators of energy usage to monitor and analyse trends in energy efficiency while, at the same time, paying ample attention to the detection of empirical regularities in the determinants of energy-saving investment, including uncertainty, energy price volatility and subsidies. This volume also contains several chapters analysing the role of energy modelling in policy design and the potential effect of energy policies on technology diffusion in energy-extensive sectors. We hope that this unique collection of research papers will serve as a useful guide and reference for academics as well as policy-makers, government officials and other professional experts in the area of energy and sustainable development.

We gratefully acknowledge the financial support of the Netherlands Organisation for Scientific Research and SenterNovem, which made it possible for a large multi-institutional research group to focus on energy efficiency from a multidisciplinary perspective during a period of four years. This edited volume is but one of the scientific outputs of this project, which in addition comprises numerous journal articles, book chapters, dissertations and conference presentations. Independent expert reviewers have been used to perform a quality assessment for all chapters of this book in order to ensure and increase the scientific quality of our work. We would like to thank the reviewers for their important suggestions, comments and independent advice. In addition we would like to thank Bert Daniëls, Ymkje de Boer, Carmen Difiglio, Ton van Dril, Hans Elzenga, Casper van Ewijk, Marga de Feber, Klaas-Jan Koops, Peter Nijkamp, Piet Rietveld, Clement Roos, Jan Ros, Erik Schmieman, Jan van de Velde, Marije Verschuur, Arno van der Vlist, Margot Weijnen and Aart de Zeeuw for their support, contributions, comments and guidance throughout various stages of the research process. Finally, we would like to thank the editorial staff at Edward Elgar for their patience and support with the editorial and typesetting process, and Julia Beckhusen and Angie Flack for their indispensable help during the stage in which the various contributions had to be turned into a complete manuscript.

Raymond J.G.M. Florax
Henri L.F. de Groot
Peter Mulder
Amsterdam, West Lafayette, February 2011

Introduction

1. Energy efficiency and technological change

Raymond J.G.M. Florax, Henri L.F. de Groot and Peter Mulder

1 INTRODUCTION

Energy consumption is essentially determined by the intimate relationship between technological and economic change. Within a relatively short period of time in the history of mankind, societies and economies have dramatically changed under the influence of economically generated technological innovations such as steam engines, railroads, internal combustion engines, electricity, airplanes, semi-conductors and satellites. This development not only implied spectacularly increasing levels of welfare, but also tremendously growing levels of energy consumption. As a result, modern societies became dependent on fossil fuels to meet their energy demand, with unfortunate consequences for energy security and the natural environment – the latter most notably in the form of an enhanced greenhouse effect that may cause considerable changes in the earth's climate system. Somewhat paradoxically, technological change is a principal source for alleviating these problems as well, not in the least because technological change facilitates a decreasing ratio of energy consumption per unit of output. Technologies often emerge as 'hopeful monstrosities' (Mokyr, 1990). Cugnot's steam truck built in 1769 had to stop every 15 minutes to have its boiler refilled, and the ENIAC computer from 1946 weighed 30000 kg. Over time, technologies became more energy efficient under the influence of the marriage between economic and technical forces. The issue is clearly illustrated in Figure 1.1, which shows for the USA that although (total primary) energy consumption per capita has increased dramatically during the last century, the per capita gross domestic product (GDP) has increased far more rapidly. In other words, it seems as if ever more output can be produced per unit of energy consumption. Figure 1.1 also shows that this rise in energy efficiency occurs at an ever increasing pace. Improving energy efficiency in the sense of using less

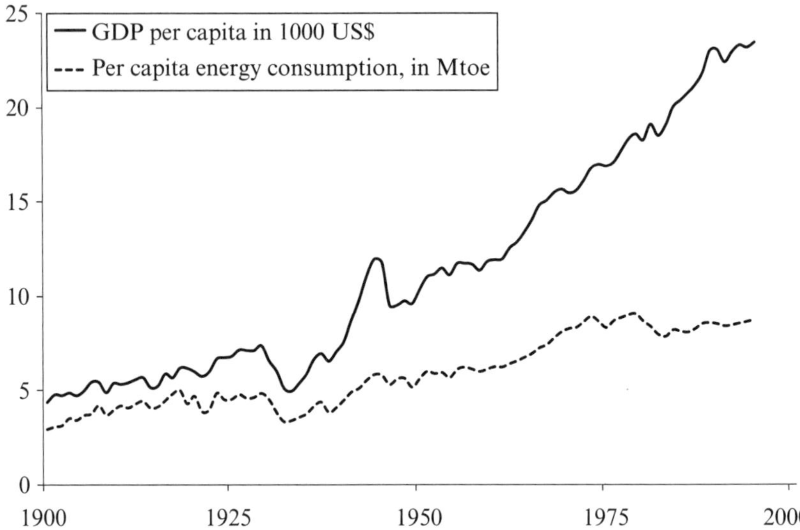

Source: Grübler (1998).

Figure 1.1 GDP in 1990 US dollars and per capita energy consumption in the USA in million tonnes of oil equivalent

energy per unit of output thus presents a powerful, although not exclusive strategy to combine economic growth, energy security and mitigation of climate change.

This book combines theoretical and empirical approaches to offer original insights into energy efficiency trends across countries and sectors, and into various aspects of the complex interaction between technological change, energy policies, markets and investment behaviour, underlying the apparent trend of energy efficiency improvements. Distinguishing features of this book are the attention paid to energy-extensive sectors such as the service sector,[1] the role of firm size and uncertainty in determining energy-saving investment behaviour, and spatial interdependencies in energy efficiency across countries. In addition this book includes critical reviews of recent theory and practice of policy-making in this area. The individual chapters in this book are grouped into three main parts. They are summarized in some detail below in order to guide the reader through this volume.

Part I presents a collection of studies of energy efficiency developments across countries and across sectors, unfolding trends and unravelling its determinants. The individual chapters put special emphasis on spatial interdependence in energy efficiency trends across countries (Mulder, Florax and de Groot, Chapter 2), the role of sectoral composition and the potential to catch up with countries using frontier technologies (Mulder and de Groot, Chapter 3), the emergence of energy-extensive industries including the service sector, as well as detailed physical indicators of energy efficiency (Ramírez, Patel and Blok, Chapters 4 and 5).

Part II provides several studies of investment behaviour in view of the availability of energy-saving technologies. Potential determinants of investment decisions considered include the role of firm size and strategic behaviour (Hunter, Chapter 6), and the role of uncertainty. The latter is studied by means of a meta-analysis scrutinizing a large sample of results from empirical studies addressing the investment–uncertainty nexus (Koetse, de Groot and Florax, Chapter 7), and an individual analysis against the background of the US gas market liberalization (van Soest and de Groot, Chapter 8).

Finally, Part III offers a collection of studies that critically assess the possibilities and impossibilities of effective policy design to stimulate energy efficiency. This part of the book re-examines the rationale for subsidy schemes as compared to other policy instruments, and draws new lessons for their effective design (van Soest and Vollebergh, Chapter 9). This analysis is followed by a theory-based evaluation of actual energy efficiency policies in a specific sector, namely the Dutch service sector (Blok, Joosen and Harmelink, Chapter 10). In addition the applicability and usefulness of utilizing applied policy models in the policy design phase is addressed by means of a critical comparison of three competing models that are simultaneously used in the Netherlands (van Groenendaal, Chapter 11).

2 TRENDS IN ENERGY EFFICIENCY

During the past decades, energy efficiency in the countries making up the Organisation for Economic Co-operation and Development (OECD) has improved with approximately 1–1.5 per cent per year (Ang, 2006; EIA, 2007; Kaufmann, 2004; Schipper and Meyers, 1992; WEC 2008). If the energy intensities of each world region had remained at their 1980 level, world energy consumption would have been 40 per cent higher in 2006 (WEC, 2008). The most rapid decline of energy intensity took place prior to the oil price collapse of 1985–86, while since 1990 the rate of energy efficiency improvements in OECD countries has been less than 1 per cent per year (IEA, 2007b). Key determinants of long-run energy efficiency improvements are changes in the production structure as well as technological change (Howarth et al., 1991; Jorgenson, 1984, 1986; Miketa, 2001; Rosenberg, 1994; Unander et al., 1999). Changes in the sectoral composition of economies matter because sectors differ considerably in terms of value added per unit of energy input due to the fact that some economic processes require more capital, skills and/or technology. Understanding aggregate trends in energy efficiency therefore requires an assessment of productivity developments in individual sectors and a careful selection and comparison of indicators (Dollar and Wolff, 1993; Jorgenson, 1984; Worrell et al., 1997).

Most of the studies that analyse energy efficiency trends focus on industry in general and on energy-intensive industrial sub-sectors, which are largely responsible for the bulk of industrial energy use. The latter include sectors such as iron and steel, chemicals and paper (see Ang and Zhang, 2000, for a survey). However, while industry traditionally accounted for about 50 per cent of the world's energy use, economic development typically has led to an increasing importance of firms and sectors with relatively low levels of energy intensity, in particular light industry (such as food processing) and the service sector. Nowadays, these energy-extensive sectors account for about 20 per cent of the economy-wide energy usage (WEC, 2008), and energy consumption in these sectors is expected to grow relatively fast. Moreover, in a globalizing world, energy efficiency improvements are unlikely to materialize in isolation. International trade flows, foreign direct investment and transnational policy making within the frameworks of the European Union and the United Nations (for example the Kyoto protocol) all profoundly influence (changes in) the energy efficiency levels of individual economies through their influence on technological development and (changes in) the production structure. In contrast to the empirical economic growth literature, which increasingly pays attention to various types of geographical interdependencies to explain cross-country productivity differences (Keller, 2002; Nijkamp and Poot, 1998), empirical studies of energy productivity or energy intensity still largely assume the wonderland of a spaceless economy. It is only recently that the hypothesized existence of an 'advantage of backwardness' that may carry a potential for rapid advantage (Abramovitz, 1986; Gerschenkron, 1952) has been empirically tested with respect to energy productivity (Mulder and de Groot, 2007; Miketa and Mulder, 2005). The latter studies confirmed that, in most sectors, countries lagging in terms of energy productivity tend to catch up with countries using frontier technologies, but also that convergence patterns are local rather than global while considerable cross-country differences in energy productivity performance persist. Part I of this book adds a collection of studies to this literature that takes into account the role of spatial effects and the shift towards energy-extensive sectors.

Chapter 2, by Mulder, Florax and de Groot, moves the analysis of international dynamics of energy productivity forward by exploring spatial heterogeneity as well as spatial dependence across OECD and non-OECD countries using modern spatial data analysis techniques. Their study considers the period 1971–2001 and two levels of sectoral aggregation: the macroeconomic level as well as the sectors agriculture, industry and services. The authors report evidence of decreasing cross-country differences in energy productivity in combination with increasing spatial clustering. The chapter shows that the European countries form a particularly strong spatial cluster over time. The econometric results generated by the authors, provide clear evidence for spatial dependence working through omitted variables with a spatial dimension, causing the regression errors from different countries to be spatially autocorrelated. This study underlines the importance of further exploring the role of individual country similarities and dissimilarities that drive energy efficiency patterns.

Chapter 3, by Mulder and de Groot, studies the international dynamics of energy productivity for a relatively small and homogenous group of 14 OECD countries during a period of about 27 years. This allows for a relatively high degree of sectoral detail: 13 sectors are considered with special attention being paid to the role of energy-extensive sectors such as services and light industry. The data reveal that within this relatively homogenous group of countries remarkable cross-country differences in energy productivity persist. The chapter presents a cross-country decomposition analysis that shows aggregate energy productivity to be mainly driven by intra-sectoral efficiency improvements, which to a large extent are realized within the energy-extensive industrial sectors. This chapter also shows that the impact on aggregate energy productivity growth of changes in the sectoral composition of economies, predominantly from industry to services, and from energy-intensive to energy-extensive industrial sectors, is mixed: in some countries they enhance and in other countries they partly offset energy productivity improvements. Mulder and de Groot reach the conclusion that sectoral energy productivity growth is positively affected by the potential to catch up, energy prices and changes in fuel mix, whereas economies of scale, investment share, openness and specialization play only a modest role in explaining cross-country variation in energy productivity growth at the sectoral level.

Chapter 4, by Ramírez, Patel and Blok, provides a study of energy efficiency developments in the Dutch service sector during a period of approximately 15 years. They report a strong increase in energy consumption, especially in electricity, while they cannot find evidence of a significant decrease in energy intensity during the past decade. A decomposition analysis of aggregate energy use in the Dutch service sector shows that the activity effect (value added) has been the main driving force behind the observed increase of energy consumption, further enhanced by an increase in labour intensity, while energy intensity (in terms of energy per employee) did not play a significant role in decreasing energy usage. Subsequently, Ramírez and colleagues analyse the developments in five sub-sectors: wholesale and retail trade, hotels and restaurants, financial installations, real estate renting and business activities, and education and hospitals. Although the picture is somewhat blurred due to the lack of data of sufficient quality, they report evidence of a very limited decrease in energy intensity at this sub-sectoral level, except for the sector real estate renting and business activities. The chapter concludes that energy policies so far have largely failed to address the service sector, notwithstanding the importance of this sector with regard to carbon dioxide reduction goals caused by its high share of electricity consumption and high growth rate.

Chapter 5 by the same group of authors completes the first part of this book by arguing that physical indicators provide a fair and feasible way of comparing energy efficiency, in spite of their inherent limitations in terms of data availability and data quality (see Chapter 2 of this book). After making this case, they illustrate their methodology by again exploring energy efficiency trends in energy-extensive sectors, but this time they focus on the meat and dairy industry of four European countries (France, Germany, the Netherlands and the UK). Changes in energy efficiency are monitored in two different ways. First, they examine trends in energy use by unit of raw material processed (that is, animal slaughtered or raw milk treated) and, second, they compare the actual energy use with the energy that would have been used if no changes in energy efficiency would have taken place. The latter indicator corrects for differences in product mix across countries and over time. The authors report that changes in product mix are significant in the period considered and therefore the second indicator should be preferred when comparing levels of energy efficiency among countries. Once changes in product mix had been taken into account they found that over the past decade this indicator increased in the meat sector, while it decreased in the dairy sector. In order to understand the observed changes in the indicators, the chapter examines factors such as different system boundaries, increased demand for food security and the differences in the values across countries.

If history provides any guide for the future, the chapters in Part I of this book do not convey an unequivocally optimistic message. Although considerable energy efficiency improvements have been realized, the rate of energy efficiency improvements in OECD countries since 1990 has in general slowed down and energy efficiency improvements have not nearly been enough to halt the growth of CO_2 emissions (IEA, 2007b). Moreover, in spite of evidence of energy productivity convergence, cross-country differences in energy consumption per unit of output remain so considerable and persistent that we apparently cannot anticipate lagging countries to catch up to forerunners as rapidly, globally and easy as we wish or maybe expect in a globalizing world. This conclusion is further strengthened by the observation that changes in the sectoral composition of economies towards energy-extensive sectors do not unambiguously contribute to aggregate energy efficiency improvements, also because energy efficiency growth in energy-extensive sectors seems generally limited. Arguably, there are reasons to expect the rate of energy efficiency to pick up again in the near future. First, in Western Europe and in the USA an increasing sense of urgency can be observed with respect to the need to decrease dependence on imported energy resources. Second, the energy and climate strategy of the EU aims for a 20 per cent reduction in greenhouse gases by 2020 as compared to 1990. Third, after nearly two decades of low oil prices (except for some incidental, mainly geopolitically motivated, spikes) it is not unlikely that after the global financial crisis and recession the world will again be entering a period with sustained high oil prices, which are not in the last place caused by structurally increasing demand from emerging economies such as China and India. This implies great challenges for the adoption and diffusion of energy-saving technologies.

3 TECHNOLOGY AND INVESTMENT BEHAVIOUR

The scope for technology-driven change is vast, as is illustrated by the economic history of the last two centuries. Economic theorizing about technological change goes at least back to the work of Schumpeter (1934) and can roughly be grouped in three major traditions of research: a growth theoretic, an evolutionary and a microeconomic approach. Studies of the economy–technology–energy nexus have, with some delay, followed these developments in the literature. Below we briefly highlight the main developments (Grübler et al., 2002; Ruttan, 2002).

The theoretical approach to economic growth has been largely inspired by Robert Solow's (1956, 1957) finding that the only way to explain the observed long-run growth in output per person is through technological change that continually offsets the dampening effect of diminishing returns. This approach takes on a relatively abstract notion of technology by defining technological development in terms of accumulation of knowledge or human capital, a process that is assumed to generate positive externalities at the economy-wide level, thus causing long-run economic growth. For a long time, economic models in this tradition necessarily relied on some form of exogenous technological change (Arrow, 1962; Nordhaus, 1969), but since the 1980s this unsatisfactory feature has been overcome in a new class of so-called endogenous growth models, in which technological change is represented as an internally dynamic, endogenous variable (Grossman and Helpman, 1991; Lucas, 1988; Romer, 1986; see Aghion and Howitt, 1998, for an overview). Within economic growth theory, many economy-environment models have been developed that explicitly address the (long-term) evolution of energy consumption (and CO_2 emissions) under the influence of technological change and environmental policies. These models can be classified in two groups: macroeconomic or 'top-down' models with few economic sectors and an aggregate stylized specification of technology (Bovenberg and Smulders, 1995; Carraro et al., 2003; Goulder and Mathai, 2000; Hart, 2008; Manne and Richels, 2004; Nordhaus, 2002; Popp, 2004; Schwoon and Tol, 2006; Zon and Hakan, 2003) and techno-economic or 'bottom-up' models with detailed, often database-driven specifications of technology options (Grübler and Messner, 1998; Oda et al., 2007; Energy Journal, 2006a, 2006b). Despite differences, these models put emphasis on innovation through research and development (R&D) and learning-by-doing processes, the latter usually being a (costless) by-product of cumulative investment, installed capacity or abatement.

The evolutionary approach emerged in the mid-1970s out of dissatisfaction with the neoclassical framework, which dominated the other two approaches, and was mainly instigated by the work of Nelson and Winter (1974, 1982). The distinguishing feature of this approach is to view technological change as an evolutionary phenomenon that is driven by microeconomic selection and search processes, which are inherently uncertain, often irreversible and path-dependent, as well as guided by 'routines' and imperfectly rational agents (see Freeman, 1994, for a survey). As such, the evolutionary framework attempts to reconcile processes of technical change at the micro-level (Dosi, 1988) with aggregate patterns of technology-driven economic growth (Silverberg and Verspagen, 2005). The evolutionary economic framework has been applied to study the transition towards sustainability, including sustainable energy systems (van den Bergh et al., 2007; Kemp, 1997; Loorbach and Kemp, 2008; Oltra and Jean, 2006). Research in this area, however, is so far mainly engaged in conceptual explorations rather than in formal theorizing or empirical analyses, except for some case studies (van den Bergh et al., 2007).

Finally, the microeconomic, approach is often considered as originating from the original observation by Hicks (1932) that innovation is driven by relative factor prices and substitution possibilities. This approach gained momentum in the 1960s and 1970s and focuses on the microeconomics of innovation and diffusion of new technologies with much attention paid to the role of relative prices, factor endowments and demand (Ahmad, 1966; Binswanger, 1974a, 1974b; David, 1975; Griliches, 1957; Kennedy, 1962; Schmookler, 1966; Stoneman, 1983; see Ruttan, 2002, for a survey). Economy-energy models within the microeconomic tradition do pay explicit attention to microeconomic characteristics of the decision to adopt energy-saving technologies. Since the classic study by Berndt and Wood (1975), many scholars have examined the role of relative factor prices and substitution possibilities on the derived demand for energy, subject to technological change. Specifically, various studies have explored a price- or product-standard induced bias towards energy-saving technological change (Newell et al., 1999; Popp, 2000; Smulders and de Nooij, 2003; Taheri and Stevenson, 2002). An important observation in this respect is the existence of a considerable gap between the most energy-efficient technologies available at some point in time and the technologies that are actually in use. This phenomenon has become known as the 'energy-efficiency paradox' since new technologies are often not adopted despite their cost-effectiveness (Jaffe and Stavins, 1994). Explanations for this phenomenon are to be sought in the complexity of actual investment behaviour. Inspired by insights from the evolutionary approach a range of investment barriers and market failures has been identified that help explain why significant cost-effective energy efficiency potential is not being tapped. They include insufficient and inaccurate information, inadequate access to capital, the need for complementary investments in technologies and skills, strategic considerations, principal–agent problems, vested interests, learning and spillover effects, and the combination of uncertainty with some degree of irreversibility in investment (Blok et al., 2004; IEA, 2007a; Jaffe et al., 2002; Mulder et al., 2003; Schleich and Gruber, 2008). Part II of this book adds three original studies on the adoption and diffusion of energy-saving investments to the existing literature, concurrently highlighting the role of firm size, strategic behaviour and uncertainty.

Chapter 6, by Hunter, combines elements of existing approaches to argue that firm size is an important determinant of the timing of investment decisions. The chapter presents a model of technology adoption incorporating three conclusions from the empirical literature regarding firms' decisions to embrace technological advances:

1. firms tend to delay the opportunity to invest in a new technology beyond what one would predict using the net present value criterion,
2. new technologies tend to be adopted in a more-or-less sequential manner or 'diffuse' into use, and
3. larger firms tend to adopt new technologies before smaller firms.

The model allows firms to learn from each others' adoption decisions and to form coalitions in order to reduce the free-rider problem and to ensure that there is adequate provision. Hunter then shows that new technologies can induce firms to merge and can also explain industry structure independently of strategic interactions in the output market. He argues that the weight of the empirical evidence on adoption timing and diffusion reassures his main conclusion that firm size is driven by technological opportunities, a result that is sometimes referred to as the 'Schumpeterian hypothesis'. Finally, the chapter lays the groundwork for future empirical study and provides valuable guidance toward a more accurate interpretation of the behaviour of the firm in existing work.

Chapter 7, by Koetse, de Groot and Florax, addresses the heterogeneity in outcomes of empirical studies that examine the effect of uncertainty on investment. While many theoretical insights have been put forward regarding the investment–uncertainty relationship, the empirical literature yields an ambiguous picture with respect to the direction and the significance of this relationship. After a review of the key theoretical contributions in this area, they apply meta-analysis to systematically and quantitatively analyse the investment–uncertainty nexus, using data from a large number of empirical studies. Their model is capable of explaining a substantial part of the observed heterogeneity in the literature on the direction and significance of the investment–uncertainty relationship. Differences in measurement of investment and uncertainty as well as empirical differences between primary studies appear to be important factors. Furthermore, the influence of differences in model specification and data on study outcomes provides evidence on important theoretical issues. For example, the authors can identify fairly precisely the impact of omitted variable bias in investment models. The analysis also shows that including Tobin's *Q*, the shadow value of capital, in investment models is important, but it does not capture the impact of uncertainty on investment spending. The chapter concludes with a call for further research that should pay attention to micro-level data, short-run versus long-run effects of uncertainty and temporary versus more permanent changes in uncertainty.

Chapter 8, by van Soest and de Groot, empirically tests the validity of the predictions of a well-known model of investment behaviour due to Dixit and Pindyck (1994) that emphasizes the combined importance of uncertainty and irreversibility of investment decisions. The model set-up assures that it might be rational to postpone technology adoption in the face of uncertainty about future development. More specifically, the model predicts that a firms' propensity to postpone investment is higher:

1. the larger the (sunk) cost involved when investing,
2. the higher the uncertainty the firms face, and
3. the higher the degree of persistence in the uncertain variable.

Van Soest and de Groot investigate the quantitative importance of such predictions by assessing the impact of the liberalization of the US natural gas markets on an energy-saving technology in the paper production industry. Since the liberalization of the US gas market started some 15 years before liberalization policies were introduced in Western Europe, the experiences in the USA can be used to predict developments in Europe. To that effect, they assess how the investment behaviour of the Dutch paper industry would react if the impact on energy prices (in terms of levels as well as volatility) of energy market liberalization in the Netherlands would be similar to the US. The authors show that the liberalization of the natural gas market will have a substantial detrimental impact on the willingness to invest in more energy-efficient technologies in the Dutch paper industry, due to a lower level as well as an increased volatility of natural gas prices. This implies the existence of a trade-off between energy market liberalization and environmental protection, suggesting that policy-makers concerned with curbing energy use are wise to consider the impact of policy measures on uncertainty of economic actors since the quantitative impacts of increases in uncertainty can be substantial.

4 ENERGY POLICY DESIGN

Policies to stimulate energy efficiency improvements are motivated by various market failures, of which the negative externalities or unpriced costs resulting from energy-related greenhouse gas emissions are probably the most fundamental. In addition, positive externalities or unpriced benefits arise from various types of knowledge spillovers associated with the adoption of new technologies, since technological development is equivalent to the creation and diffusion of knowledge. Hence, in various ways social gains can be realized from the adoption and diffusion of energy-saving technologies by individual firms. Over the past few decades various policy instruments have been developed aimed at the encouragement of firms to undertake action in order to meet environmental policy goals (Anderson and Newell, 2004; Grubb and Ulph, 2002; Kverndokk and Rosendahl, 2007). Policy instruments can be characterized as either command-and-control or market-based approaches. Command-and-control instruments intend to influence economic actors by means of force, usually through setting uniform performance- or technology-based standards. Market-based instruments aim to influence firms by means of changing their financial incentive structure, such that their own interests align with those of society at large, for example by means of taxes, subsidies or tradable permits. In general, the economics profession tends to prefer market-based instruments because of their cost-effectiveness: in some form or another, market-based instruments allow for flexibility among heterogeneous firms, and they provide economic rationale to spur innovation and diffusion as opposed to technology standards, which tend to freeze technological development. Moreover, within the group of market-based instruments economists discriminate in general against the use of subsidies for a number of reasons. As compared with taxes, subsidies do not generate government revenues that can be used to reduce other distortionary taxes and increase welfare, they often imply higher administration costs, they may increase energy consumption and emissions because of the inherent income transfer, and they may involve free-rider behaviour (Blok et al., 2004; Jaffe et al., 2002; Requate, 2005; Requate and Unold, 2003).

The fairly high degree of theoretical consensus contrasts with the diversity and complexity of the actual practice of policy-making. Effective policy design is apparently very difficult, and this is understandable for a number of reasons. An obvious elementary cause is firm heterogeneity. Different firms embody different characteristics while policies are often uniform in their design, the combination of which effectively leads to inefficient situations like free-rider behaviour and windfall gains. Moreover, technologies are heterogeneous: energy efficiency is only one of many technology characteristics. Another difficulty for proper policy design lies in the heterogeneity of results from applied research, leading to differences in policy advice. Moreover, measuring the effectiveness of past policies and programmes is often hampered by lack of data and/or an adequate framework for evaluation. Part III of this book addresses exactly these issues. Rather than presenting stylized theoretical analyses about optimal policy schemes, the studies collected in the final part of this book critically examine actual policies to stimulate energy efficiency, and they present new empirical material.

Chapter 9, by van Soest and Vollebergh, analyses how government subsidy schemes should be designed to effectively induce firms to adopt new energy-saving technologies. Although widely applied in practice, environmental subsidies are not very popular among economists. The authors present a review of the theoretical pros and cons of subsidy schemes as compared to other policy instruments (such as environmental taxes), to conclude that subsidies deserve a second look based on insights from the recent theoretical and empirical literature. Subsequently, the authors analyse the implications of subsidy design for the variation in effectiveness and efficiency of subsidies given to firms, evaluating a subsidy programme currently in place in the Netherlands as well as an economic experiment in a controlled environment. They argue that proper subsidy design, meaning the choice of a subsidy's base and rate, is of critical importance in realizing efficient and effective policies. Key issues to consider are the level of subsidy, whether eligibility is directly associated with the investment itself or with the actual amounts of emissions abated, and whether a subsidized technology directly contributes to the firm's competitiveness. Finally, on the basis of the experimental evidence the authors conclude that subsidies matter, both because of their financial incentive and because of their help in limiting the firm managers' search for new technologies by eliminating inefficient technologies.

Chapter 10, by Blok, Joosen and Harmelink, examines the effectiveness of a range of policy instruments for energy efficiency improvement directed at the service sector in the Netherlands. Using theory-based evaluation, they find evidence for accelerated energy efficiency on top of the unknown autonomous energy efficiency improvement. The improvements in energy efficiency are partly caused economic policy instruments (subsidies and taxes), but normative policy instruments (standards and environmental permits) contributed to the results as well. However, the authors also conclude that the uncertainties in the results are substantial, due to an incomplete and inconsistent system of monitoring. This may help explain why their main conclusions contradict the conclusion of Chapter 5, where Ramírez and colleagues report failing energy policies in the same Dutch service sector, as illustrated by a limited decrease in energy intensity. The chapter concludes with a call to have the necessary monitoring already in place while policy instruments are in force in order to bring target setting and evaluation more in line with each other, and to reconsider the role of regulation, for example in the form of a building labelling system.

Chapter 11, by van Groenendaal, reviews three energy models that were for some time simultaneously used for policy-making in the Netherlands. Although the three models differ in structure and are equipped with particular diffusion mechanisms, they have in common that they utilize a vintage structure and they are also sectoral models using the technology options and their direct costs contained in the same technology database to evaluate investment options. Furthermore, the policy models are partial models using the same exogenously determined energy price data and economic growth scenarios. After discussing the models and their pros and cons in some detail, van Groenendaal presents the results of a modelling exercise with the three models. The author compares the way in which the three models deal with the same three policies (CO_2-tax, investment subsidy and investing in all technologies with an internal rate of return of 10 per cent or more) for four different energy-extensive sectors (services, other metal, food, and textiles and beverages) and two forms of energy (electricity and heat). The modelling exercise shows remarkable differences between the models, implying that the Dutch government is likely to receive different advice when posing the same question to each of the institutes employing the three different models. The fact that the three research institutes work together on energy issues, makes the differences even more astounding. Van Groenendaal concludes that more research is required for the formulation of good policy advice on energy and energy technology matters for the Netherlands. All three models are partial models with limited or no endogenous behaviour, thus limiting their relevance for long-term assessments of policy impacts on behaviour. One suggested avenue for further research is therefore the modelling of endogenous behaviour accounting for the feedback between energy use and the rest of the economy.

NOTE

1. The terminology in the energy literature is not consistent but we will use the adjective 'energy-extensive' to refer to firms, sectors or countries with a relatively low energy usage per unit of output or per capita.

REFERENCES

Abramovitz, M. (1986), 'Catching up, forging ahead, and falling behind', *The Journal of Economic History*, **47**, 385–406.

Aghion, P. and P. Howitt (1998), *Endogenous Growth Theory*, Cambridge, MA: MIT Press.

Ahmad, S. (1966), 'On the theory of induced innovation', *Economic Journal*, **76**, 344–57.

Anderson, S.T. and R.G. Newell (2004), 'Information programs for technology adoption: the case of energy-efficiency audits', *Resource and Energy Economics*, **26**, 27–50.

Ang, B.W. (2006), 'Monitoring changes in economy-wide energy efficiency: from energy-GDP ratio to composite efficiency index', *Energy Policy*, **34**, 574–82.

Ang, B.W. and F.Q. Zhang (2000), 'A survey of index decomposition analysis in energy and environmental studies', *Energy*, **25**, 1149–76.

Arrow, K.J. (1962), 'The economic implications of learning-by-doing', *Review of Economic Studies*, **29**, 155–73.

Berndt, E.R. and D.O. Wood (1975), 'Technology, prices, and the derived demand for energy', *Review of Economics and Statistics*, **57**, 259–68.

Binswanger, H.P. (1974a), 'The measurement of technical change biases with many factors of production', *American Economic Review*, **64**, 964–76.

Binswanger, H.P. (1974b), 'A microeconomic approach to induced innovation', *Economic Journal*, **84**, 940–58.

Blok, K., H.L.F. de Groot, E.E.M. Luiten and M.G. Rietbergen (2004), *The Effectiveness of Policy Instruments for Energy Efficiency Improvements in Firms*, Dordrecht: Kluwer.

Bovenberg, A.L. and S. Smulders (1995), 'Environmental quality and pollution-augmenting technological change in a two-sector endogenous growth model', *Journal of Public Economics*, **57**, 369–91.

Carraro, C., R. Gerlagh and B. van der Zwaan (2003), 'Endogenous technical change in environmental macroeconomics', *Resource and Energy Economics*, **25**, 1–10.

David, P.A. (1975), *Technical Choice, Innovation and Economic Growth*, Cambridge: Cambridge University Press.

Dixit, A.K. and R.S. Pindyck (1994), *Investment under Uncertainty*, Princeton, NJ: Princeton University Press.

Dollar, D. and E.N. Wolff (1993), *Competitiveness, Convergence, and International Specialization*, Cambridge, MA: MIT Press.

Dosi, G. (1988), 'Sources, procedures and microeconomic effects of innovation', *Journal of Economic Literature*, **26**, 1120–71.

Energy Information Administration (EIA) (2007), *International Energy Outlook 2007*, Washington, DC: EIA.

Energy Journal (2006a), 'Endogenous technological change and the economics of atmospheric stabilisation', *The Energy Journal*, **28**, special issue, April.

Energy Journal (2006b), 'Hybrid modeling of energy-environment policies: reconciling bottom-up and topdown', *The Energy Journal*, **28**, special issue, November.

Freeman, C. (1994), 'The economics of technological change', *Cambridge Journal of Economics*, **18**, 463–514.

Gerschenkron, A. (1952), 'Economic backwardness in historical perspective', in B. Hoselitz (ed.), *The Progress of Underdeveloped Areas*, Chicago, IL: University of Chicago Press.

Goulder, L.W. and K. Mathai (2000), 'Optimal CO_2 abatement in the presence of induced technological change', *Journal of Environmental Economics and Management*, **39**, 1–38.

Griliches, Z. (1957), 'Hybrid corn: an exploration of the economics of technical change', *Econometrica*, **25**, 501–22.

Grossman, G.M. and E. Helpman (1991), *Innovation and Growth in the Global Economy*, Cambridge, MA: MIT Press.

Grubb, M. and D. Ulph (2002), 'Energy, the environment, and innovation', *Oxford Review of Economic Policy*, **18**, 92–106.

Grübler, A. (1998), *Technology and Global Change*, Cambridge: Cambridge University Press.

Grübler, A. and S. Messner (1998), 'Technological change and the timing of abatement measures', *Energy Economics*, **20**, 495–512.

Grübler, A., N. Nakicenovic and W.D. Nordhaus (2002), *Technological Change and the Environment*, Washington, DC: Resources for the Future.

Hart, R. (2008), 'The timing of taxes on CO_2 emissions when technological change is endogenous', *Journal of Environmental Economics and Management*, **55**, 194–212.

Hicks, J.R. (1932), *The Theory of Wages* (2nd edn, 1966), London: Macmillan.

Howarth, R.B., L. Schipper, P.A. Duerr, and S. Strøm (1991), 'Manufacturing energy use in 8 OECD countries', *Energy Economics*, **13**, 135–42.

International Energy Agency (IEA) (2007a), *Mind the Gap – Quantifying Principal–Agent Problems in Energy Efficiency*, Paris: IEA.

International Energy Agency (IEA) (2007b), *Energy Use in the New Millennium – Trends in IEA Countries*, Paris: IEA.

Jaffe, A.B. and R.N. Stavins (1994), 'The energy paradox and the diffusion of conservation technology', *Resource and Energy Economics*, **16**, 91–122.

Jaffe, A.B., R.G. Newell and R.N. Stavins (2002), 'Environmental policy and technological change', *Environmental and Resource Economics*, **22**, 41–69.

Jorgenson, D.W. (1984), 'The role of energy in productivity growth', *Energy Journal*, **5**, 11–26.

Jorgenson, D.W. (1986), 'The great transition: energy and economic change', *Energy Journal*, **7**, 1–13.

Kaufmann, R.K. (2004), 'The mechanisms for autonomous energy efficiency increases: a co-integration analysis of the US Energy/GDP ratio', *The Energy Journal*, **25**, 63–86.

Keller, W. (2002), 'Geographic localization of international technology diffusion', *American Economic Review*, **92**, 120–42.

Kemp, R. (1997), *Environmental Policy and Technical Change: A Comparison of the Technological Impact of Policy Instruments*, Cheltenham, UK and Brookfield, VT, USA: Edward Elgar Publishing.

Kennedy, C. (1962), 'The character of improvements and technical progress', *Economic Journal*, **72**, 899–911.

Kverndokk, S. and K.E. Rosendahl (2007), 'Climate policies and learning by doing: impacts and timing of technology subsidies', *Resource and Energy Economics*, **29**, 58–82.

Loorbach, D. and R. Kemp (2008), 'Transition management for the Dutch energy transition: multilevel governance aspects', in J. van den Bergh and F. Bruinsma (eds), *Managing the Transition to Renewable Energy Sources: Theory and Practice from Local, Regional and Macro Perspectives*, Cheltenham, UK and Northampton, MA, USA: Edward Elgar Publishing, pp. 245–66.

Lucas, R.E. (1988), 'On the mechanics of economic development', *Journal of Monetary Economics*, **22**, 3–42.

Manne, A. and R. Richels (2004), 'The impact of learning-by-doing on the timing and costs of CO_2 abatement', *Energy Economics*, **26**, 603–19.

Miketa, A. (2001), 'Analysis of energy intensity developments in manufacturing sectors in industrialized and developing countries', *Energy Policy*, **29**, 769–75.

Miketa, A. and P. Mulder (2005), 'Energy productivity across developed and developing countries in 10 manufacturing sectors: patterns of growth and convergence', *Energy Economics*, **27**, 429–53.

Mokyr, J. (1990), *The Level of Riches: Technological Creativity and Economic Progress*, Oxford: Oxford University Press.

Mulder, P. and H.L.F. de Groot (2007), 'Sectoral energy- and labour-productivity convergence', *Environmental and Resource Economics*, **36**, 85–112.

Mulder, P., H.L.F. de Groot and M.W. Hofkes (2003), 'Explaining slow diffusion of energy-saving technologies: a vintage model with returns to diversity and learning-by-using', *Resource and Energy Economics*, **25**, 105–26.

Nelson, R.R. and S.G. Winter (1974), 'Neo-classical vs. evolutionary theories of economic growth', *Economic Journal*, **84**, 886–905.

Nelson, R.R. and S.G. Winter (1982), *An Evolutionary Theory of Economic Change*, Cambridge, MA: Harvard University Press.

Newell, R.G., A.B. Jaffe and R.N. Stavins (1999), 'The induced innovation hypothesis and energy-saving technological change', *Quarterly Journal of Economics*, **114**, 941–75.

Nijkamp, P. and J. Poot (1998), 'Spatial perspectives on new theories of economic growth', *Annals of Regional Science*, **32**, 7–37.

Nordhaus, W.D. (1969), *Invention, Growth and Welfare*, Cambridge, MA: MIT Press.

Nordhaus, W.D. (2002), 'Modeling induced innovation in climate change policy', in A. Grübler, N. Nakićenović and W.D. Nordhaus (eds), *Modelling Induced Innovation in Climate Change Policy*, Washington, DC: Resources for the Future.

Oda, J., K. Akimoto, F. Sano and T. Tomoda (2007), 'Diffusion of energy efficient technologies and CO_2 emission reductions in iron and steel sector', *Energy Economics*, **29**, 868–88.

Oltra, V. and M.S. Jean (2006), 'Environmental innovation and clean technology: an evolutionary framework', *International Journal of Sustainable Development*, **8**, 153–72.

Popp, D.C. (2000), 'Induced innovation and energy prices', *American Economic Review*, **92**, 160–80.

Popp , D.C. (2004), 'ENTICE: Endogenous technological change in the DICE model of global warming', *Journal of Environmental Economics and Management*, **48**, 742–68.

Requate T. (2005), 'Dynamic incentives by environmental policy instruments – a survey', *Ecological Economics*, **54**, 175–95.

Requate T. and W. Unold (2003), 'Environmental policy incentives to adopt advanced abatement technology: will the true ranking please stand up?', *European Economic Review*, **47**, 125–46.

Romer, P.M. (1986), 'Increasing returns and long-run growth', *Journal of Political Economy*, **94**, 1002–37.

Rosenberg, N. (1994), 'Energy-efficient technologies: past and future perspectives', in N. Rosenberg (ed.), *Exploring the Black Box: Technology, Economics and History*, Cambridge: Cambridge University Press.

Ruttan, V.W. (2002), 'Source of technical change: induced innovation, evolutionary theory, and path dependence', in A. Grübler, N. Nakićenović and W.D. Nordhaus (eds), *Technological Change and the Environment*, Washington, DC: Resource for the Future.

Schipper, L. and S. Meyers (1992), *Energy Efficiency and Human Activity: Past Trends and Future Prospects*, Cambridge: Cambridge University Press.

Schleich, J. and E. Gruber (2008), 'Beyond case studies: barriers to energy efficiency in commerce and services sector', *Energy Economics*, **30**, 449–64.

Schmookler, K. (1966), *Invention and Economic Growth*, Cambridge, MA: Harvard University Press.

Schumpeter, J.A. (1934), *The Theory of Economic Development*, Cambridge, MA: Harvard University Press.

Schwoon, M. and R.S.J. Tol (2006), 'Optimal CO_2-abatement with socio-economic inertia and induced technological change', *The Energy Journal*, **27**, 25–59.

Silverberg, G. and B. Verspagen (2005), 'Evolutionary theorizing on economic growth', in K. Dopfer (ed.), *The Evolutionary Foundations of Economics*, Cambridge: Cambridge University Press.

Smulders, S. and M. de Nooij (2003), 'The impact of energy conservation on technology and economic growth', *Resource and Energy Economics*, **25**, 59–79.

Solow, R.M. (1956), 'A contribution to the theory of economic growth', *Quarterly Journal of Economics*, **70**, 65–94.
Solow, R.M. (1957), 'Technical change and the aggregate production function', *Review of Economics and Statistics*, **39**, 312–20.
Stoneman, P. (1983), *The Economic Analysis of Technological Change*, Oxford: Oxford University Press.
Taheri, A.A. and R. Stevenson (2002), 'Energy price, environmental policy, and technological bias', *Energy Journal*, **32**, 85–107.
Unander, F., S. Karbuz, L. Schipper, M. Khrushch and M. Ting (1999), 'Manufacturing energy use in OECD countries: decomposition of long term trends', *Energy Policy*, **27**, 769–78.
van den Bergh, J.C.J.M., A. Faber, A.M. Idenburg and F.H. Oosterhuis (2007), *Evolutionary Economics and Environmental Policy: Survival of the Greenest*, Cheltenham, UK and Northampton, MA, USA: Edward Elgar Publishing.
World Energy Council (WEC) (2008), *Energy Efficiency Policies around the World: Review and Evaluation*, London: WEC.
Worrell, E., L. Price, N. Martin, J. Farla and R. Schaeffer (1997), 'Energy intensity in the iron and steel industry: a comparison of physical and economic indices', *Energy Policy*, **25**, 727–44.
Zon, A. van and Y. Hakan (2003), 'An endogenous growth model with embodied energy-saving technical change', *Resource and Energy Economics*, **25**, 81–103.

PART I

Trends in Energy Efficiency

2. A spatial perspective on global energy productivity trends

Peter Mulder, Raymond J.G.M. Florax and Henri L.F. de Groot

1 INTRODUCTION

By way of further introducing the theme of this book, this chapter presents a bird's eye view on cross-country differences in energy productivity. We define energy productivity as the ratio of output measured as gross domestic product (GDP) to energy input, which is by definition equivalent to the inverse of energy intensity. Differences in energy productivity are caused by differences across countries in the level of technological development, the sectoral composition of an economy, factor input ratios, (relative) factor prices or the fuel mix (Berndt, 1978; Jorgenson, 1984; Schipper and Meyers, 1992). Over time these differences may change as the result of, among others, technological change, factor accumulation and (relative) factor price changes, which in turn are facilitated by, for example, trade flows, foreign direct investment (FDI), market conditions and government policies. Obviously, many of these economic indicators have a clear spatial or geographical dimension, which implies that spatial effects are potentially important in explaining spatial and temporal energy productivity dynamics across countries.

The aim of this chapter is twofold. First, we outline the main historical trends in energy productivity dynamics across countries as a way to further introduce several key issues addressed in this book. What do we know about cross-country energy consumption, in per capita terms as well as per unit of output? What have been historical rates of energy productivity improvement in different parts of the world? To what extent do world regions differ in terms of energy productivity levels? Do these differences in energy productivity levels decrease over time? Do energy-inefficient countries catch up to energy efficient countries? Second, we explore the geographical dimension of energy productivity dynamics across countries. The questions we address are to what extent are energy productivity dynamics characterized by spatial effects, such as the rise and fall of regional clusters, and does geographical proximity play a role in explaining cross-country variation in energy productivity differences over time?

The study of productivity differences across countries has a long tradition in the economic growth literature. A key notion in this respect is the concept of convergence, defined as a long-run tendency towards equalization of productivity or income per capita across countries. Theoretically, the concept of productivity convergence has its roots in the neoclassical Solow–Swan growth model that postulates convergence of per capita income, driven by the assumption of diminishing returns to capital accumulation at the economy-wide level (Solow, 1956; Swan, 1956). New or endogenous growth theory has presented a more diverse picture, with convergence mainly depending on international knowledge spillovers while learning effects, externalities and market imperfections may lead to patterns of divergence (Lucas, 1988; Romer, 1986, 1990). Following the empirical work of Baumol (1986) and Abramovitz (1986), numerous empirical convergence studies have been published, typically presenting evidence of convergence of per capita income, labour productivity or total factor productivity.[1] These studies usually ignore the role of spatial effects, in spite of recognizing the importance of 'spatial phenomena' such as technology diffusion, trade and factor mobility in driving convergence patterns. Various authors have presented evidence of misspecification in testing the convergence hypothesis owing to ignored spatial dependence (Florax and Nijkamp, 2005; Rey, 2001; Rey and Montouri, 1999).

It is only recently that evidence of energy productivity convergence has been presented (Miketa and Mulder, 2005; Mulder and de Groot, 2007; Ezcurra 2007). Overall, these authors found clear patterns of energy productivity convergence for countries participating in the Organisation for Economic Co-operation and Development (OECD) and non-OECD countries, with energy productivity levels converging to country-specific steady states and exact results depending on the level of sectoral detail. Mulder and de Groot (2007) report that cross-country variation of productivity levels is typically larger for energy than for labour. In this chapter we extend these analyses by explicitly controlling for spatial effects in testing for convergence of energy productivity across countries, using exploratory spatial data analysis and spatial econometric techniques. Throughout the chapter we distinguish two levels of aggregation: the macroeconomic level and the sectoral level, specifically the sectors agriculture, manufacturing and services.

This chapter proceeds as follows. In section 2 we provide a brief description of the data used. In section 3 we present a global perspective on key trends in energy productivity dynamics across countries and regions during the last decades. Section 4 offers an exploratory spatial perspective on energy productivity growth, including the identification of spatial autocorrelation, regional clusters and the tendency towards equalization of cross-country variation in energy productivity levels. In section 5, we provide a spatial econometric perspective by testing the hypothesis that countries with relatively low initial productivity levels grow relatively fast, thus implying that relatively backward countries catch up to more advanced countries. In doing so, we explicitly control for the role of spatial effects. Section 6 concludes.

2 DATA

For the analysis presented in this chapter we use data on energy consumption and GDP for OECD and non-OECD countries. The source of the energy data are the Energy Balances 1971–2001 for OECD and non-OECD countries as published by the International Energy Agency (IEA). We use data on total final energy consumption, thus excluding transformation and transport losses. When analysing trends and growth rates we exclude end-use consumption of renewable energy types, including biomass, due to incomplete recording of these data in the IEA Energy Balances.[2] Whereas over the last decades wind or solar energy account for a very small fraction of final energy consumption, the exclusion of biomass admittedly is a serious drawback because in many developing countries biomass is still the principle component in the energy mix. Hence, unless stated otherwise, in this chapter energy (productivity) must be interpreted as commercial energy (productivity). Nevertheless it should be kept in mind that the reported energy productivity growth rates of developed and developing countries are somewhat difficult to compare.

We obtained our GDP data also from OECD/IEA, who used the World Development Indicators (2003) as the main source for non-OECD countries, and the OECD publication National Accounts, Volume 1 (2003) for OECD countries. Data for the Czech Republic and Poland prior to 1990, for Hungary prior to 1991, and for the Slovak Republic prior to 1992 are IEA Secretariat estimates based on GDP growth rates from the World Bank. The GDP data have been compiled by the IEA for individual countries at market prices in local currency and annual rates. These data have been scaled up/down to the price levels of 1995 and then converted to US dollars using the yearly average 1995 exchange rates.[3]

Throughout this chapter we distinguish two levels of analysis: the macroeconomic level and the sectoral level, distinguishing the sectors agriculture, manufacturing, and services. At the macroeconomic level we use data for the period 1971–2001 while the analyses for the three sectors are confined to the period 1981–99 due to limited data availability. Tables 2.1 and 2.2 present the key descriptive statistics of our dataset. At the macroeconomic level our dataset includes 109 countries representing 5.9 billion people in 2001, which is about 95 per cent of the total world population. At the sectoral level our samples are smaller because of data restrictions: for the sectors agriculture, manufacturing and services our dataset includes 46, 71 and 57 countries, respectively. A list of countries per sector is provided in Table 2A.1 in the Appendix. In addition, Table 2A.2 in the Appendix provides a more detailed picture by differentiating the indicators presented in Table 2.1 across nine world regions.

From Table 2.1 it can be seen that final (commercial) energy consumption per capita has increased substantially over the last decades, on average with about 55 per cent between 1971 and 2001. Roughly the same applies for average GDP per capita, thus implying that average total energy productivity, measured as GDP per unit of energy, has been approximately constant during this 30-year period. However, Table 2.1 also indicates substantial variation across countries. This is true with respect to energy consumption and income per capita, which necessarily implies that energy productivity levels vary substantially across countries. In our dataset in 2001 the gap between best practice and the country with the lowest total energy productivity level is about 25 per cent. Table 2.1 also shows that the picture is not much different at the sectoral level. In Table 2.2 we present descriptive statistics of average annual growth rates of energy productivity, as obtained from our dataset.[4]

Table 2.2 shows that, on average, annual energy productivity growth rates have been negative during the last decade, except for the manufacturing sector. Again, however, these aggregate figures mask substantial variation across countries. For example, at the macroeconomic level average annual growth rates over the period 1971–2001 range from −9.2 to 4.6 per cent. A more complete picture of these results will be presented in the next section.

3 A GLOBAL PERSPECTIVE

The provision and use of (commercial) energy bears a close relationship to economic development. This has two straightforward but important implications: like economic development, energy consumption patterns vary greatly across countries, and they are inherently dynamic. To

Table 2.1 Descriptive statistics for energy productivity

	1971				2001				Number of countries
	Mean	Min.	Max.	St. dev.	Mean	Min.	Max.	St. dev.	
Population (million)	33.4	0.1	841.1	100.4	54.2	0.3	1271.9	160.1	109
Final energy consumption per capita (toe)	1.07	0.00	7.33	1.44	1.66	0.02	15.29	2.21	109
Final energy consumption per capita, including renewable (toe)	1.15	0.01	7.33	1.44	1.81	0.13	15.29	2.19	109
GDP per capita (US$)	5967	114	37467	8297	9530	85	57201	12856	109
	1971				2001				
Energy productivity (ln)[a]									
Macroeconomic level	15.4	11.9	17.9	0.880	15.3	13.1	16.6	0.659	109
	1981				1999				
Energy productivity (ln)[a]									
Services	17.9	16.0	19.6	0.904	17.6	15.7	19.3	0.789	57
Manufacturing	15.3	12.8	18.1	0.882	15.4	13.4	17.1	0.720	71
Agriculture	16.5	14.5	18.6	1.029	16.2	14.6	18.3	0.812	47

Note: [a] Energy productivity is defined as the ratio of output (US$) to energy input (ktoe).

Table 2.2 Descriptive statistics for energy productivity growth rates[a]

	Mean	Median	Min.	Max.	St. dev.	Number of countries
Macroeconomic						
1971–2001	−0.4	−0.3	−9.2	4.6	2.4	109
1971–86	−0.6	0.3	−11.7	6.3	3.3	109
1986–2001	−0.2	−0.5	−7.8	6.8	2.3	109
Services						
1981–99	−1.7	−2.0	−8.2	8.1	3.2	57
Manufacturing						
1981–99	0.2	0.5	−10.0	8.7	2.8	71
Agriculture						
1981–99	−1.3	−1.1	−9.6	6.2	3.8	47

Note: [a] Percentages, except for the last column.

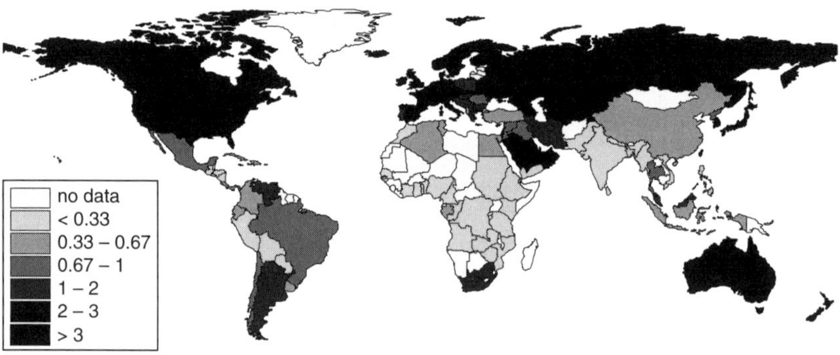

Figure 2.1 Per capita total final energy consumption in ton oil equivalent, 2001

illustrate the first point we depict in Figure 2.1 the spatial distribution of total final energy consumption per capita (including biomass and other renewables) in the world in 2001.[5]

Figure 2.1 presents an unequal distribution of per capita energy consumption, with highest levels of energy consumption per capita being concentrated in North America, Europe, the Middle East, Japan and Oceania, whereas the lowest levels of per capita energy consumption are concentrated in Latin America, Asia and particularly sub-Saharan Africa. Figure 2.1 thus resembles the strong positive relationship between energy

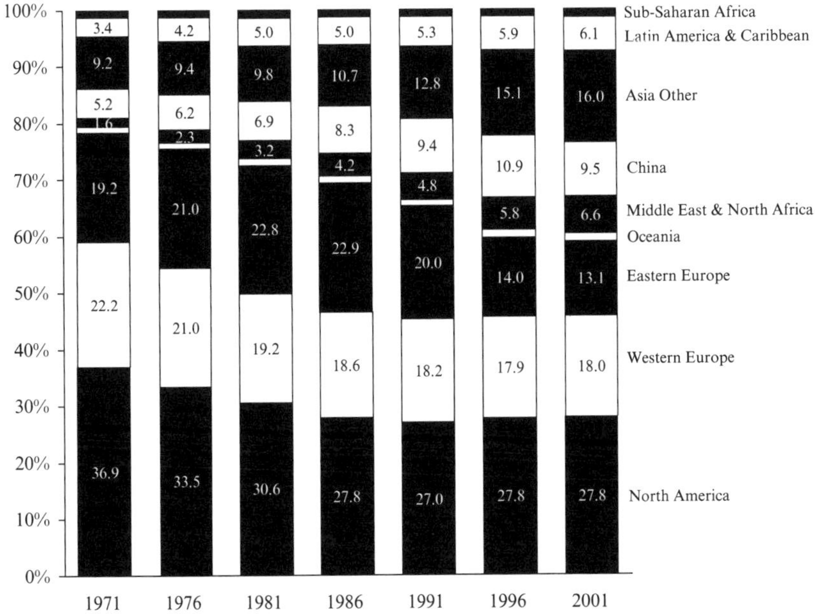

Figure 2.2 Share of regions in the world's final energy consumption, percentage

consumption and the level of economic advancement and prosperity. Moreover, Figure 2.1 shows that levels of energy consumption in various Middle East countries are very high given their income per capita, indicating a highly energy-intensive economic structure, as we will also see below.

To illustrate the dynamic nature of energy consumption patterns around the world, we provide in Figure 2.2 an overview of the share of various world regions in the world's total energy consumption between 1971 and 2001. The figure shows that whereas North America and Western Europe together used to be responsible for about 60 per cent of total energy consumption in 1971, in 2001 this share has dropped to about 45 per cent. The collapse of Eastern European economies following the disintegration of the former Soviet Union is reflected by a substantial decrease of their energy consumption share after 1991. These declining shares are mirrored by increasing shares in the world's total energy use for China and other parts of Asia, reflecting the emerging role of Asia in the world economy. In 2001, Asia accounted for about 25 per cent of total energy consumption in the world, as opposed to about 14 per cent in 1971.

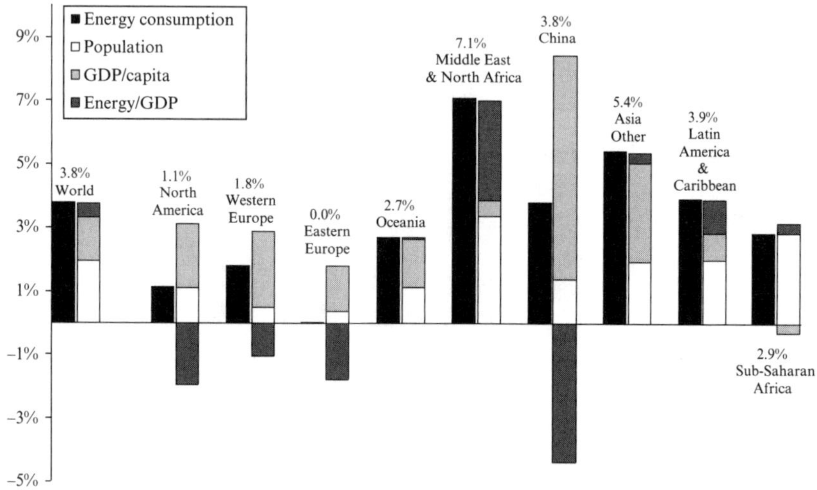

Figure 2.3 Total final energy consumption growth and its determinants, percentage

Energy consumption is driven by three principal factors: population growth, economic growth and technological change. This relationship can be captured in a simple definition, as follows:

$$Energy \equiv Population \cdot \frac{GDP}{Population} \cdot \frac{Energy}{GDP} \qquad (2.1)$$

An increase in the first and second term on the right-hand side is often referred to as a scale effect, equivalent to the product of population growth and growth of GDP per capita (economic growth). A change in the third term on the right-hand side (energy intensity) is primarily driven by technological change, either directly in the form of energy-saving technologies or indirectly through technology-driven or demand-driven changes in the sectoral composition of an economy and/or changes in the energy mix. This factor is therefore often called an efficiency effect. To illustrate the relative importance of these different factors in determining changing energy consumption patterns across the world, Figure 2.3 shows the relative contribution of population growth, economic growth and changes in energy intensity to the (unweighted) average annual growth in energy consumption during the period 1971–2001. We do this for all 109 countries in our sample, using the label 'World', as well as for nine different world regions.[6]

Figure 2.3 shows that for the total sample ('World') energy consumption grew at an average of 3.8 per cent per year, resulting mainly from population growth (about 2 per cent) and income growth (about 1.5 per cent), while increasing energy intensity has been responsible for only approximately 0.3 per cent. Figure 2.3 shows a very different picture for the various world regions. Average annual energy consumption growth has been modest in the western world (North America, Western Europe, Oceania) and sub-Saharan Africa, virtually zero in Eastern Europe, strong in Asia and Latin America and very strong in the Middle East and North Africa. In most regions, increasing per capita income has been the key driving force behind growing energy consumption, except for the regions Middle East and North Africa and sub-Saharan Africa. In the Middle East and North Africa region energy consumption growth has been on average the highest of all regions (7.1 per cent), most of which is to be explained by population growth and the increasing energy intensity of its economies. In sub-Saharan Africa population growth explains virtually all growth in energy consumption, while it is the only region where the average annual change in per capita income has been negative. Finally, Figure 2.3 also shows a mixed picture in terms of energy intensity changes. Whereas energy intensity has decreased on average in North America, Europe, and particularly China, it has increased in the other regions, and especially in the Middle East and North Africa. For the world as a whole this results in slightly increasing energy intensity, and thus in a negative average annual growth rate of energy productivity as reported in Table 2.2.

If we divide both sides of equation (2.1) by population it follows that per capita energy consumption is driven by the product of per capita income and energy intensity:

$$\frac{Energy}{Population} \equiv \frac{GDP}{Population} \cdot \frac{Energy}{GDP} \tag{2.2}$$

Accordingly we present in Figure 2.4 a decomposition of the average annual growth rate of per capita energy consumption, similar to the analysis provided in Figure 2.3. Figure 2.4 shows that for the total sample ('World') per capita energy consumption grew each year with 1.8 per cent on average, resulting mainly from income growth (approximately 1.4 per cent) while increasing energy intensity has been responsible for the remaining part (approximately 0.4 per cent). The figure shows again a very different picture for the various world regions, and clearly illustrates that increasing energy intensities have contributed to per capita energy consumption, particularly in the region Middle East and North Africa, while decreasing energy intensities have slowed down growth in per capita energy consumption in North America, Europe and China. In order to get an idea of how much energy has been saved as a result of decreasing energy

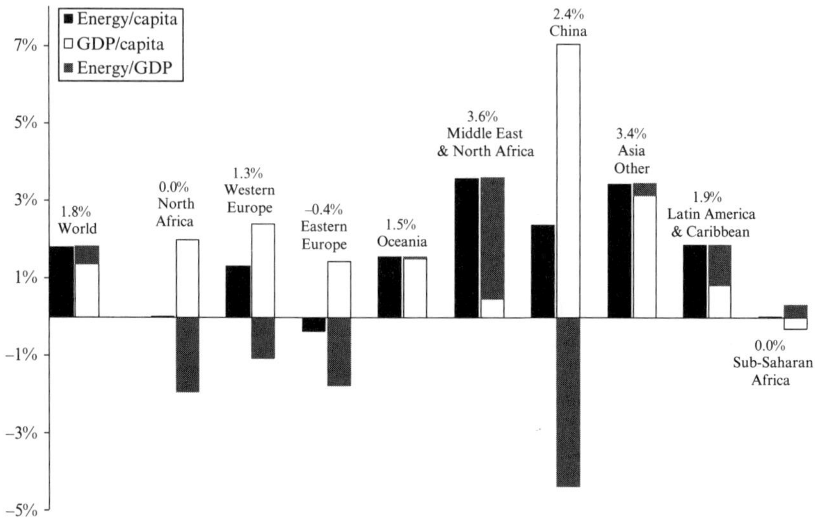

Figure 2.4 Per capita energy consumption growth and its determinants, percentage

intensity levels we have calculated energy consumption in the various world regions according to equation (2.1), keeping the energy intensity ratio (the third term on the right-hand side) constant at its 1971 level.

In Figure 2.5 we compare the energy consumption level obtained in this manner with actual energy consumption. From Figure 2.5 it can be concluded that if energy intensity levels had remained at their 1971 levels, total energy consumption in the world would have been about 60 per cent higher in 2001. Energy productivity improvements played a particularly dramatic role in China, where energy consumption would have been more than 260 per cent higher in 2001 if no energy productivity improvements had been realized since 1971. This large improvement in China's energy productivity results from a combination of factors: more efficient use of coal, switch from coal to oil, industry restructuring (rapid growth of equipment manufacturing industries) and higher energy prices (WEC, 2008, p. 17). Summarizing, in most parts of the world energy productivity improvements have clearly played an essential role in combining economic growth, energy security and mitigation of climate change.

In order to provide a more detailed picture of the cross-country variation in energy productivity trends, Figure 2.6 provides a geographical image of the macroeconomic average annual growth rates per country, over the period 1971–2001.[7] It can be seen that growth rates differ substantially, ranging from −9.2 per cent for Iraq to 4.6 per cent for China.

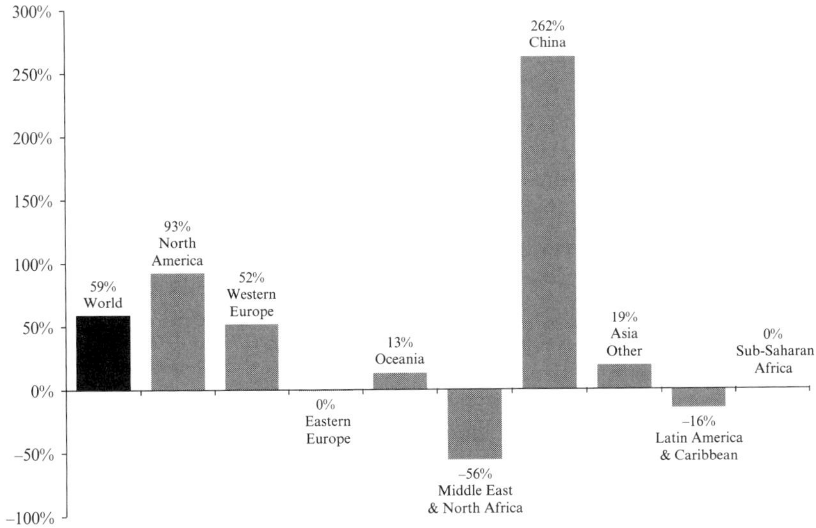

Figure 2.5 *Energy savings from energy productivity increase, percentage, 1971–2001*

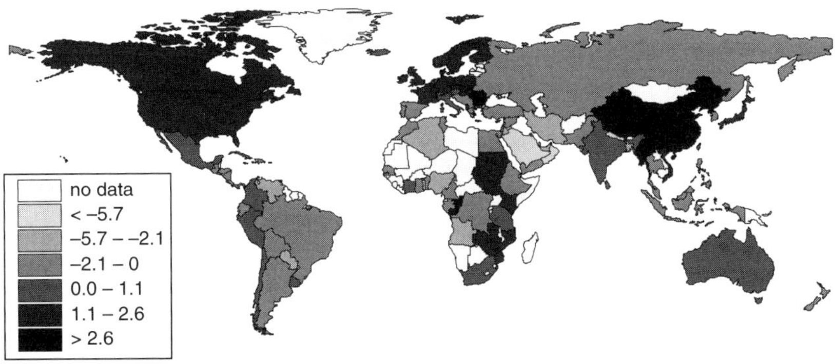

Figure 2.6 *Average annual growth rates energy productivity, percentage, 1971–2001*

In general, the highest average growth rates are to be found in China, followed by Japan, Europe and North America, as well as in several African countries, although the latter results should be interpreted with caution due to the exclusion of biomass data. In addition one should note that various countries display negative growth rates, especially in the Middle East.

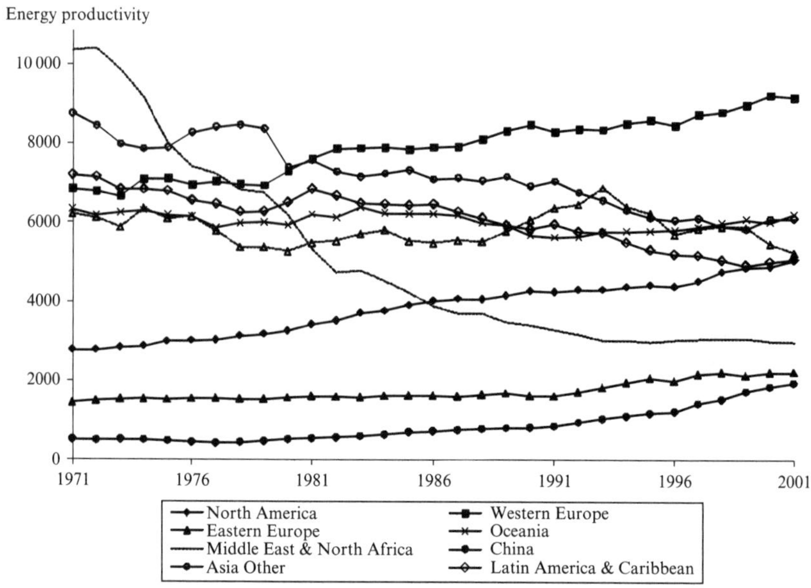

Figure 2.7 Levels of energy productivity in 1000 ktoe/US$ per region, 1971–2001

High growth rates are not equivalent to high levels of energy productivity, and therefore we plot in Figure 2.7 the evolution of energy productivity levels during the period 1971–2001 for each region. The figure clearly shows that the highest levels of energy productivity can be found in the countries of Western Europe, while China is still located at the bottom end of the distribution despite its impressive growth in energy productivity performance during the past 15 years. These results clearly indicate that China is catching up to more energy efficient countries. For most regions, however, identifying such patterns is less obvious, with Western Europe as an important counter example in that it combines a high productivity level with a relatively high growth rate. The latter result suggests the existence of a European cluster consisting of a concentration of countries showing evidence of high performance in terms of energy productivity, may be stimulated by spillover effects. In the following sections we explore these issues in more detail. Figure 2.7 also clearly illustrates the dramatically declining performance of the Middle East and North Africa in terms of energy productivity.

As noted before, energy productivity improvements are to a large extent driven by technological change, either directly in the form of energy-saving technologies, or indirectly through technology-driven or demand-driven changes in the sectoral composition of an economy and/or changes in the energy mix. The latter effects are often referred to as composition effects. They arise from the fact that value added per unit of (energy) input differs substantially across different economic activities and thus across sectors, and/or from the fact that some energy carriers (for example, natural gas) are inherently more efficient than others (for example, coal) (Berndt, 1978; Cleveland et al., 2000; Kaufmann, 2004). The past decades are in general characterized by a shift away from the use of coal and oil towards the use of natural gas and electricity. With respect to the sectoral composition of production we note that over the past few decades the share of agriculture and manufacturing in total GDP has been declining in favour of an increasing share for services in most countries, with African countries being the main exception. These issues are explored further in Chapter 3 of this volume, where the results of a decomposition analysis for 14 OECD countries are presented, which identify the impact of sectoral shifts on aggregate energy productivity performance. Concurrent regression analyses identify the significant impact of changes in fuel mix on sectoral energy productivity growth, controlling for other potential relevant factors such as energy prices, economies of scale, investment share, openness and specialization.

Finally, our data on energy productivity growth rates for the sectors agriculture, manufacturing and services, averaged over the period 1981–99, reveal that also at the sectoral level China's energy productivity growth rates are among the highest in the world. Moreover, several African countries demonstrate high growth rates at the sectoral level, although these results should again be interpreted with caution due to the exclusion of biomass data. The suggestion of a well-performing European cluster is confirmed for the manufacturing sector only; in the agricultural and service sectors, cross-country variation of growth rates is substantial among European countries. In general, studying energy productivity developments in the service sector is seriously hampered by lack of data. This issue is addressed further in Chapter 4, where the results of an in-depth study of energy efficiency developments in the Dutch service sector during a period of about 15 years are presented.

4 AN EXPLORATORY SPATIAL PERSPECTIVE

In this section we provide an exploratory spatial analysis of energy productivity growth rates across countries. To explore the existence of clusters or spatial non-stationarity in a systematic way we calculate the so-called univariate Moran's I statistic, a frequently used measure to assess the degree to which data are spatially autocorrelated (Cliff and Ord, 1981). It is defined as:

$$I_t = \frac{N}{S_0} \frac{\sum_{i=1}^{N} \sum_{j=1}^{N} w_{ij}(x_{it} - \bar{x}_t)(x_{jt} - \bar{x}_t)}{\sum_{i=1}^{n} (x_{it} - \bar{x}_t)^2} \tag{2.3}$$

where N is the number of countries, S_0 a scaling factor equal to the sum of the elements of the spatial weight matrix, x_{it} a variable for country i at time t, $\bar{x}_t$ the mean of variable x_{it} at time t, and w_{ij} the element in row i and column j of a spatial weights matrix. A statistically significant positive value of Moran's I statistic indicates the occurrence of similar (positive or negative) values of a variable over space, and hence spatial clustering, which could not have been caused by chance. A statistically significant negative value indicates the joint occurrence of high and low values of a variable in close-by locations that could not have been caused by chance.

The spatial weight matrix contains information on the spatial configuration of the countries in the sample. Following Abreu et al. (2004) the matrix has been constructed as follows:

$$\begin{cases} w_{ij} = d_{ij}^{-1} & \text{if } d_{ij} < 6000 \text{ miles} \\ w_{ij} = 0 & \text{otherwise} \end{cases}$$

where d_{ij} is the great-circle distance between the centroids of countries i and j in miles. It is assumed that the diagonal elements of the spatial weights matrix are zero, implying absence of self-association. The elements of the spatial weight matrix have been row-standardized, so that for each i the sum of all elements over j equals unity, in order to ensure that the parameter space is compact and the parameters are comparable across spatial models. The critical cut-off distance of 6000 miles is imposed to limit the extent to which new data points change the connectedness of existing data points, motivated by compliance with the regularity conditions, a requirement to ensure consistency of the estimator in spatial models (Anselin, 2001, 2002). In addition, this cut-off distance is motivated by the location of the world's major technological centres: the distances between the USA, Europe, Japan, Brazil and South Africa all fall within the range of 6000 miles.

Table 2.3 lists the Moran's I statistics for energy productivity growth rates and their associated z-values and probability values. It can be seen that the values of Moran's I statistic are positive, except for the agricul-

Table 2.3 Moran's I *statistics of energy productivity growth rates*

	Moran's *I*	St. dev.	*z*-value	Probability
Macroeconomic				
1971–2001	0.1412	0.0213	7.0801	0.0000
1971–86	0.1254	0.0213	6.3375	0.0000
1986–2001	0.1439	0.0213	7.2037	0.0000
Services				
1981–99	0.0289	0.0416	1.1239	0.2610
Manufacturing				
1981–99	0.0089	0.0327	0.7081	0.4789
Agriculture				
1981–99	−0.0042	0.0439	0.3988	0.6901

tural sector, indicating positive spatial autocorrelation. At the macroeconomic level the value is significant at the 1 per cent level. However, the reported values for the manufacturing, agricultural and service sector are not significant at the 10 per cent level. In other words, at the macroeconomic level we find similar energy productivity growth rates in nearby countries, which could not have been caused by chance, while at the sectoral level the possibility of a random allocation of energy productivity growth rates over space cannot be rejected.

A Moran scatterplot can help to identify the existence of spatial clusters, by plotting the (standardized) energy productivity growth rate of a country against its (standardized) spatial lag. A country's spatial lag is a weighted average of the energy productivity growth rates of its neighbouring countries, with the weights taken from the spatial weight matrix. The four quadrants in the scatterplot identify four types of local spatial association: quadrant I (upper-right) includes countries with high values (growth rates) for themselves and their neighbours, quadrant II (lower-right) includes countries with high values that have neighbours with low values, quadrant III (lower-left) includes countries with low values for themselves and their neighbours, and quadrant IV (upper-left) includes countries with low values that have neighbours with high values. Therefore, quadrant I and III represent positive local spatial dependence, while the other two quadrants represent a negative form of spatial autocorrelation.

Figure 2.8 is a Moran scatterplot of energy productivity growth rates at the macroeconomic level. The figure confirms the relatively high degree of spatial autocorrelation reported in Table 2.3: the observations are clustered around an upward sloping trend line. More specifically, the figure shows a regional cluster of European countries in the high-high quadrant as well as cluster of countries in the Middle East in the low-low quadrant.

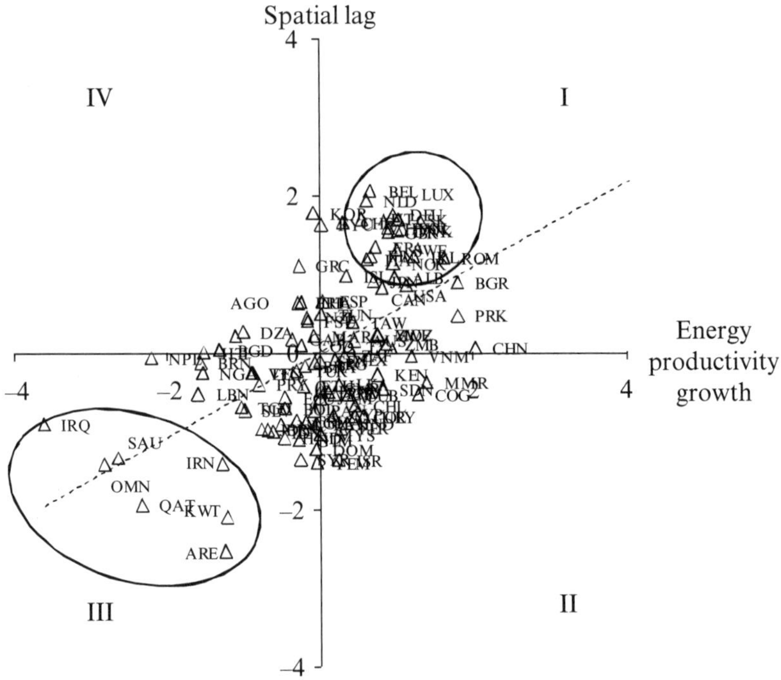

Note: Both variables are expressed in deviations from the mean and divided by the standard deviation of the variable on the *x*-axis.

Figure 2.8 Moran scatterplot of average annual energy productivity growth, 1971–2001

In other words, most European countries are characterized by relatively high positive energy productivity growth rates for themselves and their neighbours, while the opposite is true for countries in the Middle East. At the same time a large group of countries is concentrated in the lower-right quadrant (quadrant II), having (relatively high) positive energy productivity growth rates whereas their neighbours exhibit relatively low energy productivity growth rates. These are mainly middle income countries, both from Latin America as well as Asia.

With respect to the underlying determinants of regional clusters it should be noted that in the macroeconomic literature it has been suggested that technology diffusion and knowledge spillovers are local rather than global phenomena (Caniëls and Verspagen, 2001; Keller, 2002). Moreover, cross-country analyses of productivity dynamics have yielded the insight that technology diffusion and trade might lead to convergence

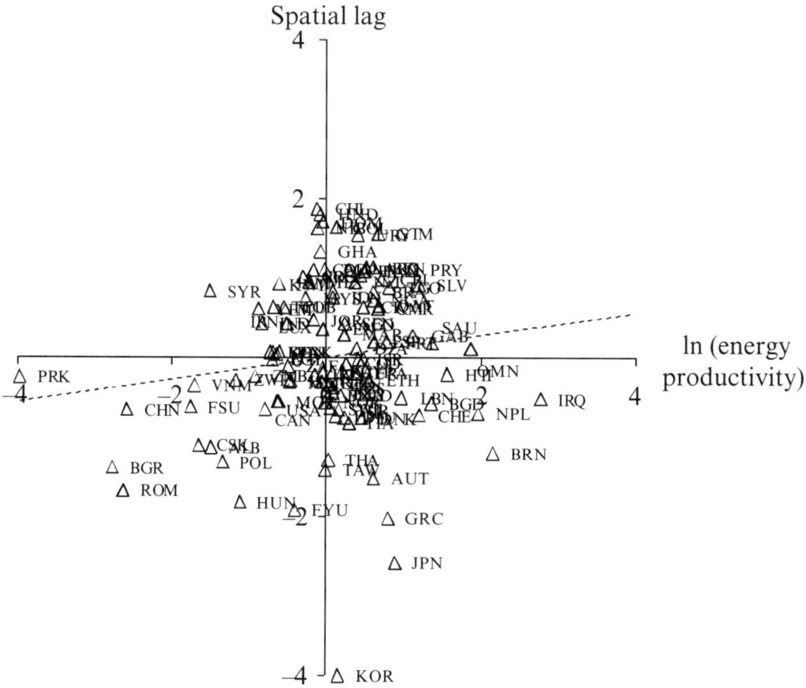

Figure 2.9 Moran scatterplot of the logarithm of energy productivity in 1971

clubs: groups of countries converging to different steady states (see, for example, Chatterji, 1992; Chatterji et al., 1993; Durlauf and Johnson, 1992; Quah, 1997). These phenomena might thus explain the formation of geographically concentrated clusters of countries with a similar energy productivity performance. The observed spatial autocorrelation of energy productivity levels among European countries suggests that relatively energy inefficient countries have been able to catch up to more energy efficient countries. Concurrently, the poor performing cluster of countries in the Middle East suggests the absence of spatially motivated interaction directed towards energy productivity improvements, such as the diffusion of energy-saving technologies. In the next section we further explore the issue of convergence and catching up by means of regression analyses.

The use of standardized variables allows us to compare Moran scatterplots over time. Figures 2.9 and 2.10 present Moran scatterplots of macroeconomic energy productivity levels in 1971 and 2001, respectively. Comparing the two figures yields a somewhat paradoxical result: while

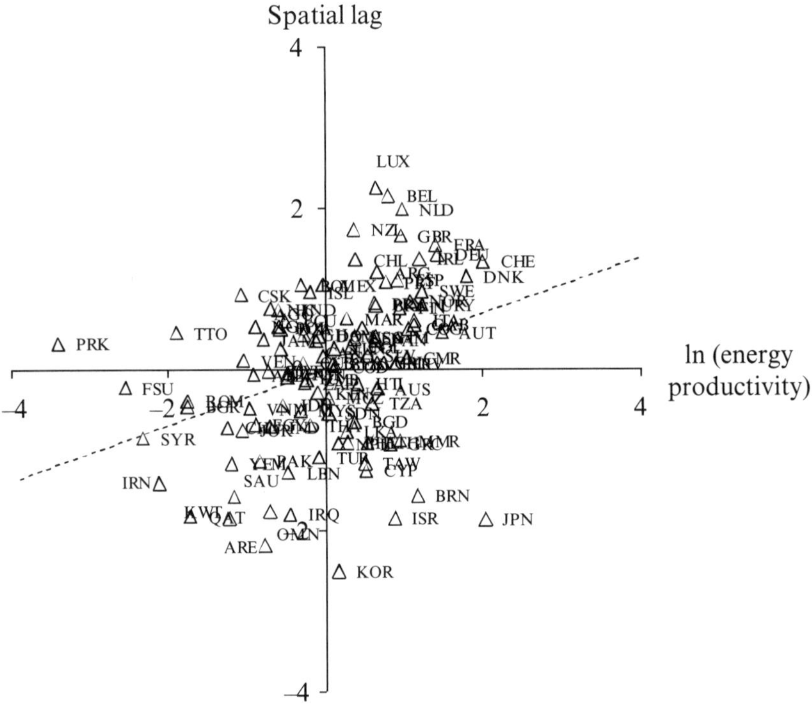

Figure 2.10 Moran scatterplot of the logarithm of energy productivity in 2001

overall spatial autocorrelation has increased over time, the spatial distribution of energy productivity levels is more clustered in 1971 than in 2001. In 1971 several European countries formed a cluster in the lower-right quadrant and several Latin America countries formed a cluster in the high-high quadrant. In 2001 the number of local outliers has decreased in favour of concentration of European and Middle Eastern countries at the two extremes of the spatial distribution, with most other countries scattered in between. Together with the results shown in Figure 2.8, this leads to a number of observations. First, most Western European countries have increased their energy productivity over time and have become more homogeneous as a group. Second, various Eastern European countries like Czechoslovakia, Poland, Hungary, the former Soviet Union and former Yugoslavia as well as their direct neighbours are characterized by low levels of energy productivity in 1971 in combination with relatively high energy productivity growth rates. This suggests that Eastern European countries have been able to catch up to Western European nations. Third, the positive spatial dependence of 1971 energy productivity levels in various Latin American countries has not been preserved over time: in 2001, Latin American energy productivity levels are typically relatively low and have diverged.

However, our data reveal that the picture is rather different at the sectoral level (not shown here). Within the service sector, the distribution of energy productivity growth rates does show some degree of spatial autocorrelation, but in terms of energy productivity levels the degree of spatial autocorrelation is close to zero, both in 1971 and in 2001. For the manufacturing and agricultural sectors the picture is exactly the opposite: although energy productivity growth rates are more or less randomly distributed over space, energy productivity levels do exhibit some degree of spatial autocorrelation, both in 1971 and in 2001. The sectoral results also reveal that in the manufacturing sector most European countries have increased their energy productivity over time and have become more homogeneous as a group.

The observed spatial autocorrelation in energy productivity growth rates and energy productivity levels across countries, particularly at the macroeconomic level, raises the question as to whether our data support the occurrence of a long-run tendency towards equalization of energy productivity levels across countries. If so, to what extent does such a convergence process feature a geographical dimension? To answer these questions we measure cross-country variation in energy productivity over time (σ) defined in terms of the unweighted cross-country standard deviation of the log of energy productivity. In addition we calculated the Moran's *I* statistic of the log of energy productivity over time. For each sample, both indicators are shown in Figure 2.11.

Figure 2.11 shows a decreasing standard deviation of the log of energy productivity performance at the macroeconomic level as well as at the sectoral level, indicating a clear pattern of σ-convergence. For the sectors services and agriculture these results corroborate the findings of Mulder and de Groot (2007) who for the period 1975–95 also reported a pattern of σ-convergence in these sectors, albeit for a sample of 14 OECD countries. Moreover, the observed σ-convergence in the industry sector confirms the results reported by Miketa and Mulder (2005), who found for a sample of 56 countries and 10 manufacturing sectors that, except for the non-ferrous metals sector, cross-country differences in absolute energy productivity levels tend to decline over the period 1971 to 1995, particularly in the less energy-intensive industries. The observed decline in cross-country differences at the industry and macroeconomic level is, however, in contrast to the findings of Mulder and de Groot (2007) for a group of 14 OECD countries, which is likely due to the substantial difference in sample used in their study and in the analysis reported here.

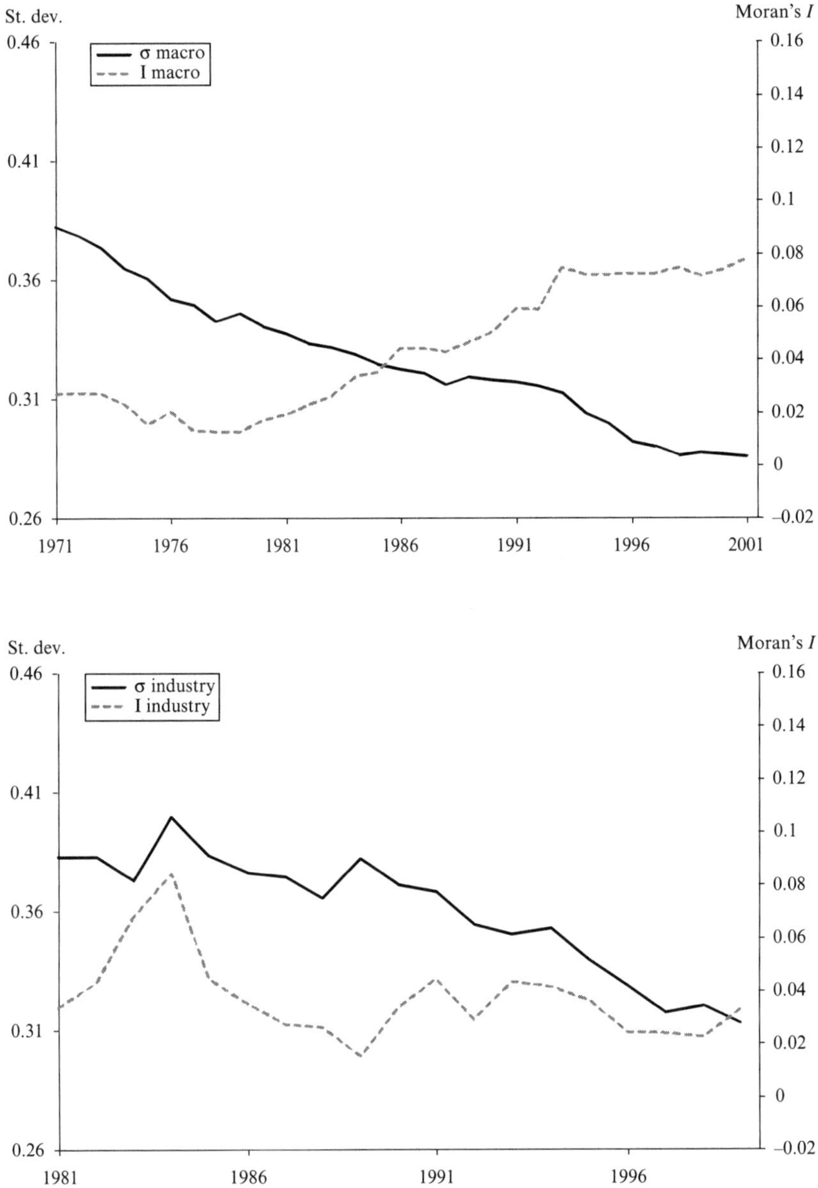

Figure 2.11 *σ-convergence energy productivity (st. dev.) and spatial autocorrelation (Moran's* I*) by sector*

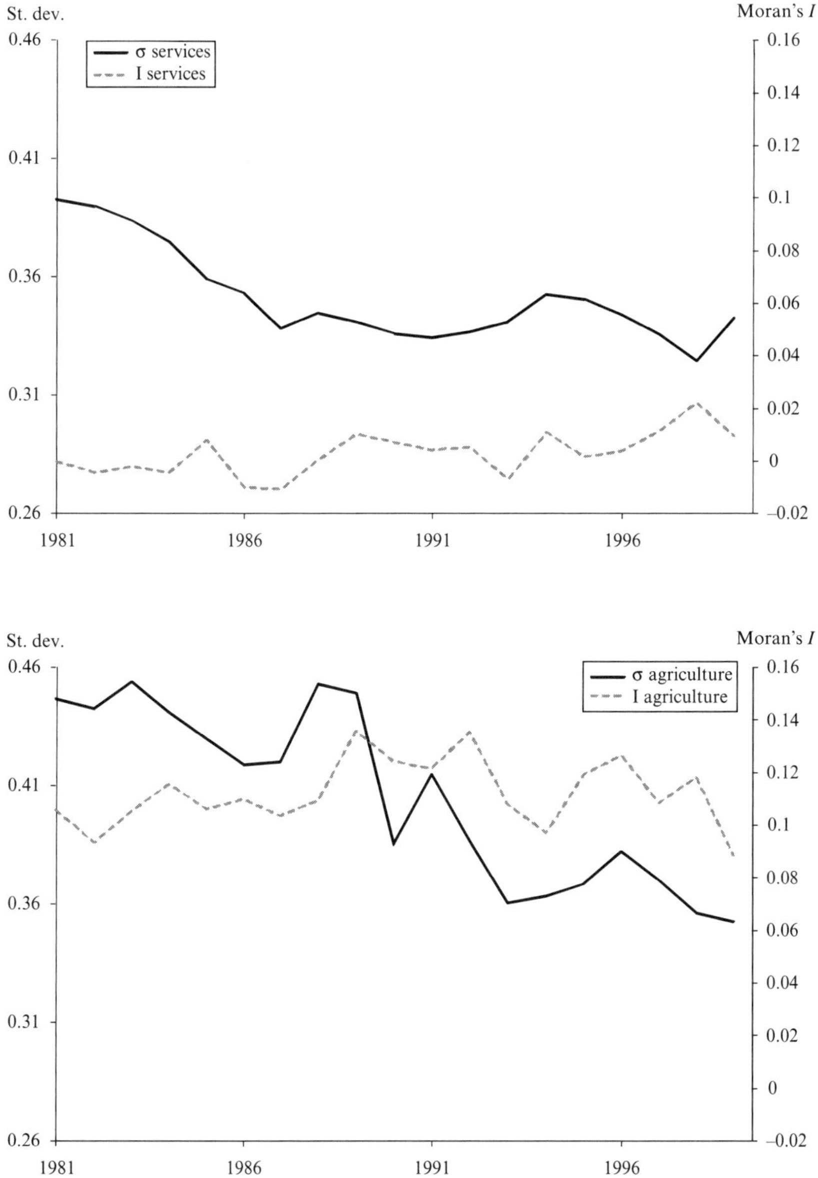

Figure 2.11 (continued)

In addition to the evolution of the standard deviation of the log of energy productivity, Figure 2.11 also shows Moran's *I* statistic of the log of energy productivity over time. It can be seen that at the macroeconomic level our data display positive and (strongly) increasing values of the Moran's *I* statistic, indicating a (strong) increase in spatial autocorrelation over time. In addition, the results indicate that in the service and agricultural sectors spatial autocorrelation for our sample of countries tends to remain fairly constant over time, despite annual fluctuations. Note, however, that for the sector services the Moran's *I* statistic is statistically insignificant at the 10 per cent level for the whole period under consideration. Finally, the manufacturing sectors exhibit a sharp increase of spatial autocorrelation in the beginning of the 1980s, after which the value of the Moran's *I* decreases gradually to its 1981 level.

In the next section we further explore the nature of convergence patterns by examining whether countries with relatively low initial energy productivity levels tend to grow relatively fast, assuming we adequately control for possible spatial effects.

5 A SPATIAL ECONOMETRIC PERSPECTIVE

The empirical macroeconomic growth literature has presented substantial evidence that productivity growth rates are inversely correlated with initial productivity levels: countries with relatively low initial productivity levels tend to grow relatively fast, and vice versa. This concept is usually referred to as β-convergence, suggesting that countries with relatively low initial energy productivity levels catch up to more advanced countries. It builds upon the proposition that growth rates tend to decline as countries approach their steady state and hence this concept can be refined by assuming either convergence towards a unique steady state (unconditional or absolute convergence) or towards multiple country-specific or region-specific steady states (conditional or relative convergence).[8]

The standard approach to test for β-convergence consists of regressing the energy productivity growth rate on the natural logarithm of its initial level, and a constant. A statistically significant negative relationship is then taken as evidence in favour of β-convergence. It is to be noted that the OLS estimation method is valid only under the assumption that the error term is independent of the explanatory variable to prevent the estimation from being biased and inconsistent (Hsiao, 1986). This requirement, however, is unlikely to be met in this case since the error term in a simple cross-country regression does contain various (unobserved) country-specific factors, such as technological change, that are most likely to be correlated with the initial level energy productivity. This problem can be dealt with by using a random effects model within our panel data structure and leads to the following specification:

$$g_{it} = \mu + \beta \ln y_{it-1} + \delta_{it}, \text{ with } \delta_{it} = \alpha_i + \varepsilon_{it} \tag{2.4}$$

where i (= 1, 2, . . ., N) and t (= 1, 2, . . ., T) denote the cross-country and the time series dimension respectively, while α_i and ε_{it} represent the time-invariant cross-sectional component and the combined error component, respectively. The inclusion of time-invariant individual country effects (α_i) allows for cross-country differences in steady states in the form of unobservable individual country effects, thus diminishing the omitted variables problem and relaxing the otherwise strong implicit assumption of convergence to a uniform steady state. However, equation (2.4) assumes zero correlation between the individual country effects (α_i) and the observed explanatory variable y_{t-1}, which is problematic as we noted above. We can solve this issue by using a fixed effects panel data model to test for conditional or relative convergence according to:

$$g_{it} = \alpha_i + \beta \ln y_{it-1} + \varepsilon_{it} \tag{2.5}$$

again with α_i representing unspecified country-specific (fixed) effects. Hence, we may assume ε_{it} to be an independently and identically distributed random variable with mean 0 and variance σ^2_ε.

In Section 2.4 we found growth rates of energy productivity to be positive spatially autocorrelated, particularly at the macroeconomic level. However, it remains to be seen whether spatial effects still play a role in explaining cross-country differences in energy productivity growth rates after controlling for initial energy productivity levels. To test for spatial autocorrelation in a convergence framework we can estimate a standard cross-section model using the following specification where T sub-vectors with N cross-sectional observations are stacked:

$$g_t = \alpha + \beta \ln y_{t-1} + \varepsilon_t \tag{2.6}$$

with the ordinary least squares estimator, and subsequently use the results to identify spatial dependence by checking several diagnostic statistics. If this exercise confirms the existence of spatial dependence, the next question can then deal with the question as to which transmission channel is responsible for creating spatial effects in energy productivity growth. The exploratory spatial analysis in section 2.4 indicated that, particularly at the macroeconomic level, countries with high energy productivity growth rates tend to have neighbours with similar high energy productivity growth rates, presumably because of the existence of some sort of spatial interaction such as knowledge spillovers. The relevance of this potential transmission channel can be tested by means of a so-called spatial lag model. The spatial lag model reads as:

$$g_t = \alpha + \beta \ln y_{t-1} + \rho W g_t + \varepsilon_t \tag{2.7a}$$

where ρ is a parameter indicating the extent of spatial correlation between the energy productivity growth rates (g_t) of countries, and ε_t a well-behaved error term. In other words, a country's energy productivity growth rate depends not only on its own initial energy productivity level, but also on the spatially weighted average of the energy productivity growth rates of its neighbouring countries. This can be illustrated by using $(1 - \rho W) g_t = \alpha + \beta \ln y_{t-1} + \varepsilon_t$ to rewrite equation (2.7a) such that:

$$g_t = (1 - \rho W)^{-1}(\alpha + \beta \ln y_{t-1} + \varepsilon_t) \tag{2.7b}$$

Spatial dependence can also work through omitted variables with a spatial dimension, suggesting the existence of a 'spatial measure of our ignorance'. We can test this hypothesis by estimating the relationship between a country's initial level of energy productivity and its subsequent growth rate by means of a so-called spatial error model according to:

$$g_t = \alpha + \beta \ln y_{t-1} + \mu_t, \text{ with } \mu_t = \lambda W \mu_t + \varepsilon_t \tag{2.8a}$$

where λ is a parameter indicating the extent of the spatial correlation between the errors, W is the exogenously determined spatial weight matrix and ε_t a well-behaved error term. Note that since $\mu_t = (I - \lambda W)^{-1} \varepsilon_t$ equation (2.8a) can be rewritten as:

$$g_t = \alpha + \beta \ln y_{t-1} + (1 - \lambda W)^{-1} \varepsilon_t \tag{2.8b}$$

which immediately demonstrates the role of spatial dependence in the error term.

In sum, our estimation strategy is the following. First, we test for β-convergence without controlling for spatial effects, using a panel data framework and distinguishing between a random and a fixed effect model according to equations (2.4) and (2.5). We use three-year time intervals in order to reduce the influence of business cycle fluctuations and serial correlation of the error term. We estimate this specification using a generalized method of moments (GMM) estimator and a least squares dummy vari-

Table 2.4a Estimates for β-convergence, macroeconomic level, 1971–2001[a]

	Panel Random effects (1)	Panel Fixed effects (2)	Cross-section OLS (3)	Cross-section Spatial lag (4)
Energy productivity 1971 (ln)	–0.028 (0.003)	–0.071 (0.005)	–0.019 (0.002)	–0.018 (0.002)
$W \times$ growth energy productivity				0.899 (0.063)
AIC			–554.971	–580.146
R^2	0.16	0.16	0.49	0.54
F-statistic (χ^2)	89.07	3.83	98.073	
$\tilde{\rho}$	0.145	0.527		
Hausman	102.580			
Jarque-Bera			3.836	
(Spatial) Breusch-Pagan			4.745	4.419
White			9.932	
Moran's *I*			8.157	
LM-error			41.604	
Robust LM-error			3.331	
LM-lag			43.897	
Robust LM-lag			5.625	
LR-lag				27.175
Number of countries	106	106	106	106
Number of observations	1060	1060	106	106

Note: [a] Standard errors in parentheses.

ables (LSDV) estimator are for the random and the fixed effects models, respectively. Next, we control for spatial effects, using the cross-section model given in equation (2.6) estimated with the ordinary least squares estimator as a point of reference, and subsequently discriminating between various types of spatial dependence, following equations (2.7a) and (2.8a). The results for the macroeconomic and the sectoral level are presented in Tables 2.4a–c and 2.5a–c, respectively.

With respect to the macroeconomic level, Tables 2.4a–c show the estimate of β for energy productivity growth to be negative and statistically significant at the 1per cent significance level for all models and in all three time periods, indicating a robust pattern of β-convergence. The higher

Table 2.4b Estimates for β-convergence, macroeconomic level, 1971–86[a]

	Panel Random effects (1)	Panel Fixed effects (2)	Cross-section OLS (3)	Cross-section Spatial lag (4)
Energy productivity 1971 (ln)	−0.028	−0.114	−0.021	−0.020
	(0.004)	(0.011)	(0.003)	(0.003)
$W \times$ growth energy productivity				0.872
				(0.079)
AIC			−461.226	−478.950
R^2	0.21	0.21	0.33	0.37
F-statistic (χ^2)	43.88	3.25	49.981	
$\tilde{\rho}$	0.231	0.760		
Hausman	76.430			
Jarque-Bera			15.291	
(Spatial) Breusch-Pagan			5.988	5.790
White			7.002	
Moran's *I*			7.109	
LM-error			31.035	
Robust LM-error			1.747	
LM-lag			31.143	
Robust LM-lag			1.855	
LR-lag				19.725
Number of countries	106	106	106	106
Number of observations	530	530	106	106

Note: [a] Standard errors in parentheses.

estimates of β in the fixed effects models, reported in columns labelled (2), as compared to the random effects model, reported in columns labelled (1), suggests that energy productivity convergence depends to a large extent on individual country effects. This is confirmed by the reported values of $\tilde{\rho}$, which indicate that individual country effects explain about 52 per cent of the total unexplained variance for the entire period under analysis and up to 85 per cent in the period 1986–2001, implying that energy productivity convergence is conditional rather than absolute. Moreover, the Hausman specification test comparing the fixed and random effects slope parameters signals that the random effects model is indeed estimated inconsistently, due to correlation between the explanatory variable and the error component.[9]

In the cross-section OLS model, reported in columns labelled (3), the

Table 2.4c Estimates for β-convergence, macroeconomic level, 1986–2001[a]

	Panel Random effects (1)	Panel Fixed effects (2)	Cross-section OLS (3)	Cross-section Spatial lag (4)
Energy productivity 1986 (ln)	−0.017 (0.003)	−0.151 (0.012)	−0.014 (0.003)	−0.014 (0.002)
$W \times$ growth energy productivity				0.841 (0.095)
AIC			−523.223	−540.446
R^2	0.26	0.26	0.21	0.28
F-statistic (χ^2)	25.170	2.540	27.727	
$\tilde{\rho}$	0.069	0.852		
Hausman	123.860			
Jarque-Bera			0.879	
(Spatial) Breusch-Pagan			0.023	0.028
White			5.086	
Moran's *I*			8.806	
LM-error			48.362	
Robust LM-error			10.471	
LM-lag			38.108	
Robust LM-lag			0.217	
LR-lag				19.223
Number of countries	106	106	106	106
Number of observations	530	530	106	106

Note: [a] Standard errors in parentheses.

estimated convergence rate is typically slightly smaller than the estimate based on the fixed and random effects estimators. In addition, a number of diagnostics for the presence of spatial effects are reported in this column. The value of Moran's *I* test provides statistically significant evidence of the existence of spatial autocorrelation. Also, the Lagrange Multiplier (LM) test statistics for both omitted spatial lag and spatial error autocorrelation are significant at the 1 per cent level. Subsequently, the robust LM tests allow us to discriminate between the different types of misspecification: the lower value of the robust LM test on spatial error autocorrelation as compared to omitted spatial lag signals the existence of spatial interaction in the spatial lag (weighted average of initial energy productivity growth rates of neighbouring countries).[10] In addition, we have estimated

Table 2.5a Estimates for β-convergence, services, 1981–99[a]

	Panel Random effects (1)	Panel Fixed effects (2)	Cross-section OLS (3)	Cross-section Spatial lag (4)
Energy productivity 1981 (ln)	−0.025 (0.006)	−0.126 (0.014)	−0.018 (0.004)	−0.018 (0.004)
$W \times$ growth energy productivity				0.171 (0.311)
AIC			−244.347	−242.651
R^2	0.22	0.22	0.26	0.26
F-statistic (χ^2)	20.66	2.28	18.906	
$\tilde{\rho}$	0.031	0.644		
Hausman	62.220			
Jarque-Bera			1.646	
(Spatial) Breusch-Pagan			1.302	1.281
White			1.315	
Moran's *I*			1.237	
LM-error			0.472	
Robust LM-error			0.145	
LM-lag			0.362	
Robust LM-lag			0.035	
LR-lag				0.304
Number of countries	57	57	57	57
Number of observations	342	342	57	57

Note: [a] Standard errors in parentheses.

a so-called cross-regressive model (Florax and Folmer, 1992; Rey and Montouri, 1999) where the spatial lag of the initial energy productivity level is added to the original specification such that a country's energy productivity growth rate depends not only on its own initial energy productivity level, but also on the initial energy productivity levels of its neighbouring countries. However, the results indicate that this model still suffers from misspecification due to ignored spatial effects. We therefore present only the results of the spatial lag model in columns labelled (4).[11]

With respect to the various sectors, the regression results presented in Table 2.5 again show statistically significant negative estimates of β for energy productivity growth at the 1 per cent significance level for all models and all three sectors, indicating a robust pattern of β-convergence. Individual country effects explain between 49 per cent (agriculture) and 86

Table 2.5b Estimates for β*-convergence, manufacturing, 1981–99*[a]

	Panel Random effects (1)	Panel Fixed effects (2)	Cross-section OLS (3)	Cross-section Spatial lag (4)
Energy productivity 1981 (ln)	−0.027 (0.005)	−0.231 (0.016)	−0.018 (0.003)	−0.018 (0.003)
$W \times$ growth energy productivity				0.012 (0.321)
AIC			−332.932	−330.934
R^2	0.38	0.38	0.34	0.34
F-statistic (χ^2)	29.36	3.43	35.491	
$\tilde{\rho}$	0.000	0.857		
Hausman	191.94			
Jarque-Bera			66.246	
(Spatial) Breusch-Pagan			0.035	0.110
White			0.095	
Moran's *I*			0.722	
LM-error			0.048	
Robust LM-error			0.152	
LM-lag			0.002	
Robust LM-lag			0.106	
LR-lag				0.002
Number of countries	71	71	71	71
Number of observations	426	426	71	71

Note: [a] Standard errors in parentheses.

per cent (industry) of the total variance, implying that energy productivity convergence is conditional rather than absolute. Contrary to what we found for the macroeconomic level, the value of Moran's *I* test in models reported in column (3) provides statistically insignificant evidence of spatial autocorrelation at the sectoral level. In addition, the estimates for the (robust) LM and Breusch-Pagan tests are all statistically insignificant for the various sectors, thus suggesting that, after controlling for initial energy productivity levels, spatial effects do not play a significant role in explaining cross-country differences in energy productivity growth rates at the sector level. This conclusion is in line with the results of the exploratory spatial analysis documented in section 2.4, where we reported relatively weak indications of spatial autocorrelation at the sectoral level.

Table 2.5c Estimates for β*-convergence, agriculture, 1981–99*[a]

	Panel Random effects (1)	Panel Fixed effects (2)	Cross-section OLS (3)	Cross-section Spatial lag (4)
Energy productivity 1981 (ln)	−0.041 (0.007)	−0.120 (0.016)	−0.023 (0.004)	−0.023 (0.004)
$W \times$ growth energy productivity				−0.432 (0.393)
AIC			−193.289	−192.206
R^2	0.22	0.22	0.37	0.38
F-statistic (χ^2)	45.35	1.67	26.915	
$\tilde{\rho}$	0.000	0.489		
Hausman	45.350			
Jarque-Bera			2.624	
(Spatial) Breusch-Pagan			0.951	1.628
White			2.486	
Moran's *I*			0.247	
LM-error			0.066	
Robust LM-error			0.498	
LM-lag			0.535	
Robust LM-lag			0.967	
LR-lag				0.917
Number of countries	47	47	47	47
Number of observations	282	282	47	47

Note: [a] Standard errors in parentheses.

As mentioned before, this is likely to have its roots in the relatively poor quality and scope of the sectoral data.

6 CONCLUSIONS

In this chapter we studied the evolution of cross-country differences in energy productivity over time and in space. More specifically, we studied cross-country energy productivity developments over the period 1971–2001 for the macroeconomic level as well as for the sectors services, industry and agriculture. We outlined the main historical trends in energy productivity dynamics across countries, and explored the geographical dimension of energy productivity dynamics across countries. We have built upon the existing empirical convergence literature, and extended it by explicitly controlling for spatial effects in testing for convergence of energy productivity across countries, using exploratory spatial data analysis and spatial econometric techniques. In most regions increasing per capita income has been the key driver behind growing energy consumption, except for the regions Middle East and North Africa and sub-Saharan Africa. Increasing energy intensities have contributed to per capita energy consumption in various parts of the world, particular in the region Middle East and North Africa, while decreasing energy intensities have slowed down growth in per capita energy consumption in North America, Europe, and particularly in China. We found that if energy intensity levels had remained at their 1971 levels total energy consumption in the world would have been about 60 per cent higher in 2001. Energy productivity improvements played a particular dramatic role in China, where energy consumption would have been more than 260 per cent higher in 2006 in the case where no energy productivity improvements would have been realized since 1971.

In sum, in most parts of the world energy productivity improvements have clearly played an essential role in combining economic growth, energy security and mitigation of climate change. We document substantial cross-country variation in levels and growth rates of energy productivity. However, we find evidence of decreasing cross-country differences in energy productivity in combination with increasing spatial autocorrelation, particularly at the macroeconomic level with the evidence being less clear at the sectoral level. Our results regarding the international dynamics of energy productivity development can be summarized as follows. Most Western European countries have increased their energy productivity over time and have become more homogenous as a group. Second, various Eastern European countries such as Czechoslovakia, Poland, Hungary, the former Soviet Union and former Yugoslavia as well as their direct neighbours have been able to catch up to Western European nations. Third, the positive spatial dependence of 1971 energy productivity levels in various Latin American countries has not been preserved over time: in 2001 Latin American energy productivity levels are typically relatively low and have diverged.

Finally, we find robust evidence of β-convergence, suggesting that countries with relatively low initial energy productivity levels catch up to more advanced countries. This convergence process, however, is conditional: countries converge to multiple, country- or region-specific, steady states rather than to a global standard. Furthermore, we could establish that spatial effects do play a role in explaining cross-country differences in energy productivity growth rates, even after controlling for initial energy productivity levels. The transmission channel is that of interaction with neighbouring countries: a country's energy productivity growth rate depends not only on its own initial energy productivity level, but also on the spatially weighted average of the energy productivity growth rates of its neighbouring countries. Arguably, this raises the question as to what exactly are these cross-country spatial interactions that contribute to energy productivity improvements within spatial clusters of countries. What are the mechanisms facilitating these spatial interactions, and to what extent can they be influenced by policy-makers? These are important questions to guide future research. For now we conclude with the observation that, given the close relationship between energy consumption and technological change, knowledge spillovers and technology diffusion across borders are likely to play an important role and that they do not materialize randomly across space. The combined evidence of the existence of spatial clusters, conditional convergence and spatial interaction lends support to the view that there is no such thing as a 'death of distance' when it comes to explaining and stimulating energy productivity improvements.

NOTES

1. Seminal contributions can be found in Barro and Sala-i-Martin (1992), Mankiw et al. (1992) and Islam (1995). For good surveys we refer to Abreu et al. (2005), Barro and Sala-i-Martin (1995), Broadberry (1996), Durlauf and Quah (1999), Economic Journal (1996), Fagerberg (1994) and Islam (2003).
2. Consistent IEA biomass statistics are available as of 1994 only.
3. The use of market exchange rates instead of purchasing power parity (PPP) data is motivated by the lack of PPPs at the sectoral level. For the macroeconomic level we also conducted the analysis measuring GDP in PPP's, which had only a very minor quantitative impact on the results and left the qualitative conclusions unaffected. Details are available upon request.
4. In Table 2A.3 in the Appendix we provide a more detailed picture by differentiating these indicators across nine world regions.
5. Because of a required match between our data and available polygon maps we present the countries according to a '1990 worldview', including the former Soviet Union, Yugoslavia, Czechoslovakia as well as a united Germany, resulting in a total 106 countries with data.
6. We refer to Table 2A.1 in the Appendix for the list of countries per region.
7. Additional information is provided in Table 2A.3 in the Appendix in the form of descriptive statistics of the average annual growth rates of energy productivity for the different world regions.
8. Obviously, β-convergence is a necessary condition for σ-convergence: without some form of catching up of backward countries there will be no decline in cross-country vari-

ation. However, β-convergence is not a sufficient condition for σ-convergence due to a statistical result known as Galton's fallacy of regression towards the mean (Quah, 1993).

9. We also extended the models labelled (1) with time dummies to control for time-specific variation in the data, but this ('mixed effect') model did not yield significantly different results while the time effects proved to be insignificant except for the 1977 and 1980 coefficients. Owing to space limitations these results are not reported.
10. This is also confirmed when comparing the results of the spatial lag model in equation (2.7a) with the spatial error model of equation (2.8a): the relatively low value of the Akaike Information Criterion (AIC) and the comparatively high Breusch-Pagan (BP) test value in model (2.8a) as well as the higher LM-test value on spatial lag dependence as compared to the LM-test on spatial error dependence in model (2.7a) all indicate the spatial lag model to be preferred over the spatial error model.
11. We also extended the spatial error model with time dummies to control for time-specific variation in the data. However, in comparison to the results for the spatial error model of equation (2.8a), this ('mixed effect') model yielded substantially lower estimates of the autoregressive error parameter while the time effects turned out to be statistically insignificant.

REFERENCES

Abramovitz, M. (1986), 'Catching up, forging ahead, and falling behind', *The Journal of Economic History*, **47**, 385–406.

Abreu, M., H.L.F. de Groot and R.J.G.M. Florax (2004), 'Spatial patterns of technology diffusion: an empirical analysis using TFP', Discussion Paper 04-079/3, Amsterdam-Rotterdam: Tinbergen Institute.

Abreu, M., H.L.F. de Groot and R.J.G.M. Florax (2005), 'A meta-analysis of β-convergence: the legendary two percent', *Journal of Economic Surveys*, **19**, 389–420.

Anselin, L. (2001), 'Spatial effects in econometric practice in environmental and resource economics', *American Journal of Agricultural Economics*, **83** (3), 705–10.

Anselin, L. (2002), 'Under the hood: issues in the specification and interpretation of spatial regression models', *Agricultural Economics*, **27** (3), 247–67.

Barro, R.J. and X. Sala-i-Martin (1992), 'Convergence', *Journal of Political Economy*, **100**, 223–51.

Barro, R.J. and X. Sala-i-Martin (1995), *Economic Growth*, New York: McGraw-Hill.

Baumol, W.J. (1986), 'Productivity growth, convergence and welfare: what the long-run data show', *American Economic Review*, **76**, 1072–85.

Berndt, E.R. (1978), 'Aggregate energy, efficiency, and productivity measurement', *Annual Review of Energy*, **3**, 225–73.

Broadberry, S.N. (1996), 'Convergence: what the historical record shows', in B. van Ark and N. Crafts (eds), *Quantitative Aspects of Post-War European Growth*, Cambridge: Cambridge University Press.

Caniëls, M.C.J. and B. Verspagen (2001), 'Barriers to knowledge spillovers and regional convergence in an evolutionary model', *Journal of Evolutionary Economics*, **11**, 307–29.

Chatterji, M. (1992), 'Convergence clubs and endogenous growth', *Oxford Review of Economic Policy*, **8**, 57–69.

Chatterji, M., B. Gilmore, K. Strunk and J. Vanasin (1993), 'Political economy, growth and convergence in less-developed countries', *World Development*, **21**, 2029–38.

Cleveland C.J., R.K. Kaufmann and D.I. Stern (2000), 'Aggregation and the role of energy in the economy', *Ecological Economics*, **32**, 301–18.

Cliff A. and J.K. Ord (1981), *Spatial Processes: Models and Applications*, London: Pion.

Durlauf, S.N. and P.A. Johnson (1992), 'Local versus global convergence across national economies', Working Paper 3996, Cambridge, MA: National Bureau of Economic Research.

Durlauf, S.N. and D. Quah (1999), 'New empirics of economic growth', in J.B. Taylor and M. Woodford (eds), *Handbook of Macroeconomics*, vol. 1, Amsterdam: North Holland Elsevier Science.

Economic Journal (1996), 'Controversy on the convergence and divergence of growth rates', *Economic Journal*, special issue, **106**, 1016–69.

Ezcurra, R. (2007), 'Distribution dynamics of energy intensities: a cross-country analysis', *Energy Policy*, **35**, 5254–59.

Fagerberg, J. (1994), 'Technology and international differences in growth rates', *Journal of Economic Literature*, **32**, 1147–75.

Florax, R.J.G.M. and H. Folmer (1992), 'Specification and estimation of spatial linear regression models: Monte Carlo evaluation of pre-test estimators', *Regional Science and Urban Economics*, **22**, 405–32.

Florax, R.J.G.M. and P. Nijkamp (2005), 'Misspecification in linear spatial regression models', in K. Kempf-Leonard (ed.), *Encyclopedia of Social Measurement*, San Diego, CA: Academic Press.

Hsiao, C. (1986), *Analysis of Panel Data*, Cambridge: Cambridge University Press.

Islam, N. (1995), 'Growth empirics: a panel data approach', *Quarterly Journal of Economics*, **4**, 1127–70.

Islam, N. (2003), 'What have we learnt from the convergence debate?', *Journal of Economic Surveys*, **17**, 309–62.

Jorgenson, D.W. (1984), 'The role of energy in productivity growth', *Energy Journal*, **5**, 11–26.

Kaufmann, R.K. (2004), 'The mechanisms for autonomous energy efficiency increases: a cointegration analysis of the US energy/GDP ratio', *Energy Journal*, **25**, 63–86.

Keller, W. (2002), 'Geographic localization of international technology diffusion', *American Economic Review*, **92**, 120–42.

Lucas, R.E. (1988), 'On the mechanics of economic development', *Journal of Monetary Economics*, **22**, 3–42.

Mankiw, N.G., D. Romer and D.N. Weil (1992), 'A contribution to the empirics of economic growth', *Quarterly Journal of Economics*, **107**, 407–37.

Miketa, A. and P. Mulder (2005), 'Energy productivity across developed and developing countries in 10 manufacturing sectors: patterns of growth and convergence', *Energy Economics*, **27**, 429–53.

Mulder, P. and H.L.F. de Groot (2007), 'Sectoral energy- and labour-productivity convergence', *Environmental and Resource Economics*, **36**, 85–112.

Quah, D. (1993), 'Galton's fallacy and test of the convergence hypothesis', *Scandinavian Journal of Economics*, **95**, 427–43.

Quah, D. (1997), 'Empirics for growth and distribution: polarization, stratification, and convergence clubs', *Journal of Economic Growth*, **2**, 27–59.

Rey, S.J. (2001), 'Spatial empirics for economic growth and convergence', *Geographical Analysis*, **33**, 195–214.

Rey, S.J. and B.D. Montouri (1999), 'US regional income convergence: a spatial econometric perspective', *Regional Studies*, **33**, 143–56.

Romer, P.M. (1986), 'Increasing returns and long-run growth', *Journal of Political Economy*, **94**, 1002–37.

Romer, P.M. (1990), 'Endogenous technological change', *Journal of Political Economy*, **98**, S71–S102.

Schipper, L. and S. Meyers (1992), *Energy Efficiency and Human Activity: Past Trends and Future Prospects*, Cambridge: Cambridge University Press.

Solow, R.M. (1956), 'A contribution to the theory of economic growth', *Quarterly Journal of Economics*, **70**, 65–94.

Swan, T.W. (1956), 'Economic growth and capital accumulation', *Economic Record*, **32**, 334–61.

World Energy Council (2008), *Energy Efficiency Policies around the World: Review and Evaluation*, London: World Energy Council.

APPENDIX

Table 2A.1 List of countries per sector

No.	Country	Code	Region	MACRO	SRV	MAN	AGR
1	Bahrain	BHR	Asia Other	×			
2	Bangladesh	BGD	Asia Other	×	×	×	×
3	Brunei	BRN	Asia Other	×	×	×	
4	Hong Kong	HKG	Asia Other	×			
5	India	IND	Asia Other	×	×	×	×
6	Indonesia	IDN	Asia Other	×	×	×	×
7	Japan	JPN	Asia Other	×	×	×	×
8	Korea, D. P. Rep.	PRK	Asia Other	×			
9	Korea Rep.	KOR	Asia Other	×	×	×	×
10	Malaysia	MYS	Asia Other	×	×	×	
11	Myanmar	MMR	Asia Other	×			
12	Nepal	NPL	Asia Other	×	×	×	×
13	Pakistan	PAK	Asia Other	×	×	×	×
14	Philippines	PHL	Asia Other	×	×	×	×
15	Singapore	SGP	Asia Other	×			
16	Sri Lanka	LKA	Asia Other	×	×	×	
17	Taiwan	TAW	Asia Other	×			
18	Thailand	THA	Asia Other	×	×	×	×
19	Vietnam	VNM	Asia Other	×			
20	China	CHN	China	×	×	×	×
21	Albania	ALB	Europe East	×		×	
22	Bulgaria	BGR	Europe East	×	×	×	×
23	Czechoslovakia	CSK	Europe East	×			
24	Former USSR	FSU	Europe East	×			
25	Former Yugoslavia	FYU	Europe East	×			
26	Hungary	HUN	Europe East	×	×	×	×
27	Poland	POL	Europe East	×			
28	Romania	ROM	Europe East	×			
29	Austria	AUT	Europe West	×	×	×	×
30	Belgium	BEL	Europe West	×	×	×	×
31	Cyprus	CYP	Europe West	×			
32	Denmark	DNK	Europe West	×	×	×	×
33	Finland	FIN	Europe West	×	×	×	×
34	France	FRA	Europe West	×	×	×	×
35	Germany	DEU	Europe West	×			
36	Greece	GRC	Europe West	×	×	×	×

Table 2A.1 (continued)

No.	Country	Code	Region	MACRO	SRV	MAN	AGR
37	Iceland	ISL	Europe West	×			
38	Ireland	IRL	Europe West	×			
39	Italy	ITA	Europe West	×	×	×	×
40	Luxembourg	LUX	Europe West	×			
41	Netherlands	NLD	Europe West	×	×	×	×
42	Norway	NOR	Europe West	×	×	×	×
43	Portugal	PRT	Europe West	×			
44	Spain	ESP	Europe West	×			
45	Sweden	SWE	Europe West	×	×	×	×
46	Switzerland	CHE	Europe West	×			
47	United Kingdom	GBR	Europe West	×	×	×	×
48	Argentina	ARG	Latin America & Caribbean	×	×	×	×
49	Bolivia	BOL	Latin America & Caribbean	×			
50	Brazil	BRA	Latin America & Caribbean	×	×	×	×
51	Chile	CHL	Latin America & Caribbean	×	×	×	
52	Colombia	COL	Latin America & Caribbean	×	×	×	×
53	Costa Rica	CRI	Latin America & Caribbean	×		×	
54	Cuba	CUB	Latin America & Caribbean	×			
55	Dominican Rep.	DOM	Latin America & Caribbean	×		×	×
56	Ecuador	ECU	Latin America & Caribbean	×	×	×	
57	El Salvador	SLV	Latin America & Caribbean	×	×	×	
58	Guatemala	GTM	Latin America & Caribbean	×	×	×	×
59	Haiti	HTI	Latin America & Caribbean	×		×	
60	Honduras	HND	Latin America & Caribbean	×	×	×	
61	Jamaica	JAM	Latin America & Caribbean	×		×	

Table 2A.1 (continued)

No.	Country	Code	Region	MACRO	SRV	MAN	AGR
62	Mexico	MEX	Latin America & Caribbean	×	×	×	×
63	Nicaragua	NIC	Latin America & Caribbean	×	×	×	×
64	Panama	PAN	Latin America & Caribbean	×	×	×	
65	Paraguay	PRY	Latin America & Caribbean	×		×	
66	Peru	PER	Latin America & Caribbean	×	×	×	×
67	Trinidad	TTO	Latin America & Caribbean	×		×	
68	Uruguay	URY	Latin America & Caribbean	×	×	×	×
69	Venezuela	VEN	Latin America & Caribbean	×	×	×	×
70	Algeria	DZA	Middle East & North Africa	×		×	
71	Egypt	EGY	Middle East & North Africa	×		×	
72	Iran	IRN	Middle East & North Africa	×	×	×	×
73	Iraq	IRQ	Middle East & North Africa	×			
74	Israel	ISR	Middle East & North Africa	×			
75	Jordan	JOR	Middle East & North Africa	×		×	×
76	Kuwait	KWT	Middle East & North Africa	×			
77	Lebanon	LBN	Middle East & North Africa	×			
78	Morocco	MAR	Middle East & North Africa	×	×	×	
79	Oman	OMN	Middle East & North Africa	×			
80	Qatar	QAT	Middle East & North Africa	×			
81	Saudi Arabia	SAU	Middle East & North Africa	×		×	

Table 2A.1 (continued)

No.	Country	Code	Region	MACRO	SRV	MAN	AGR
82	Syria	SYR	Middle East & North Africa	×			
83	Tunisia	TUN	Middle East & North Africa	×	×	×	×
84	Turkey	TUR	Middle East & North Africa	×	×	×	×
85	United Arab Em.	ARE	Middle East & North Africa	×			
86	Yemen	YEM	Middle East & North Africa	×			
87	Canada	CAN	North America	×	×	×	×
88	United States	USA	North America	×	×	×	×
89	Australia	AUS	Oceania	×	×	×	×
90	New Zealand	NZL	Oceania	×	×	×	×
91	Angola	AGO	Sub-Saharan Africa	×			
92	Benin	BEN	Sub-Saharan Africa	×		×	
93	Cameroon	CMR	Sub-Saharan Africa	×	×	×	
94	Congo, Dem. Rep.	COD	Sub-Saharan Africa	×		×	
95	Congo, Rep.	COG	Sub-Saharan Africa	×			
96	Ethiopia	ETH	Sub-Saharan Africa	×	×		
97	Gabon	GAB	Sub-Saharan Africa	×			
98	Ghana	GHA	Sub-Saharan Africa	×	×	×	×
99	Ivory Coast	CIV	Sub-Saharan Africa	×		×	
100	Kenya	KEN	Sub-Saharan Africa	×		×	×
101	Mozambique	MOZ	Sub-Saharan Africa	×			
102	Nigeria	NGA	Sub-Saharan Africa	×	×	×	
103	Senegal	SEN	Sub-Saharan Africa	×	×	×	×

Table 2A.1 (continued)

No.	Country	Code	Region	MACRO	SRV	MAN	AGR
104	South Africa	ZAF	Sub-Saharan Africa	×	×	×	×
105	Sudan	SDN	Sub-Saharan Africa	×			
106	Tanzania	TZA	Sub-Saharan Africa	×			
107	Togo	TGO	Sub-Saharan Africa	×	×	×	
108	Zambia	ZMB	Sub-Saharan Africa	×	×	×	×

Table 2A.2 Descriptive statistics population, energy and GDP

	1971				2001				Number of countries
	Mean	Min.	Max.	St. dev.	Mean	Min.	Max.	St. dev.	
Population (million)									
World	33.4	0.1	841.1	100.4	54.2	0.3	1271.9	160.1	109
North America	114.8	22.0	207.7	131.3	158.5	31.1	285.9	180.2	2
Western Europe	18.7	0.2	78.3	23.7	20.6	0.3	82.3	25.5	19
Eastern Europe	44.3	2.2	245.1	81.7	51.3	3.2	289.1	96.7	8
Oceania	8.0	2.9	13.2	7.3	11.7	3.9	19.5	11.0	2
Middle East and North Africa	9.6	0.1	36.6	11.8	20.7	0.6	68.6	23.1	18
China	841.1	841.1	841.1	–	1271.9	1271.9	1271.9	–	1
Asia Other	64.8	0.1	560.3	128.3	115.6	0.3	1032.4	235.9	18
Latin America and the Caribbean	12.9	1.0	98.4	22.2	23.5	1.3	172.4	39.6	22
Sub-Saharan Africa	11.9	0.5	54.7	13.0	27.3	1.3	129.9	30.4	19
Final energy consumption per capita excluding renewable (toe)									
World	1.07	0.00	7.33	1.437	1.66	0.02	15.29	2.213	109
North America	5.41	5.05	5.78	0.518	5.43	5.22	5.64	0.296	2
Western Europe	2.58	0.55	7.33	1.563	3.49	1.75	8.46	1.845	19
Eastern Europe	1.59	0.48	3.05	0.789	1.40	0.42	2.35	0.643	8
Oceania	2.17	1.85	2.49	0.453	3.41	3.29	3.52	0.163	2
Middle East and North Africa	1.13	0.05	5.59	1.650	2.69	0.16	15.29	3.756	18
China	0.22	0.22	0.22	–	0.45	0.45	0.45	–	1
Asia Other	0.38	0.00	1.90	0.476	1.01	0.04	2.75	1.038	18
Latin America and the Caribbean	0.38	0.02	0.95	0.287	0.71	0.06	4.28	0.867	22
Sub-Saharan Africa	0.17	0.01	1.27	0.283	0.16	0.02	1.09	0.246	19

Table 2A.2 (continued)

	1971				2001				Number of countries
	Mean	Min.	Max.	St. dev.	Mean	Min.	Max.	St. dev.	
Final energy consumption per capita including renewable (toe)									
World	1.15	0.01	7.33	1.442	1.81	0.13	15.29	2.186	109
North America	5.67	5.39	5.95	0.398	5.67	5.39	5.95	0.398	2
Western Europe	2.67	0.62	7.33	1.583	3.66	1.83	8.50	1.849	19
Eastern Europe	1.66	0.65	3.05	0.749	1.46	0.45	2.38	0.632	8
Oceania	2.31	1.85	2.76	0.641	3.66	3.57	3.75	0.130	2
Middle East and North Africa	1.14	0.05	5.61	1.648	2.71	0.16	15.29	3.748	18
China	0.22	0.22	0.22	–	0.62	0.62	0.62	–	1
Asia Other	0.45	0.05	1.90	0.452	1.13	0.13	2.75	0.959	18
Latin America and the Caribbean	0.60	0.20	0.95	0.212	0.86	0.22	4.28	0.825	22
Sub-Saharan Africa	0.20	0.01	1.48	0.325	0.46	0.22	1.30	0.312	19
GDP per capita (US$)									
World	5967	114	37467	8297	9530	85	57201	12856	109
North America	15067	12882	17252	3090	27241	23080	31401	5884	2
Western Europe	15360	4894	36494	7378	29968	13061	57201	11340	19
Eastern Europe	2002	649	3578	1128	3015	1033	5543	1796	8
Oceania	13492	13204	13780	408	21254	18470	24039	3938	2

Middle East and North Africa	8766	120	37467	11950	6293	316	22057	6748	18
China	114	114	114	–	878	878	878	–	1
Asia Other	3597	143	21344	6670	8624	248	44396	12629	18
Latin America and the Caribbean	2143	503	7088	1502	2882	354	7468	1923	22
Sub-Saharan Africa	821	117	4226	1123	814	85	4377	1219	19
Energy productivity, macroeconomic level (ln)[a]									
World	15.4	11.9	17.9	0.880	15.3	13.1	16.6	0.659	109
North America	14.8	14.8	14.9	0.111	15.4	15.2	15.6	0.272	2
Western Europe	15.7	14.9	16.5	0.392	16.0	15.2	16.6	0.315	19
Eastern Europe	14.0	13.0	15.1	0.690	14.5	13.7	15.0	0.486	8
Oceania	15.7	15.5	15.8	0.179	15.6	15.5	15.7	0.138	2
Middle East and North Africa	15.7	14.1	17.9	0.909	14.8	13.8	15.9	0.525	18
China	13.1	13.1	13.1	–	14.5	14.5	14.5	–	1
Asia Other	15.5	11.9	17.3	1.223	15.4	13.1	16.6	0.803	18
Latin America and the Caribbean	15.7	14.8	16.8	0.520	15.3	14.1	16.1	0.487	22
Sub-Saharan Africa	15.5	14.4	16.4	0.585	15.4	14.7	16.0	0.392	19

Note: [a] Energy productivity is defined as the ratio of output (US$) to energy input (ktoe).

Table 2A.3 Descriptive statistics for average annual energy productivity growth rates per region, 1971–2001[a]

	Mean	Median	Min.	Max.	St. dev.	Number of countries
World	−0.4	−0.3	−9.2	4.6	2.4	109
North America	2.0	2.0	1.6	2.4	0.6	2
Western Europe	1.1	1.2	−1.1	2.8	1.1	19
Eastern Europe	1.8	2.0	−0.8	4.0	1.7	8
Oceania	0.0	0.0	−0.8	0.7	1.1	2
Middle East and North Africa	−3.0	−2.6	−9.2	0.2	2.8	18
China	4.6	4.6	4.6	4.6	–	1
Asia Other	−0.2	0.0	−5.7	4.0	2.5	18
Latin America and the Caribbean	−1.0	−1.0	−4.1	1.0	1.3	22
Sub-Saharan Africa	−0.3	−0.2	−4.1	2.7	1.9	19

Note: [a] Percentage, except for the last column.

3. Energy-productivity performance across 14 OECD countries: the role of energy-extensive sectors

Peter Mulder and Henri L.F. de Groot

1 INTRODUCTION

Economic development is typically associated with a change in the sectoral composition of economies, with emphasis shifting from agriculture, via industry towards services. In addition, economic development is characterized by increasing trade, enhanced international specialization and technological progress. At the same time, economic development has generally a strong positive relationship with energy consumption. The nature of this relationship is neither linear nor exogenous, but instead depends to a large extent on the same changes in the sectoral composition of economies and patterns of trade, specialization and technological progress that affect economic growth as well. Against this background, we explore the interplay between economic growth and energy use across 14 countries associated with the Organisation for Economic Co-operation and Development (OECD) during the period 1970–97 in this chapter. Data show that within this relatively homogenous group of countries remarkable cross-country differences in energy productivity exist. We examine trends and differences in energy-productivity performance across countries and search for their determinants, with an emphasis on the role of energy-extensive sectors, such as services and light industry, in driving aggregate trends. Specifically, we document several stylized facts and decompose country's energy-productivity growth rates into a part that is due to intra-sectoral productivity improvements and a part that is due to changes in sectoral composition. Subsequently, we identify the sector-specific fundamentals determining these energy-productivity developments.

For our empirical analysis we constructed a database that establishes a link between economic and energy data at a detailed sectoral level, distinguishing 13 sectors for a range of OECD countries. The unique feature of this database is that it is comprehensive and consistent over time as well as over space, with as many countries, years and sectors included as possible given current data availability. Previous research based on this database, but with a different focus, has been reported in Mulder and de Groot (2003, 2007) and Mulder (2005). In several respects, the research reported in this chapter differs from existing empirical analyses of productivity developments (see, for example, Eichhammer and Mannsbart, 1997; Greening et al., 1997; Howarth et al., 1991; Miketa, 2001; Park et al., 1993; Schipper and Meyers, 1992; Unander et al., 1999). First, we combine a relatively high level of sectoral detail with a wide range of countries to trace back aggregate energy productivity trends to developments at the level of individual sectors.[1] Moreover, contrary to most studies in the field of energy-intensity developments, our analysis is not confined to the manufacturing sector, but also includes a decomposition of macroeconomic energy productivity growth rates. Second, particular attention is paid to the role of the service sector and light industry. So far, most studies in this field have focused on heavy industries because of their energy-intensive production processes. Total energy use in the energy-extensive sectors, however, is clearly substantial at approximately 20 per cent of economy-wide energy use, and so are the potential savings. Moreover, within OECD countries these sectors become increasingly important as economic development progresses, with the emphasis shifting away from heavy industry to knowledge-intensive activities that are in general relatively energy extensive. Third, in our search for the sector-specific factors driving energy productivity developments we do not only include the more traditional determinants such as energy prices and investment shares. In addition, we incorporate potential determinants that have been identified in the more recent literature, such as fuel mix, economies of scale and the potential to catch up with countries using frontier technologies.

This chapter proceeds as follows. In section 2 we give a brief description of the data used in this study. Section 3 documents various stylized facts concerning energy consumption, growth of the gross domestic product (GDP) and energy productivity trends across countries, with particular attention for the role of energy-extensive sectors. In section 4 we provide a decomposition analysis of aggregate energy productivity trends, identifying the role of structural changes and intra-sectoral efficiency improvements in driving these trends. Section 5 presents the result of a regression analysis that aims to explain energy productivity growth per sector from a range of potential determinants. Section 6 concludes.

2 DATA

The analysis presented in this chapter is based on a database that merges energy data from the Energy Balances, as published by the International Energy Agency (IEA), and economic data from the International Sectoral Database (ISDB) and the Structural Analysis Database (STAN), both published by the OECD. The main idea behind the construction of this database is to establish a link between economic and energy data at a sectoral level.[2] This results in the sector classification described in Table 3.1.

Table 3.1 Sector classification

Sector	Abbreviation	ISIC rev. 2 code
Manufacturing[a]		
1 Food and tobacco (E)	FOD	31
2 Textiles and leather (E)	TEX	32
3 Wood and wood products (E)	WOD	331[b]
4 Paper, pulp and printing (I)	PAP	34
5 Chemicals (I)	CHE	351 + 352[c]
6 Non-metallic minerals (I)	NMM	36
7 Iron and steel (I)	IAS	371
8 Non-ferrous metals (I)	NFM	372
9 Machinery (E)	MAC	381 + 382 + 383[d]
10 Transport equipment (E)	MTR	384
11 Non-specified industry	NSI	332 + 355 + 356+ 385[e]
Other sectors		
12 Services	SRV	61 + 62 + 63 + 72 + 81 + 82 + 83 + 90[f]
13 Transport	TAS	71
14 Agriculture	AGR	10

Notes:

[a] I and E in parentheses are used to indicate energy-intensive and energy-extensive manufacturing sectors, respectively.

[b] WOD excludes furniture since the sector WOD in the IEA Energy Balances excludes furniture.

[c] CHE includes non-energetic energy consumption (using energy carriers as feedstock).

[d] MAC = Metal products (BMA, 381) + Agricultural and industrial machinery (MAI, 382) + Electrical goods (MEL, 383).

[e] NSI = Furniture (332) + Rubber (355) + Plastic products not classified elsewhere (356) + Professional, scientific, measuring and controlling equipment (not classified elsewhere), photographic and optical goods (385).[3]

[f] SRV = Wholesale and retail trade, restaurants and hotels (RET) + Communication (COM) + Finance, insurance, real estate and business services (FNI) + Community, social and personal services (SOC).

In principal, the database covers the period 1970–97, although the exact period of data availability varies per country, sector and variable.[4] The database includes the following countries: Australia (AUS), Belgium (BEL), Canada (CAN), Denmark (DNK), Finland (FIN), France (FRA), WestGermany (WGR), Italy (ITA), Japan (JPN), the Netherlands (NLD), Norway (NOR), Sweden (SWE), the United Kingdom (GBR) and the United States (USA).

We measure energy productivity by gross value-added per unit of final energy consumption. Value-added data have been converted to constant 1990 US$, using 1990 expenditure purchasing power parities (PPP) as given by the OECD. Following the IEA, energy use is defined as final energy consumption in kilo tonnes of oil equivalence (ktoe), with sectoral data excluding transformation losses. Moreover, the database includes data on investment, energy prices, export and import and the fuel mix, all at the sectoral level. The sector-specific energy prices are constructed by dividing sector-specific expenditures on energy over total sectoral energy consumption. The sector-specific expenditures are calculated as the product of the sectoral consumption of the four main energy carriers (coal, natural gas, electricity and oil) available from the Energy Balances, and the (annual) price of each energy carrier at the aggregate industrial sector, available from the IEA Energy Prices and Taxes series. In addition, some missing aggregate energy price data series have been constructed.

In the remainder of this chapter we use the distinction between energy-intensive and energy-extensive industries. With the former we refer to the sum of activities in the iron and steel, non-ferrous metals, non-metallic minerals, chemicals and paper industries, while the latter includes the food, textiles, wood, machinery and transport equipment industries. We think that this operational classification, albeit rudimentary is useful. It should be kept in mind, though, that some energy-intensive processes take place within energy-extensive sectors, and vice versa.

3 SOME STYLIZED FACTS

Before we examine the energy productivity trends, we document some stylized facts with respect to energy productivity developments, the role of energy-extensive sectors and the potential importance of structural differences across the 14 OECD countries. In Figure 3.1, we plot the development over time of non-residential energy productivity levels at the macroeconomic level for each country.[5]

Figure 3.1 shows a diverse picture with substantial cross-country differences in total energy productivity. One should note that the revealed cross-country variation is much larger than, for example, cross-country variation in labour productivity levels (Mulder and de Groot, 2007). According to Figure 3.1, the highest energy productivity levels are to be found in Italy and Japan while Canada, Finland, Norway and Sweden show the lowest levels of aggregate energy productivity. All other countries form a medium energy productivity group. The USA tends to leave the group with low levels of energy productivity over time, and catches up with the medium group.

In Tables 3.2 and 3.3 we present the share of the respective sectors in total final energy consumption and total value added (GDP). The tables clearly illustrate the declining role of manufacturing and agriculture, and the increasing importance of the service and transport sectors. The share of total value added generated in the service sector in 1990 has on average increased 7 per cent as compared to 1976, accompanied by a 21 per cent average increase in its share of total energy consumption. The transport sector shows a remarkable increase in energy share of 30 per cent. Moreover, Table 3.2 shows that on average the decrease in value added and energy share is higher in the energy-extensive industry than in the energy-intensive industry. Because of the rather rudimentary classification as well as the omission of various important countries in calculating the percentage change, one cannot draw firm conclusions from this analysis. However, a first look at the data leads to the conclusion that in spite of a general pattern of structural change in the OECD from manufacturing towards services (Tables 3.2 and 3.3), substantial cross-country differences in total energy productivity persist (Figure 3.1). Can this be explained by cross-country differences in technology-driven energy efficiency improvements only, or do (disaggregated) structural differences across countries play a role as well? This is the question to which we turn in the next section.

4 DECOMPOSING ENERGY PRODUCTIVITY DEVELOPMENT

Overall productivity performance is not only the result of technology-driven productivity performance in individual sectors, but also of the allocation of production factors across sectors. The underlying reason is that some sectors produce more value-added per unit of energy than others, because some activities require more capital, more high-skilled workers and/or more advanced technologies. As a result, understanding technology-driven energy productivity performance requires an assessment of productivity developments in individual sectors (see also, for example, Dollar and Wolff, 1993; Jorgenson 1984; Wagner and van Ark,

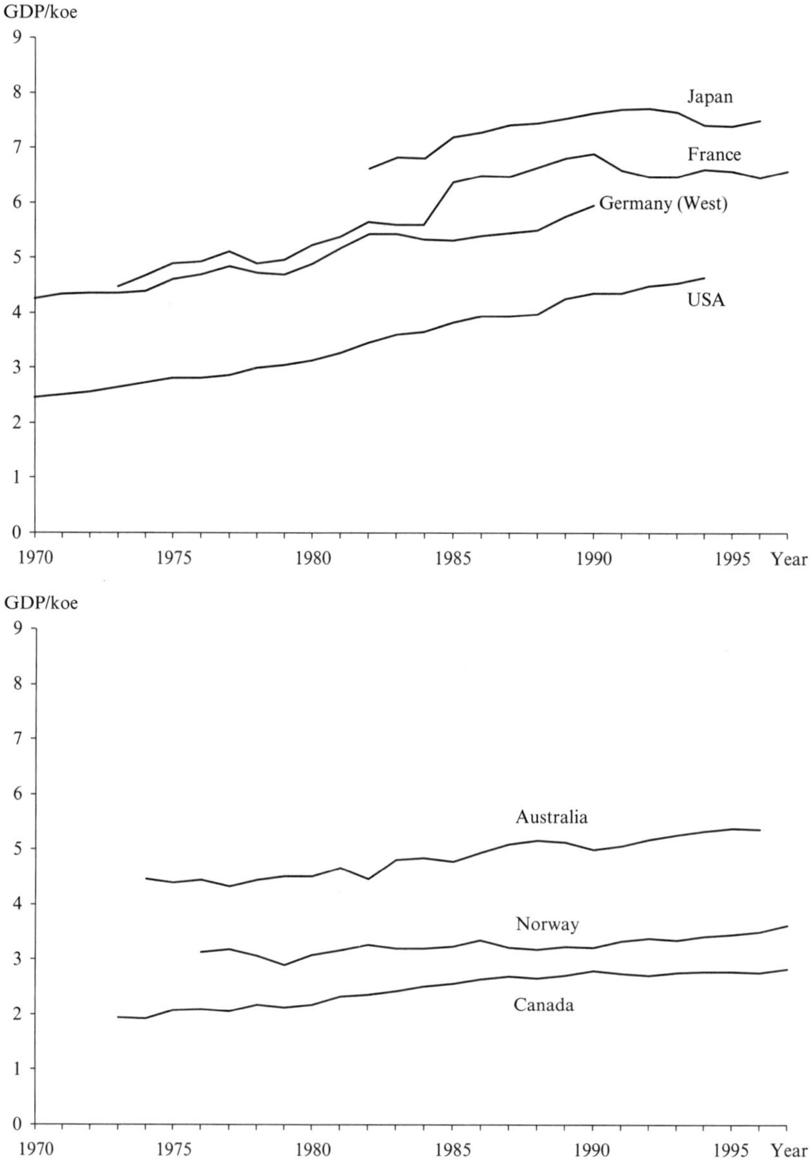

Note: Measured as GDP in 1990 US dollars per kilogram of oil equivalent (koe).

Figure 3.1 Trends in macroeconomic energy productivity development

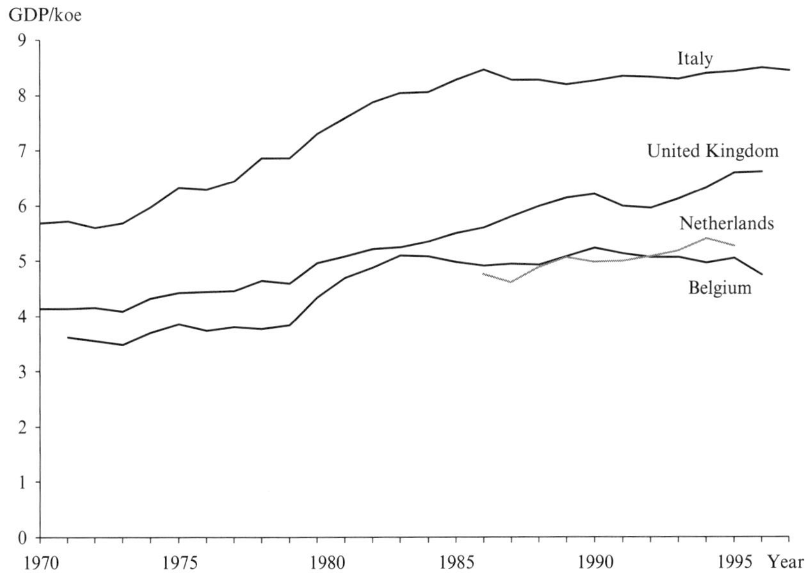

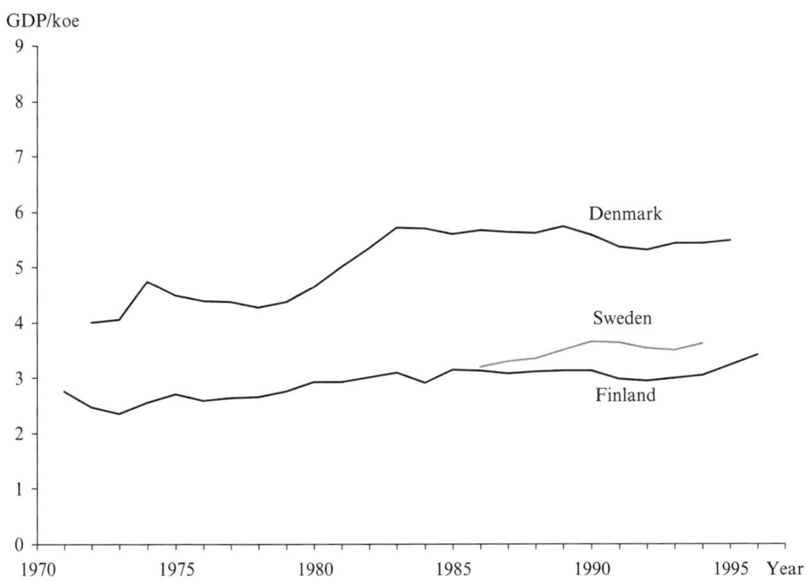

Figure 3.1 (continued)

Table 3.2 Energy and value-added shares for manufacturing[a]

	Energy-extensive manufacturing[b]				Energy-intensive manufacturing[b]			
	Energy		GDP		Energy		GDP	
	1976	1990	1976	1990	1976	1990	1976	1990
Australia	12	8	–	–	35	33	–	–
Belgium	6	7	18	16	50	41	9	10
Canada	–	–	–	14	–	–	–	8
Denmark	15	12	18	17	18	13	9	7
Finland	12	8	19	18	49	48	11	12
France	8	8	22	17	26	28	8	8
Germany (West)	9	8	30	25	42	36	12	10
Italy	14	9	20	19	47	36	7	8
Japan	–	7	–	20	–	36	–	9
Netherlands	–	7	–	13	–	48	–	8
Norway	10	7	16	11	48	44	8	8
Sweden	–	10	–	20	–	38	–	10
United Kingdom	14	10	23	18	33	24	11	9
United States	–	–	13	12	15	12	8	7
Percentage change[c]		–25		–15		–15		–5

Notes:

[a] The macroeconomic level, excluding the residential sector is set equal to 100.

[b] For classification of energy-intensive and extensive manufacturing, see Table 3.1.

[c] Percentage change of unweighted cross-country average of sector share in 1990 as compared to 1976, excluding those countries for which no 1976 and/or 1990 data were available.

1996). We will therefore correct the trends in aggregate energy productivity performance for the impact of shifts in sectoral energy shares, in order to get a better view on the role of sector-specific technology-driven productivity improvements and structural changes in driving aggregate energy productivity growth.

Many studies have measured the relative contribution of structural and technological change to aggregate productivity growth, using so-called index number decomposition or shift-share analysis. The studies differ from each other in several respects, including the number of sectors and countries included, the methodology (Laspeyres, Paasche or Divisia indices), the area of application (total factor productivity, capital, labour or energy), the type of indicator (quantity, intensity, productivity or elasticity) and the type of analysis (time series or period-wise). For a lucid exposition of the methodology and a survey of studies on (total factor) productivity we refer

Table 3.3 Energy and value-added shares for services, agriculture and transport[a]

	Services				Agriculture				Transport			
	Energy		GDP		Energy		GDP		Energy		GDP	
	1976	1990	1976	1990	1976	1990	1976	1990	1976	1990	1976	1990
Australia	6	8	68	72	3	3	5	4	45	49	5	6
Belgium	13	12	61	61	2	2	2	2	23	34	9	9
Canada	–	18	–	67	–	3	–	4	–	37	–	6
Denmark	12	17	57	57	14	7	5	6	41	49	9	10
Finland	3	6	47	50	5	6	12	9	23	27	8	8
France	–	18	56	61	3	3	5	5	30	41	5	5
Germany (West)	17	16	45	53	2	2	2	2	27	36	4	4
Italy	2	4	58	59	3	4	6	4	31	43	5	6
Japan	–	14	–	55	–	5	–	3	–	31	–	8
Netherlands	–	5	–	63	–	9	–	5	–	30	–	6
Norway	13	15	52	57	2	2	7	5	26	32	14	17
Sweden	–	17	–	56	–	2	–	4	–	31	–	7
United Kingdom	13	12	53	60	2	1	2	2	32	48	8	7
United States	16	16	66	71	1	1	2	2	44	50	5	4
Percentage change[b]		21		7		–14		–12		30		7

Notes:

[a] The macroeconomic level, excluding the residential sector is set equal to 100.

[b] Percentage change of unweighted cross-country average of sector share in 1990 as compared to 1976, excluding those countries for which no 1976 and/or 1990 data were available.

to Syrquin (1984) as well as to Ang (1995a, 1995b, 1999), and to Ang and Zhang (2000) for applications in the context of energy studies. We follow, among others, Fagerberg (2000), Timmer and Szirmai (2000) and Peneder (2003) in applying a decomposition or shift-share analysis that is based on the following definition of aggregate energy productivity (*EP*):

$$EP^t = \frac{Y_t}{E_t} = \sum_{i=1}^{n} \frac{Y_{i,t}}{E_{i,t}} \frac{E_{i,t}}{E_t} = \sum_{i=1}^{n} EP_{i,t} S_{i,t} \tag{3.1}$$

where Y is GDP, E final energy consumption, S the sectoral share in total energy consumption, and the subscripts i and t denote the sub-sector and time, respectively. Equation (3.1) implies that aggregate energy productivity is the sum of the energy productivity of each sub-sector (the first term on the right-hand side) multiplied by the energy share of each sub-sector (the second term on the right-hand side). On the basis of equation (3.1), changes in aggregate energy productivity (*EP*) between time 0 and time t can be written as:

$$EP_t - EP_0$$

$$= \underbrace{\sum_{i=1}^{n} (EP_{i,t} - EP_{i,0}) S_{i,0}}_{\text{Intra-sectoral effect}} + \underbrace{\sum_{i=1}^{n} (S_{i,t} - S_{i,0}) EP_{i,0}}_{\text{Static shift effect}} + \underbrace{\sum_{i=1}^{n} (EP_{i,t} - EP_{i,0})(S_{i,t} - S_{i,0})}_{\text{Dynamic shift effect}} \tag{3.2}$$

In equation (3.2) we decompose aggregate energy productivity growth into an intra-sectoral efficiency effect and the effects of structural change, which consist of a static and a dynamic shift effect. In other words, the intra-sectoral efficiency effect is the aggregate energy productivity growth insofar as it is caused by changes in the energy productivity performance within each individual sub-sector, keeping the sectoral energy shares constant. Vice versa, the static shift effect is the aggregate energy productivity growth insofar as it is caused by shifts in sectoral energy shares, keeping the levels of energy productivity performance for each individual sub-sector constant. The dynamic shift effect is an interaction term that captures the shift of energy towards sectors with improving energy productivity performance over time.[6]

In Figure 3.2 we present the results of the decomposition of the *macroeconomic* energy productivity growth rates into the aforementioned efficiency effect and the two structural effects, based on the role of the four main sectors (agriculture, services, manufacturing and transport). The

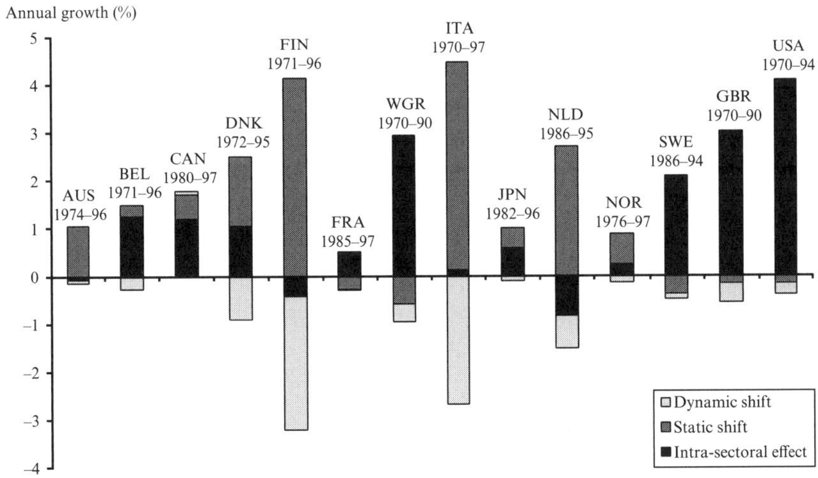

Figure 3.2 Decomposition of aggregate energy productivity per country based on four main sectors

figure shows that in most countries macroeconomic energy productivity growth is mainly driven by intra-sectoral efficiency improvements, while at the same time structural changes play a dominant role in countries such as Denmark, Finland, Italy, the Netherlands and Norway.

In order to see which sectors drive these aggregate results, Table 3.4 provides each sector's percentage contribution to the country-specific intra-sectoral efficiency effect as well as the static and dynamic shift effect. If we consider, for example, the case of Australia, the results reveal that its energy productivity growth of 0.94 per cent is largely driven by a shift towards the service sector, which has a high, although declining, energy productivity and has grown substantially in terms of its share in total energy consumption.[7] In contrast, energy productivity growth in countries such as the United Kingdom (UK) and the United States (US) is mainly driven by intra-sector (productivity) effects, predominantly observed in manufacturing and the service sector.

Table 3.4 shows that in virtually all countries efficiency improvements are mainly realized within the manufacturing sector, while the largest structural shift is the shift from manufacturing to services. Moreover, the table shows that in Finland, Italy and the Netherlands the small or even negative aggregate efficiency growth as well as the extraordinary positive effect of structural changes on macroeconomic energy productivity growth can be explained by an exceptional increasing energy share of the service sector. A closer look at the data reveals that this result is due to the combination of an exceptionally

Table 3.4 Decomposition of aggregate energy productivity per country based on four main sectors

	Annual energy productivity growth	Percentage energy productivity growth attributable to:			
		Intra-sectoral effect	Static shift	Dynamic shift	Total effect
Australia, 1974–96					
Total	0.94	−5	113	−8	100
Agriculture	0.0	0	1	0	1
Manufacturing	0.3	6	−20	−1	−15
Services	−0.3	−19	129	−7	102
Transport	−1.2	7	4	1	12
Belgium, 1971–96					
Total	1.24	102	18	−20	100
Agriculture	−3.1	−6	12	−7	0
Manufacturing	2.0	57	−19	−12	26
Services	1.0	58	10	3	71
Transport	−1.2	−8	15	−4	3
Canada, 1980–97					
Total	1.78	67	29	4	100
Agriculture	−1.9	−4	6	−2	0
Manufacturing	1.2	19	−1	0	18
Services	1.2	47	26	6	78
Transport	−1.6	5	−1	0	3
Denmark, 1972–95					
Total	1.6	66	90	−56	100
Agriculture	1.8	8	0	0	8
Manufacturing	3.1	82	−32	−33	18
Services	−1.1	−32	116	−25	59
Transport	1.1	8	6	2	15
Finland, 1971–96					
Total	0.95	−42	437	−294	100
Agriculture	−1.4	−20	−10	3	−27
Manufacturing	2	88	−17	−11	60
Services	−3.9	−113	459	−287	60
Transport	0.3	3	5	0	8
France, 1985–97					
Total	0.24	211	−105	−7	100
Agriculture	0.5	12	−27	−2	−17
Manufacturing	0.5	63	−99	−6	−42

Table 3.4 (continued)

	Annual energy productivity growth	Percentage energy productivity growth attributable to:			
		Intra-sectoral effect	Static shift	Dynamic shift	Total effect
Services	0.5	131	1	0	132
Transport	0.3	6	20	1	27
Germany (West), 1970–90					
Total	2	147	−28	−19	100
Agriculture	1.5	2	−1	0	1
Manufacturing	1.4	42	−21	−7	14
Services	3.4	104	−12	−12	80
Transport	−0.3	−1	5	0	4
Italy, 1970–97					
Total	1.8	7	240	−148	100
Agriculture	−1.8	−6	5	−2	−3
Manufacturing	3	73	−18	−23	32
Services	−2.6	−60	247	−123	64
Transport	0.2	0	6	0	7
Japan, 1982–96					
Total	0.93	63	45	−9	100
Agriculture	−3.5	−12	3	−1	−11
Manufacturing	1.7	69	−32	−9	28
Services	0.1	7	65	1	74
Transport	−0.1	−1	9	0	9
Netherlands, 1986–95					
Total	1.19	−68	227	−59	100
Agriculture	−0.9	−4	15	−1	10
Manufacturing	2.1	52	−38	−8	6
Services	−2.6	−122	242	−50	69
Transport	1.3	6	8	1	15
Norway, 1976–97					
Total	0.75	33	83	−16	100
Agriculture	−3.8	−23	50	−27	−1
Manufacturing	−0.1	−3	−32	1	−34
Services	0.4	27	44	4	75
Transport	1.5	32	21	7	60
Sweden, 1986–94					
Total	1.63	128	−23	−6	100
Agriculture	4.1	13	−10	−4	−1

Table 3.4 (continued)

	Annual energy productivity growth	Percentage energy productivity growth attributable to:			
		Intra-sectoral effect	Static shift	Dynamic shift	Total effect
Manufacturing	1.2	26	−2	0	23
Services	2.2	81	−13	−2	65
Transport	2.5	9	3	1	12
United Kingdom, 1970–90					
Total	2.47	122	−6	−16	100
Agriculture	3.4	5	−1	−1	2
Manufacturing	2.8	61	−27	−21	12
Services	2.2	58	13	7	78
Transport	−0.6	−2	10	−1	7
USA, 1970–94					
Total	3.7	110	−4	−6	100
Agriculture	3.2	4	−1	−1	2
Manufacturing	3.2	36	−9	−10	17
Services	2.8	69	4	4	76
Transport	1.1	2	2	1	4

low initial level, and a subsequent steep increase of energy consumption in these countries, ultimately converging to an average level. Lack of appropriate data prevents us from further exploring this issue through an analysis of energy consumption levels in the underlying sub-sectors.[8]

In order to attain a more detailed picture of the manufacturing sector, we repeat the previous analysis, decomposing *manufacturing* energy productivity growth rates on the basis of the role of ten sub-sectors of the manufacturing industry. The results are presented in Figure 3.3 and Table 3.5. Figure 3.3 shows that in all countries, manufacturing energy productivity growth is mainly driven by intra-sectoral efficiency improvements, while the role of structural changes is mixed. In Belgium, Denmark, France, West Germany, Italy, Japan, the UK and the USA the efficiency effect is accompanied by a shift in the relative shares of sectors towards high-productive (energy-extensive) sectors. However, in other countries, such as Finland, the Netherlands and Norway, these structural shifts between sub-sectors counteract productivity growth resulting from intra-sectoral efficiency improvements.

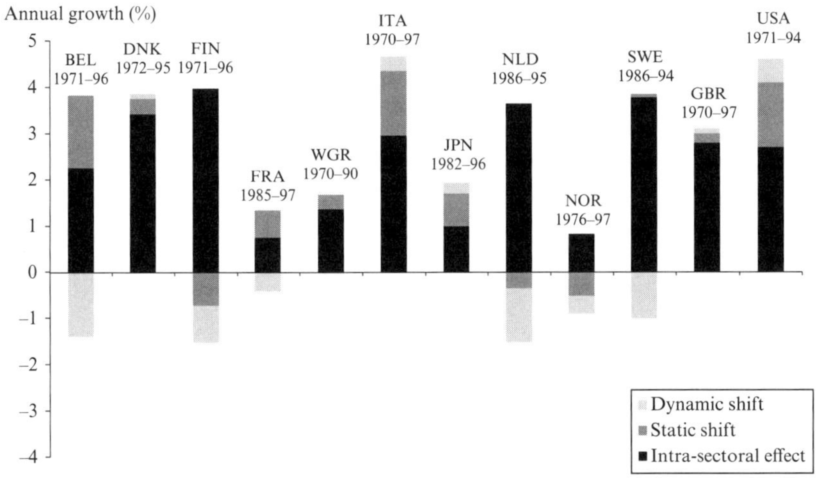

Figure 3.3 *Decomposition of manufacturing energy productivity per country based on 10 manufacturing sub-sectors*

Table 3.5 confirms that intra-sectoral efficiency improvements are the main source of manufacturing energy productivity growth, while generally the observed structural changes constitute a shift away from energy-intensive to energy-extensive sectors. The observed structural effects are mainly caused by the sectors machinery and chemicals and to a lesser extent also by the sectors food, paper, textiles and transport equipment. The intra-sectoral efficiency improvements materialize in both energy-intensive and energy-extensive sectors. Specifically, the results reveal that chemicals, food and paper production are primary sources of efficiency improvements, while the production of machinery, textiles and transport equipment contributes substantially to both aggregate efficiency improvements and structural changes. Given the overall shift towards energy-extensive industries, this implies that – in spite of their relatively small energy consumption – the energy-extensive sectors explain a relatively large part of manufacturing energy productivity growth (see the last column in Table 3.5 for the different countries). The impact of structural changes on manufacturing energy productivity growth in Belgium, Denmark, France, Italy and the USA are clearly driven by an increasing share of energy-extensive sectors in total manufacturing energy consumption. In contrast, in Finland and Norway the decreasing share of energy-extensive sectors in total manufacturing energy consumption drives the negative impact of structural changes on their manufacturing energy productivity growth. The principal source of the negative structural effect

Table 3.5 Decomposition of manufacturing energy productivity per country based on 10 manufacturing sub-sectors

Sector[a]	Annual energy productivity growth	Percentage energy productivity growth attributable to:			
		Intra-sectoral effect	Static shift	Dynamic shift	Total effect
Belgium, 1971–97					
Total	2.44	92	65	−57	100
Food (E)	−0.5	−4	18	−2	12
Machinery (E)	3.0	42	−12	−15	15
Transport equipment (E)	−0.8	−3	17	−3	11
Textiles (E)	−2.9	−10	17	−9	−2
Wood (E)	0.2	0	1	0	1
Chemicals (I)	2.4	14	8	7	28
Iron and steel (I)	1.4	5	−3	−2	0
Non-ferrous metals (I)	3.5	5	0	−1	4
Non-metallic minerals (I)	1.1	4	−1	0	2
Paper and pulp (I)	−2.8	−6	24	−12	6
Non-specified industry	7.7	45	−3	−19	23
Denmark, 1972–97					
Total	3.86	88	9	3	100
Food (E)	1.4	7	10	4	21
Machinery (E)	3.2	36	1	2	38
Transport equipment (E)	−0.2	0	−3	0	−3
Textiles (E)	3.0	8	−4	−5	0
Wood (E)	0.1	0	3	0	3
Chemicals (I)	6.0	19	0	−1	18
Iron and steel (I)	6.3	4	0	−2	2
Non-ferrous metals (I)					0
Non-metallic minerals (I)	0.6	1	−3	0	−2
Paper and pulp (I)	1.7	8	−1	0	7
Non-specified industry	2.3	5	7	5	16

Table 3.5 (continued)

Sector[a]	Annual energy productivity growth	Percentage energy productivity growth attributable to:			
		Intra-sectoral effect	Static shift	Dynamic shift	Total effect
Finland, 1971–97					
Total	2.47	161	−28	−33	100
Food (E)	2.1	18	−11	−8	0
Machinery (E)	5.5	78	−2	−5	71
Transport equipment (E)	4.0	19	−7	−12	0
Textiles (E)	1.2	5	−9	−3	−8
Wood (E)	2.0	10	−5	−4	1
Chemicals (I)	1.0	3	3	1	7
Iron and steel (I)	3.3	4	1	2	7
Non-ferrous metals (I)	1.7	1	1	0	2
Non-metallic minerals (I)	3.0	9	−4	−5	0
Paper and pulp (I)	1.2	11	3	1	15
Non-specified industry	0.6	3	1	0	4
France, 1973–97					
Total	0.94	79	64	−42	100
Food (E)	0.1	2	18	1	20
Machinery (E)	−1.1	−27	75	−17	30
Transport equipment (E)	1.0	15	−10	−3	3
Textiles (E)	1.5	20	−28	−12	−20
Wood (E)					0
Chemicals (I)	2.7	22	6	6	34
Iron and steel (I)	1.9	12	−10	−6	−3
Non-ferrous metals (I)	3.7	10	−3	−4	3
Non-metallic minerals (I)	0.4	−2	−1	0	−3
Paper and pulp (I)	−1.1	−8	19	−5	7
Non-specified industry	1.8	33	−3	−2	29
Germany (West), 1970–90					
Total	1.67	81	19	0	100
Food (E)	0.6	4	1	0	5
Machinery (E)	0.8	18	15	3	36

Table 3.5 (continued)

Sector[a]	Annual energy productivity growth	Percentage energy productivity growth attributable to:			
		Intra-sectoral effect	Static shift	Dynamic shift	Total effect
Transport equipment (E)	2.4	14	7	4	25
Textiles (E)	−0.2	−1	−5	0	−5
Wood (E)	1.5	2	−1	0	1
Chemicals (I)	0.5	2	12	1	16
Iron and steel (I)	2.3	12	−8	−4	0
Non-ferrous metals (I)	1.6	2	2	1	4
Non-metallic minerals (I)	1.9	6	−3	−1	2
Paper and pulp (I)	0.0	0	4	0	4
Non-specified industry	2.2	21	−6	−3	11
Italy, 1970–97					
Total	4.67	63	30	7	100
Food (E)	1.5	5	4	2	10
Machinery (E)	3.0	22	5	6	32
Transport equipment (E)	−1.4	−3	8	−3	3
Textiles (E)	1.9	10	3	2	14
Wood (E)	−1.9	−1	3	−1	0
Chemicals (I)	6.4	12	0	−2	10
Iron and steel (I)	0.7	1	−1	0	1
Non-ferrous metals (I)	1.0	0	0	0	1
Non-metallic minerals (I)	3.7	10	−1	−2	7
Paper and pulp (I)	3.1	5	1	1	8
Non-specified industry	1.1	2	9	3	14
Japan, 1982–97					
Total	1.94	51	36	13	100
Food (E)	−1.2	−9	−6	1	−14
Machinery (E)	2.5	38	41	19	98
Transport equipment (E)	−0.5	−3	13	−1	9
Textiles (E)	1.3	5	−14	−3	−11
Wood (E)					0
Chemicals (I)	0.2	1	8	0	9

Table 3.5 (continued)

Sector[a]	Annual energy productivity growth	Percentage energy productivity growth attributable to:			
		Intra-sectoral effect	Static shift	Dynamic shift	Total effect
Iron and steel (I)	2.4	11	−5	−2	4
Non-ferrous metals (I)	0.7	1	−1	0	0
Non-metallic minerals (I)	1.8	4	−2	−1	1
Paper and pulp (I)	0.2	1	0	0	1
Non-specified industry	0.3	2	1	0	2
Netherlands, 1982–97					
Total	2.14	170	−15	−55	100
Food (E)	1.6	16	−3	−1	12
Machinery (E)	0.9	12	14	2	28
Transport equipment (E)	2.1	7	−3	−1	2
Textiles (E)	0.0	0	−1	0	−1
Wood (E)	1.7	1	0	0	1
Chemicals (I)	4.0	32	−1	−1	30
Iron and steel (I)	−2.4	−3	4	−1	0
Non-ferrous metals (I)	0.2	0	0	0	0
Non-metallic minerals (I)	1.2	2	1	0	4
Paper and pulp (I)	2.6	17	−4	−2	11
Non-specified industry	8.0	86	−22	−51	13
Norway, 1976–97					
Total	−0.08	1068	−648	−520	−100
Food (E)	0.6	164	−319	−46	−201
Machinery (E)	3.3	1206	−393	−389	424
Transport equipment (E)	−1.3	−202	−165	40	−327
Textiles (E)	1.1	48	−102	−25	−79
Wood (E)	−1.7	−100	82	−24	−43
Chemicals (I)	1.7	105	129	54	289
Iron and steel (I)	1.0	40	−42	−9	−11
Non-ferrous metals (I)	0.8	51	15	3	69
Non-metallic minerals (I)	1.3	86	−100	−31	−46
Paper and pulp (I)	0.1	27	84	3	113
Non-specified industry	−4.2	−357	164	−96	−289

Table 3.5 (continued)

Sector[a]	Annual energy productivity growth	Percentage energy productivity growth attributable to:			
		Intra-sectoral effect	Static shift	Dynamic shift	Total effect
Sweden, 1973–97					
Total	2.84	132	3	−35	100
Food (E)	1.4	7	−1	0	5
Machinery (E)	5.7	95	−10	−27	58
Transport equipment (E)	0.1	0	5	0	5
Textiles (E)	1.1	2	−5	−1	−4
Wood (E)	−1.7	−4	8	−3	1
Chemicals (I)	3.1	10	1	1	12
Iron and steel (I)	4.2	9	−2	−3	4
Non-ferrous metals (I)	2.1	2	0	0	2
Non-metallic minerals (I)	2.3	5	−4	−3	−2
Paper and pulp (I)	1.0	7	3	1	10
Non-specified industry	0.2	1	7	0	8
United Kingdom, 1970–97					
Total	3.10	90	6	4	100
Food (E)	1.6	8	4	2	15
Machinery (E)	1.8	20	4	2	27
Transport equipment (E)	1.9	10	0	0	10
Textiles (E)	0.9	3	−2	0	1
Wood (E)					0
Chemicals (I)	3.0	11	4	5	21
Iron and steel (I)	2.1	5	−2	−2	0
Non-ferrous metals (I)	1.2	1	0	0	1
Non-metallic minerals (I)	3.4	8	−2	−3	2
Paper and pulp (I)	2.1	9	1	1	12
Non-specified industry	3.1	15	−1	−2	12
USA, 1971–94					
Total	4.60	58	30	12	100
Food (E)	1.4	4	2	1	8
Machinery (E)	1.8	8	25	13	46
Transport equipment (E)	2.6	9	−1	−1	7

Table 3.5 (continued)

Sector[a]	Annual energy productivity growth	Percentage energy productivity growth attributable to:			
		Intra-sectoral effect	Static shift	Dynamic shift	Total effect
Textiles (E)	2.4	4	1	1	5
Wood (E)					0
Chemicals (I)	2.9	6	5	4	15
Iron and steel (I)	4.4	9	−3	−5	0
Non-ferrous metals (I)	1.2	1	0	0	1
Non-metallic minerals (I)	−1.1	−1	3	−1	2
Paper and pulp (I)	1.9	6	2	1	9
Non-specified industry	1.8	12	−3	−2	7

Note: [a] In parentheses, 'E' signals an energy-extensive sector and 'I' an energy-intensive sector.

in the Netherlands is associated with the non-specified industry sector, and thus seems predominantly a data quality issue.[9]

In sum, the decomposition analysis reveals that the energy-extensive sectors play an increasingly important role in explaining aggregate energy productivity growth. We find that on the whole, aggregate energy productivity growth is mainly driven by intra-sectoral productivity improvements, which are presumably technology-driven. Concurrently, changes in the sectoral composition of the economy contribute considerably to aggregate energy productivity growth in various countries, while in some other countries changes in the sectoral composition of the economy partly offset energy-efficiency improvements. To further identify the sources of these observed cross-country differences we continue by presenting a regression analysis that tries to explain energy productivity growth per sector on the basis of a suite of potential determinants.

5 EXPLAINING ENERGY PRODUCTIVITY GROWTH

The results presented in the preceding section are useful to obtain a detailed view of the *proximate* sources of energy productivity growth. They clearly reveal the contributions of changing sectoral energy productivity on the one hand, and changing sectoral compositions on the other. Despite their usefulness, the results provide no answer to the more fundamental question regarding the *ultimate* sources of energy productivity growth (see Mulder and de Groot, 2007, for a more detailed discussion in the context of an empirical analysis on the convergence of labour and energy productivity). In order to provide an answer to the query for the ultimate sources of energy productivity growth we adopt a panel data framework in which we regress average energy productivity growth rates on a suite of explanatory variables for each sector. We use the following specification:

$$g_{it} = \mu_i + \sum_{j=1}^{9} \gamma_j x_{it}^j + \varepsilon_{it} \tag{3.4}$$

where i and t denote the cross-country and the time series dimension, respectively, and ε_{it} is the usual error term. Following Islam (1995) we use five-year time intervals in order to reduce the influence of business cycle fluctuations and serial correlation of the errors. Hence, the growth rate (g) in equation (3.4) is an average over a five-year period, so when $t = 1975$, for example, the initial year (t_0) is 1970. We assume ε_{it} to be an independently identically distributed random variable with mean 0 and variance σ_ε^2. The variables μ_i represent unspecified individual country effects. These so-called 'fixed effects' allow for cross-country differences in steady states, and effectively imply the intercepts for the different countries in equation (3.4) are allowed to be different.[10] In addition to unobservable individual country effects, we include a vector of country-specific explanatory variables x_{it}^j. These variables are defined at the sectoral level and include:

- the initial energy productivity level, $x_{it}^1 = \ln{(Y/E)}_{i,t-1}$,
- energy prices, $x_{it}^2 = (p_{E,i,t} + p_{E,i,t-1} + p_{E,i,t-2})/3$,
- the investment share, $x_{it}^3 = (I_{i,t}/Y_{i,t})$,
- an openness indicator, $x_{it}^4 = (XGS_{i,t} + MGS_{i,t}/Y_{i,t})$,
- the Balassa index,

$$x_{it}^5 = \frac{XGS_{i,t}/\sum_{i=1}^{14} XGS_{i,t}}{\sum_{s=1}^{10} XGS_{i,t,s}/\sum_{i=1}^{14}\sum_{s=1}^{10} XGS_{i,t,s}},$$

- a variable measuring economies of scale, $x_{it}^6 = Y_{i,t}/\sum_{s=1}^{13} Y_{i,t}$, and
- the fuel mix indicators, $x_{it}^7 = \ln(\text{Oil/Energy})_{i,t}$, $x_{it}^8 = \ln(\text{Electricity/Energy})_{i,t}$ and $x_{it}^9 = \ln(\text{Natural Gas/Energy})_{i,t}$.

By including the initial level of energy productivity we test for the β-convergence hypothesis, implying that countries with relatively low initial productivity levels tend to grow relatively fast. There are several potential mechanisms that may lead to this tendency: advanced economies may suffer from diminishing returns, lagging countries may benefit from knowledge spillovers and technology diffusion, or production processes may converge due to increasing competition.[11]

Concerning the second variable, we expect a positive correlation between energy prices and energy productivity growth. We took a three-year moving average for the energy price to avoid capturing the effect of short-term price fluctuations, assuming that investments in energy-efficient technologies do respond to a structural trend in energy price developments rather than to short-term fluctuations.

The third explanatory variable allows us to test for the embodiment hypothesis or a vintage effect, assuming that higher investment will contribute to increasing energy productivity growth via technological change embodied in new capital goods (Howarth et al., 1991; Mulder et al., 2003).

We assume that openness will have a positive impact on energy productivity growth, since an open sector faces relatively strong competition as well as exchange of knowledge, which we both assume to have a stimulating effect on productivity growth. The Balassa index is an indicator that measures relative specialization patterns. We assume a positive relationship for each sector between a country's degree of specialization and its level of technological advancement, with subsequent positive effects on energy productivity growth. The indicator for the relative size of a sector within a country captures the potential effect of economies of scale on productivity growth, assuming that a large sector can make substantial investments in R&D and new capital goods, and hence might be a technological leader displaying relatively high productivity growth rates.

Finally, several authors have stressed the fact that changes in energy mix are an important source of aggregate energy productivity developments, because some energy types (for example, natural gas and electricity) are more efficient than others (for example, coal and oil) in terms of available energy (Berndt, 1978; Cleveland et al., 2000; Kaufmann, 2004). We use a logarithmic transformation to capture diminishing returns and leave out coal, because the other three energy types are substitutes for coal. We expect a positive effect of all three energy type shares on average energy productivity growth.

Table 3.6 shows that the estimate of the initial energy productivity level is negative in all sectors and in most cases statistically significant.[12] The data therefore strongly support the existence of energy productivity convergence across all sectors of the economy, possibly with the exception

Table 3.6 Results for panel data regressions of determinants of energy productivity growth per sector

	Agriculture		Services		Trans.	Chemicals		Food		Iron and steel	
$\ln(Y/E)_0$	−0.862	−1.014	−0.804	−1.225	−0.512	−0.781	−0.685	−0.040	−0.629	−0.867	−1.020
	(−4.385)	(−8.323)	(−3.378)	(−6.891)	(−1.967)	(−4.975)	(−4.428)	(−0.106)	(−3.696)	(−5.163)	(−3.329)
Energy price	1.355	−2.174	1.118	−0.200	0.054	1.072	−0.058	1.035	−1.438	3.864	1.225
	(1.148)	(−2.439)	(1.754)	(−0.211)	(0.331)	(1.203)	(−0.056)	(1.345)	(−1.878)	(3.004)	(0.439)
Investment share	0.218	−0.999	−0.531	2.625	−0.017	1.885	−0.500	−0.549	1.157	−1.316	−0.761
	(0.194)	(−1.747)	(−0.678)	(0.744)	(−0.047)	(1.616)	(−0.341)	(−0.539)	(3.686)	(−2.080)	(−0.769)
Openness						0.009	0.010	−0.260	−0.177	0.005	0.033
						0.248	0.273	−2.264	−3.130	0.168	0.506
Balassa index						−0.504	0.092	−0.576	−0.315	0.933	0.592
						−1.271	0.214	−1.463	−2.648	4.079	1.443
Econ. of scale	12.366	12.872	7.179	1.160	16.839	81.712	33.084	−0.759	−1.263	−5.432	6.655
	(1.607)	(3.307)	(2.235)	(0.322)	(3.844)	(4.515)	(1.273)	(−0.069)	(−0.174)	(−0.296)	(0.235)
ln(oil share)		0.125		−0.327			0.172		0.204		−0.034
		(0.693)		(−0.609)			(1.342)		(1.211)		(−0.316)
ln(elect. share)		0.563		0.609			0.631		0.389		0.746
		(4.034)		(1.299)			(2.296)		(3.067)		(1.616)
ln(nat. gas share)		0.070					0.002		0.049		−0.276
		(1.771)					(0.472)		(3.072)		(−0.850)
R^2	0.37	0.91	0.59	0.87	0.63	0.67	0.74	0.2	0.95	0.69	0.73
Standard error	0.2	0.08	0.14	0.07	0.07	0.11	0.09	0.1	0.03	0.13	0.13
F-statistic	2.38	17.77	3.71	8.18	4.75	4.79	5.25	1.48	27.26	5.21	4.5
No. of countries	9	7	5	4	7	9	9	9	7	9	8
No. of observations	29	22	16	12	23	27	27	27	22	28	22

	Machinery		Non-ferrous metal		Non-metallic min.		Paper		Textile	Wood	Mtr.
$\ln(Y/E)_0$	−0.596	−1.418	−0.026	−0.492	−0.953	−0.863	−0.471	−0.788	−0.750	−0.568	−1.144
	(−2.183)	(−4.023)	(−0.091)	(−23.47)	(−5.419)	(−3.414)	(−2.338)	(−4.567)	(−2.774)	(−0.306)	(−3.905)
Energy price	−0.372	0.794	0.584	2.920	0.217	−0.491	1.072	0.060	2.439	−0.947	−1.671
	(−0.317)	(0.636)	(0.525)	(7.151)	(0.136)	(−0.179)	(1.891)	(0.115)	(1.899)	(−0.899)	(−2.825)
Investment share	3.854	−3.603	0.035	−1.694	1.357	1.857	−0.948	0.648	3.129	−0.175	3.420
	(1.201)	(−0.768)	(0.082)	(−7.791)	(1.060)	(1.232)	(−0.927)	(0.899)	(1.189)	(−0.025)	(2.409)
Openness	0.101	0.129	−0.049	−0.012	0.032	−0.046	−0.068	−0.040	−0.174	−0.184	0.202
	(1.039)	(1.102)	(−0.895)	(−1.266)	(0.214)	(−0.221)	(−0.615)	(−0.426)	(−3.196)	(−0.388)	(2.808)
Balassa index	0.657	−0.076	0.083	0.524	−0.178	−0.256	−0.011	0.013	0.347	0.030	0.266
	(0.801)	(−0.077)	(0.382)	(7.013)	(−0.856)	(−0.813)	(−0.126)	(0.252)	(1.186)	(0.184)	(0.514)
Econ. of scale	3.774	37.783	132.989	−110.07	−17.654	−1.337	−5.335	26.238	−49.336	68.612	57.398
	(0.643)	(2.005)	(1.658)	(−2.645)	(−0.537)	(−0.030)	(−0.278)	(1.675)	(−2.288)	(0.206)	(2.147)
ln(oil share)		0.517		−0.456		−0.307		0.195			
		(1.445)		(−9.374)		(−1.278)		(2.655)			
ln(elect. share)		1.155		1.299		0.044		1.086			
		(2.922)		(10.006)		(0.135)		(4.921)			
ln(nat. gas share)		−0.080		0.431		−0.024		0.023			
		(−1.370)		(9.305)		(−0.311)		(0.523)			
R^2	0.26	0.76	0.03	0.98	0.59	0.52	0.38	0.82	0.53	−0.20	0.6
Standard error	0.16	0.09	0.19	0.01	0.15	0.16	0.14	0.07	0.15	0.22	0.17
F-statistic	1.65	4.86	1.06	458.03	3.76	2.53	2.11	6.9	2.98	0.8	3.46
No. of countries	9	7	8	6	9	8	9	8	9	6	7
No. of observations	27	19	25	16	28	24	26	22	26	14	21

of the non-ferrous metals sector. Countries characterized by an initially low level of energy productivity exhibit a relatively high level of energy productivity growth, which indicates a process of catching up.

The energy-price effect is positive in most sectors, as expected. It is, however, statistically significant in a few sectors only, most notably in the iron and steel sector, and to some extent also in the paper and non-ferrous metals producing sectors. This result is intuitively clear in that these sectors are among the most energy intensive. We also find some evidence of energy prices having a statistically significant positive effect on energy productivity growth in the energy-extensive sectors producing services and textiles. In addition, we find that changes in the fuel mix have the expected and statistically significant effects on energy productivity growth in various sectors, most notably in iron and steel, and non-ferrous metals. The impact of the other explanatory variables on energy productivity growth is, however, both mixed and rather limited. We find hardly any statistically significant positive effect of openness, investment, economies of scale or specialization on sectoral energy productivity growth.

6 CONCLUSIONS

In this chapter we presented an empirical analysis of energy productivity trends for 14 OECD countries and 13 sectors over the period 1970–97. Much emphasis has been put on tracing back macroeconomic productivity developments to developments at the level of individual sectors, in particular for the energy-extensive service sector and light industry. A cross-country decomposition analysis reveals that aggregate energy productivity growth is mainly driven by intra-sectoral productivity growth, which is presumably technology driven. These efficiency improvements are mainly realized within the manufacturing sector.

We find that the energy-extensive sectors play an increasingly important role in explaining aggregate energy productivity growth. In addition to the energy-intensive sectors chemicals and paper, the highest percentage contribution to aggregate manufacturing efficiency improvements comes from the (relatively) energy-extensive sectors machinery, food and also textiles and transport equipment. At the same time, in some countries changes in sectoral composition contribute considerably to aggregate energy productivity growth, while in other countries these changes partly offset energy efficiency improvements. The largest structural shift is apparent in the manufacturing to services trend. Within the manufacturing sector the observed structural changes are in general a shift away from energy-intensive sectors to energy-extensive sectors. Finally, we find that the catch-up potential, energy prices, and changes in fuel mix positively affect energy productivity growth in various sectors, whereas economies of scale, investment share, openness and specialization at best seem to play a very limited role in explaining cross-country variation in energy productivity growth.

NOTES

1. There are studies with a more refined sectoral composition, but they cover fewer countries and/or a shorter time span.
2. For a detailed description of the dataset, including descriptive statistics, we refer to sections 1 and 2 of a separate document that can be downloaded from http://www.henridegroot.net.
3. The sector non-specified industries also includes manufacturing energy consumption for which no further breakdown into manufacturing sub-sectors could be given. In most countries, non-specified industries are rather unimportant with an average share of only 2 per cent of total energy consumption. Exceptions are France, Norway and the Netherlands where a substantial part of efficiency improvements materializes in the non-specified industry sector, and where the share of this sector is on average approximately 12 per cent. As a result, caution is required in interpreting the results, since an efficiency improvement and a changing energy share in non-specified industries is partly due to developments in the sectors constituting the non-specified industry sector (specifically, ISIC 332, 355, 356 and 385), but it might also partly be due to data inaccuracy.
4. For example, for the USA we had to exclude data for the manufacturing sub-sectors starting from 1995. The reason is that until 1994 the energy consumption data in the IEA Energy Balances for the different manufacturing sub-sectors in the USA include coal and electricity consumption only, with all manufacturing consumption of oil products and natural gas being collected in the sector non-specified industry. As of 1995, data on oil products and gas consumption are available by manufacturing sub-sector. However, to prevent biased results due to this inconsistency in the data series, we include the data until 1994 only. Moreover, we calculate energy productivity growth rates for the different manufacturing sub-sectors by defining final energy consumption as the sum of coal and electricity consumption only, assuming that the share of the sum of oil and gas in final energy consumption is more or less constant over time. Hence, the documented growth rates of energy productivity for the breakdown of the US manufacturing sector should be interpreted with caution.
5. The macroeconomic level is defined here as the sum of the manufacturing, transport, services and agricultural sectors.
6. In interpreting the results that follow, it is important to keep in mind that sectoral effects are based on changes in shares of energy consumption rather than shares of value added. Several previous studies have looked at energy intensity, instead of following the productivity approach that we use. In those studies, sectoral effects are based on effects related to shares of value added. We also performed a decomposition analysis using this approach, and the qualitative results are very similar to those presented here. Details are available upon request from the authors.
7. Note that the intra-sector effect, the static shift effect and the dynamic shift effect horizontally sum up to the total effect per sector, and that the effects for the four sectors vertically sum up to the total of the effects (that is, the sum of the intra-sector, the static shift, the dynamic shift and the total effect).
8. We note, however, that this particular result might be due to poor quality of the data.

See, for example, Ramírez et al. (2002) who in a detailed analysis of the Dutch service sector for the period 1984–98 only found a minor increase in energy productivity that was hardly affected by structural changes.

9. We also performed a decomposition that explicitly distinguishes between the energy-intensive and the energy-extensive sector as well as the non-specified industries. This variant obviously provides less detailed results, but the qualitative implications of the results are similar to those described here. Detailed results are available from the authors upon request.
10. For a more detailed discussion of the most appropriate estimation technique and specification tests in this particular context, we refer to Mulder and de Groot (2007). There are compelling theoretical as well as econometric arguments to choose a fixed effects approach, and we present only the fixed effects results in Table 3.5. Results for other specifications, specifically pooled ordinary least squares, random effects and a Mundlak specification, are available upon request from the authors.
11. The concept of productivity convergence has its roots in traditional neoclassical growth theory, with its central notion of a transitional growth path to a steady state. A discussion of the empirical macroeconomic growth literature is beyond the scope of this chapter. For good surveys we refer to Abreu et al. (2005), Barro and Sala-i-Martin (1995), Broadberry (1996), Durlauf and Quah (1999), Fagerberg (1994), Economic Journal (1996) and Islam (2003). Most of the literature focuses on convergence of per capita income, labour productivity or total factor productivity. For and analysis of energy productivity convergence see Mulder and de Groot (2007), Mulder (2005) and Miketa and Mulder (2005).
12. We omit the aggregate manufacturing sector, because of limited data availability for various explanatory variables, due to the fact that the exact period of data availability varies per sub-sector. We also controlled for different specifications of various explanatory variables. For energy prices we also included: current prices, a five-year moving average, as well as the logarithm of three- and five-year moving averages. For investment share we also included $(I/Y)_{t-1}$, (I/K), $(I/K)_{t-1}$ and $\ln(I/K)_{t-1}$, as well as an interaction term of investment share and log initial energy productivity, $\ln(Y/E)_0 \times (I/Y)$. For fuel mix we also included the energy shares without applying the logarithmic transformation. All these specifications did not substantially alter the estimates. Details are available from the authors upon request.

REFERENCES

Abreu, M., R.J.G.M. Florax and H.L.F. de Groot (2005), 'A meta-analysis of β-convergence: the legendary two percent', *Journal of Economic Surveys*, **19**, 389–420.

Ang, B.W. (1995a), 'Decomposition methodology in industrial energy demand analysis', *Energy*, **20**, 1081–95.

Ang, B.W. (1995b), 'Multilevel decomposition of industrial energy consumption', *Energy Economics*, **17**, 39–51.

Ang, B.W. (1999), 'Decomposition methodology in energy demand and environmental analysis', in J.C.J.M. van den Bergh (ed.), *Handbook of Environmental and Resource Economics*, Cheltenham, UK and Northampton, MA, USA: Edward Elgar Publishing.

Ang, B.W. and F.Q. Zhang (2000), 'A survey of index decomposition analysis in energy and environmental studies', *Energy*, **25**, 1149–76.

Barro, R.J. and X. Sala-i-Martin (1995), *Economic Growth*, New York: McGraw Hill.

Berndt, E.R. (1978), 'Aggregate energy, efficiency, and productivity measurement', *Annual Review of Energy*, **3**, 225–73.

Broadberry, S.N. (1996), 'Convergence: what the historical record shows', in B. van Ark and N. Crafts (eds), *Quantitative Aspects of Post-War European Growth*, Cambridge: Cambridge University Press.

Cleveland C.J., R.K. Kaufmann and D.I. Stern (2000), 'Aggregation and the role of energy in the economy', *Ecological Economics*, **32**, 301–18.

Dollar, D. and E.N. Wolff (1993), *Competitiveness, Convergence, and International Specialization*, Cambridge, MA: MIT Press.

Durlauf, S.N. and D. Quah (1999), 'New empirics of economic growth', in J.B. Taylor and M. Woodford (eds), *Handbook of Macroeconomics*, vol. 1, Amsterdam: North Holland Elsevier Science.

Economic Journal (1996), 'Controversy on the convergence and divergence of growth rates', *Economic Journal*, special issue, **106**, 1016–69.

Eichhammer, W. and W. Mannsbart (1997), 'Industrial energy efficiency: indicators for a European cross-country comparison of energy efficiency in the manufacturing industry', *Energy Policy*, **25**, 759–72.

Fagerberg, J. (1994), 'Technology and international differences in growth rates', *Journal of Economic Literature*, **32**, 1147–75.

Fagerberg, J. (2000), 'Technological progress, structural change and productivity growth', *Structural Change and Economic Dynamics*, **11**, 393–411.

Greening, L.A., W.B. Davis and L. Schipper (1997), 'Comparison of six decomposition methods: application to aggregate energy intensity for manufacturing in ten OECD countries', *Energy Economics*, **19**, 375–90.

Howarth, R.B., L. Schipper, P.A. Duerr and S. Strøm (1991), 'Manufacturing energy use in eight OECD countries', *Energy Economics*, **13**, 135–42.

Islam, N. (1995), 'Growth empirics: a panel data approach', *Quarterly Journal of Economics*, **4**, 1127–70.

Islam, N. (2003), 'What have we learnt from the convergence debate?', *Journal of Economic Surveys*, **17**, 309–62.

Jorgenson, D.W. (1984), 'The role of energy in productivity growth', *Energy Journal*, **5**, 11–26.

Kaufmann R.K. (2004), 'The mechanisms for autonomous energy efficiency increases: a cointegration analysis of the US energy/GDP ratio', *Energy Journal*, **25**, 63–86.

Miketa, A. (2001), 'Analysis of energy intensity developments in manufacturing sectors in industrialized and developing countries', *Energy Policy*, **29**, 769–75.

Miketa, A. and P. Mulder (2005), 'Energy-productivity convergence across developed and developing countries in ten manufacturing sectors', *Energy Economics*, **27**, 429–53.

Mulder, P. (2005), *The Economics of Technology Diffusion and Energy Efficiency*, Cheltenham, UK and Northampton, MA, USA: Edward Elgar Publishing.

Mulder, P. and H.L.F. de Groot (2003), 'International comparisons of sectoral energy- and labour-productivity performance: stylised facts and decomposition of trends', CPB Discussion Paper, no. 22, The Hague: Netherlands Bureau for Economic Policy Analysis.

Mulder, P. and H.L.F. de Groot (2007), 'Sectoral energy- and labour-productivity convergence', *Environmental and Resource Economics*, **36**, 85–112.

Mulder, P., H.L.F. de Groot and M.W. Hofkes (2003), 'Explaining slow diffusion of energy-saving technologies: a vintage model with returns to diversity and learning-by-using', *Resource and Energy Economics*, **25**, 105–26.

Park, S.H., B. Dissmann and K.Y. Nam (1993), 'A cross-country decomposition analysis of manufacturing energy consumption', *Energy*, **18**, 843–58.

Peneder, M. (2003), 'Industrial structure and aggregate growth', *Structural Change and Economic Dynamics*, **14**, 427–48.

Ramírez, A., M. Patel and K. Blok (2002), 'An analysis of the Dutch service sector: energy consumption and trends', Report NWS E-2002-47, Utrecht: Copernicus Institute, Department of Science, Technology and Society, Utrecht University.

Schipper, L. and S. Meyers (1992), *Energy Efficiency and Human Activity: Past Trends and Future Prospects*, Cambridge: Cambridge University Press.

Syrquin, M. (1984), 'Resource allocation and productivity growth', in M. Syrquin, L. Taylor and L.E. Westphal (eds), *Economic Structure Performance: Essays in Honor of Hollis B. Chenery*, Orlando, FL: Academic Press.

Timmer, M.P. and A. Szirmai (2000), 'Productivity growth in Asian manufacturing: the structural bonus hypothesis re-examined', *Structural Change and Economic Dynamics*, **11**, 371–91.

Unander, F., S. Karbuz, L. Schipper, M. Khrushch and M. Ting (1999), 'Manufacturing energy use in OECD countries: decomposition of long term trends', *Energy Policy*, **27**, 769–78.

Wagner, K. and B. van Ark (1996), *International Productivity Differences: Measurement and Explanation*, Contributions to Economic Analysis vol. 233, Amsterdam: Elsevier.

4. Using physical indicators to monitor energy efficiency in energy-extensive sectors

Andrea Ramírez, Martin K. Patel and Kornelis Blok

1 INTRODUCTION

Defining and measuring energy efficiency as well as devising specific programmes encouraging energy efficiency constitute challenging tasks. Energy efficiency improvement refers to using less energy for producing the same amount of services or useful output. The measurement of (technical) energy efficiency at the lowest level of aggregation, for instance a machine, is simple and straightforward. However, policy-makers are generally interested in higher levels of aggregation, such as energy efficiency of an industrial sector or a country. In this case, energy efficiency cannot be measured directly and it therefore has to be analysed by the use of surrogate measures or indicators. An energy efficiency indicator should be easily observable with little or no time lag, close to the policy actions in the sense that it is quickly affected by the policy undertaken, and related to the target and goal variables. There is general consensus that an energy efficiency indicator should relate the amount of energy use to the useful output or activity by means of a ratio.[1] Specifically, an efficiency indicator equals the amount of energy use divided by the amount of activity.

Activity can be measured in economic or physical terms (for example, value-added or tonnes of product, respectively). In the first case, the indicator is referred to as economic energy intensity, in the second case as physical energy intensity. In the literature, energy efficiency is often analysed by examining changes in economic energy intensity. Energy per unit of gross domestic product (*E*/*GDP*) is the most common indicator of energy intensity. The value of the measure lies in its simplicity and therefore the relative ease of cross-country comparisons. However, it has long been recognized that *E*/*GDP* is a highly aggregate measure that provides little insight into the reasons for differences between countries as well as the dynamics over time. For instance, changes in *E/GDP* are due not only to changes in energy efficiency, but also to structural changes within the economy, changes in goods traded with other countries and changes in fuel mix. The fact that there is not a direct relationship between the economic output of an activity and energy use constitutes the main drawback of economic energy intensity indicators.

On the contrary, changes in physical energy intensity can be directly associated with changes in energy efficiency, because physical measures relate energy use to the unit of physical service provided. In comparisons of monetary-based and physical-based energy indicators it has been recognized that physical energy intensity is a better indicator of energy efficiency than economic energy intensity. For instance, Freeman et al. (1997) examined different measures of industrial output for use in constructing estimates of industrial energy efficiency. They found that economic energy intensities seem to be a poor source of information for policy-makers. Farla (2000) concluded that physical indicators are preferable to indicators based on economic values, because the latter do not properly correct for the differences in output mix between different countries (that is, structural differences). Nanduri et al. (2002), Nyboer et al. (1996), Ross and Hwang (1992) and Williams et al. (1987) report similar results.

Physical energy intensity has become the preferred way of analysing energy efficiency developments in energy-intensive industries, such as iron and steel (Ozawa et al., 2002), pulp and paper (Farla et al., 1997), ammonia (Rafiqul et al., 2005), the transport sector (He et al., 2005) and buildings/households (Haas, 1997). The main drawbacks of the development of physical indicators is not so much the lack of strength of the indicator in measuring changes in energy efficiency, but rather the larger amount of data needed, the availability and reliability of the data, and in the case of heterogeneous sectors/processes such as energy-extensive sectors, the incommensurability of physical units (for example, tonnes of potatoes versus litres of milk; Golove and Schipper, 1997; Nanduri et al., 2002; Sinton and Levine, 1994).

Against this background, and taking into account new developments such as the European Union (EU) draft directive on energy end-use efficiency and energy services which has as one of its main target groups non-energy intensive sectors (European Commission, 2003a), the main goal of this chapter is to examine whether physical energy-intensity indicators provide a feasible way of analysing changes in energy efficiency in energy-extensive sectors. We use the meat and dairy sectors as case studies of the energy-extensive sector.

The structure of this chapter is as follows. Section 2 covers the main aspects of the methodology used, and section 3 describes the data we use.

Section 4 provides a brief description of the developments in energy use in the meat and dairy sectors. In section 5, physical energy efficiency indicators are presented, and a discussion of the results is provided in section 6. Conclusions are given in section 7.

2 METHODOLOGY

The development of physical energy intensity for processes that generate a single product (or use a main raw material) is straightforward: the physical energy intensity indicator is defined as the ratio of the energy used in a sector (E_N) to the physical amount of the product or raw material (RM_N). We refer to this indicator as PEI_1:

$$PEI_1 = \frac{E_N}{RM_N} \tag{4.1}$$

However, if more than one product is generated, which is generally the case at higher levels of aggregation, for instance at the level of an industrial sector, the physical energy intensity indicator needs to account for heterogeneous products. Heterogeneous products will have different energy requirements (for instance, tonnes of butter versus tonnes of milk powder), and hence results provided by PEI_1 may be partly due to changes in the relative share of the different products (product mix) and partly due to changes in energy efficiency. It is possible, however, to correct equation (4.1) for differences in product mix, while at the same time dealing with the problem of incommensurability of physical units. Following the methodology developed by Worrell (1994), Phylipsen et al. (1998) and Farla (2000), the physical energy intensity indicator PEI_2 of a sector N that produces products i (= 1, 2, . . ., n) in year j can be expressed as:

$$PEI_{2,N,j} = \frac{E_{N,j}}{\sum_i^n m_{i,j} \cdot SEC_{i,0}} = \frac{E_{N,j}}{\sum_i^n m_{i,j} \cdot \left(\frac{E_{i,0}}{m_{i,0}}\right)} \tag{4.2}$$

where E is the energy consumption, m the production quantity, $SEC_{i,0}$ a specific energy consumption of product i in a reference year defined as the amount of energy needed in the reference year to produce one unit of physical product i, for instance GJ/tonne product. $SEC_{i,0}$ reflects the type of process, the technology and the efficiency level used to produce each product i in the reference year. Equation (4.2) estimates energy efficiency as the ratio between the real energy use $E_{N,j}$ and a 'frozen' energy efficiency development: the energy that would have been used if there were no increases in energy efficiency, in other words, if energy efficiency had remained equal to the reference year. The use of $SEC_{i,0}$-values as multiplier allows physical amounts of products to be transformed into energy amounts, which can then be directly added up. Note that since the frozen efficiency development accounts for yearly changes in the structure of production, PEI_2 is an indicator that corrects for structural changes. In international comparisons of energy efficiency one common set of references should be used for all countries. Depending on data availability this can be best practice, or typical or reference values for one of the countries. This method is in essence identical to the one used by the Dutch Energy and Environmental Agency (SenterNovem) to monitor improvements in energy efficiency as a consequence of the so-called Long-Term Agreements (LTAs).[2]

3 DATA

In this section we present results for the meat (NACE 151) and the dairy industry (NACE 155) of four European countries (France, Germany, the Netherlands and the UK) following the classification of economic activities in the European Union (NACE). The availability of energy and production data determined the time periods for which the analysis was carried out. These time periods are 1986–2000 for France, 1989–2000 for the Netherlands, 1993–2000 for Germany, and 1990–2000 for the UK. A detailed account of the data is given below.

3.1 Energy Data

Energy data were taken from the energy supply statistics of the Netherlands (CBS), France (Agreste), Germany (Statistisches Bundesamt) and the UK (Department of Trade and Industry). System boundaries for energy data are shown in Figure 4.1. Data availability depends on the level of aggregation. At the three-digit level of statistical aggregation, which is the relevant degree of detail for the dairy and meat industry, only net available energy data are available. The main implication is that the role of cogeneration of heat and power (CHP) in the development of the energy efficiency indicators cannot be assessed at the three-digit level. We will discuss this point further when analysing the results.

In this chapter we express energy use in terms of net calorific values.[3] We correct the British and German energy figures, which are based on gross calorific values, with the following net/gross ratios: 0.95 for coal and oil,

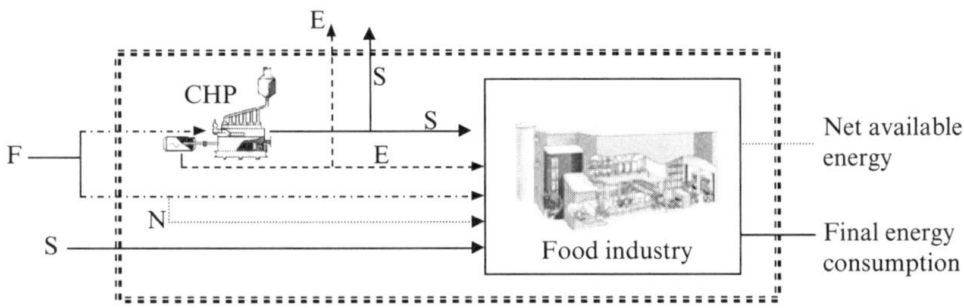

Note: CHP = cogeneration of heat and power, E = electricity, F = fossil fuels, S = steam and N = non-energy use.

Figure 4.1 System boundaries

and 0.9 for natural gas. Primary energy is calculated using the following conversion factors from final to primary:[4] 1.05 for oil products, 1.01 for natural gas, 2.5 for electricity bought from the grid (reflecting a 40 per cent conversion efficiency)[5] and 1.11 for heat from boilers (reflecting a 90 per cent conversion efficiency). For steam produced from a CHP unit, steam has been valued at 83 per cent of its energy content.[6] We have corrected energy data for climate influences using the Eurostat temperature method, assuming an average heating share of 20 per cent based on data published in European Commission (2003b) and RIVM (1995).[7]

3.2 Reference Levels for Energy Consumption Data

As specified in section 2, we use the amounts of energy needed to produce each product as weight factors in the calculation of the physical energy intensity indicator PEI_2. We use the label *SEC* to refer to the reference level energy consumption. There are several options in choosing *SEC* reference values. The reference values can reflect best available technologies, best practice, the best process or typical values for a country. Due to the large number of processes and products and the limited availability of *SEC* reference values, we use values that reflect average technologies in use in the mid-1990s. *SEC* reference values should then be obtained for all key products. In the case of the dairy industry, the key products selected are: fluid milk, fermented milk products (yogurt, cream, desserts, buttermilk), cheese (fresh, processed cheese and quark), butter, milk powder (whole milk powder, non-fat milk powder, semi-skimmed milk powder, buttermilk powder, cream powder), condensate milk (sweetened and unsweetened condensed milk and coffee milk), whey products (whey powder, whey protein concentrates), caseins and lactose. In the case of the meat industry, we identified as key

Table 4.1 Specific energy consumption values in the dairy and meat industry[a]

Product	Unit	*SEC*[b]
Dairy sector		
Liquid milk & fermented products	MJ/tonne final product	1100
Cheese	MJ/tonne final product	5100
Butter	MJ/tonne final product	2200
Milk powder	MJ/tonne final product	12 100
Condensed milk	MJ/tonne final product	2700
Whey products (powder)	MJ/tonne final product	12 700
Caseins & lactose	MJ/tonne final product	6400
Meat sector		
Beef, veal and sheep	MJ/tonne dress carcass weight	1390
Poultry	MJ/tonne dress carcass weight	3096
Pork	MJ/tonne dress carcass weight	2097
Meat products	MJ/tonne final product	5500
Rendering	MJ/tonne of raw material processed	1625

Notes:

[a] In primary energy, typical technologies for the mid-1990s.

[b] Data sources: Arcadis (2000), Suijkerbuijk et al. (1995), Pontopiddan and Hansen (2000, 2001), Nielsen (2004) and Schreurs (2004).

products: beef/veal meat, goat meat, sheep and lamb meat, poultry meat, pig meat, rendered products and processed products. Despite extensive efforts to gather information on *SEC* reference levels for processed meat products, only aggregate figures were obtained. An overview of the *SEC* reference values used for the meat and dairy sector is shown in Table 4.1.

3.3 Production Data

Since the goal of this chapter is not only to study the development in energy efficiency for the last few years but also to develop a methodology that can be applied in the future, data sources for physical production were selected according to three conditions: the data must be compiled annually, they must be easily available, and the source should be reliable. In this chapter, we rely primarily on sources from industrial associations of the individual countries: Agreste, Zentrale Markt- und Preisberichtstelle (ZMP), Onilait, La maison du lait, Prodzuivel, Productschappen Vee, Vlees and Eieren, and Defra. We also use international data sources (for instance, FAOSTAT, and UN Comtrade) for closing data gaps, and interviews for cross-checks.

A special point to be taken into account is that in the meat sector, production data associated with the processing of meat by-products, known as rendering, are not complete. The reason is that most of the products are not processed for sale because they constitute fallen stock or offal that is considered a danger to public health, and therefore by-products do not appear in most production statistics. To account for this factor, we use the amount of raw material processed, which avoids underestimating the contribution of rendering in the energy consumption of the sector. As there is a lack of statistical data on the total annual amount of raw material to be rendered, we assume that: the total weight of raw material to be rendered per beast is typically 198 kg for cattle, 21 kg for pigs, 14 kg for sheep and 0.7 kg for poultry, and fallen stock represents 10 per cent of the tonnage processed annually (Meat Livestock Commission, 1998). The total amount of raw material to be rendered is calculated using data on the total number of slaughtered animals which are available from national statistics.

4 ENERGY USED BY THE MEAT AND DAIRY INDUSTRY

Figure 4.2 shows the primary energy demand of the food sector compared to the total Dutch manufacturing industry, and its distribution by food sub-sector. The shares shown in the figure for the Dutch industry are representative of the situation in the other countries. In the Netherlands, the meat and dairy sector[8] accounts for 24 per cent of the primary energy used by the food industry; the shares in the other countries are between 22 and 28 per cent. Interestingly, and despite the strong role that energy plays in the meat and dairy industry because bacterial growth is controlled by heat treatment, energy accounts only for a limited share (1–3 per cent) of total production costs (Ramírez, 2005). The low share of energy costs in the total production cost is a typical characteristic of energy-extensive sectors.

Figures 4.3 and 4.4 show the historical development of net energy consumption in the dairy and meat industry by type of fuel and by country. The breakdown shows that, in general, fuel use in the dairy and meat industry have shifted in the period under consideration: the share of natural gas has increased at the expense of coal and petroleum products. The situation in the Netherlands is an exception since it already has a large share of natural gas use at the beginning of the time period. This large share can be explained not only because the country has the second largest gas reserves in Europe after Norway, but also because gas penetration in

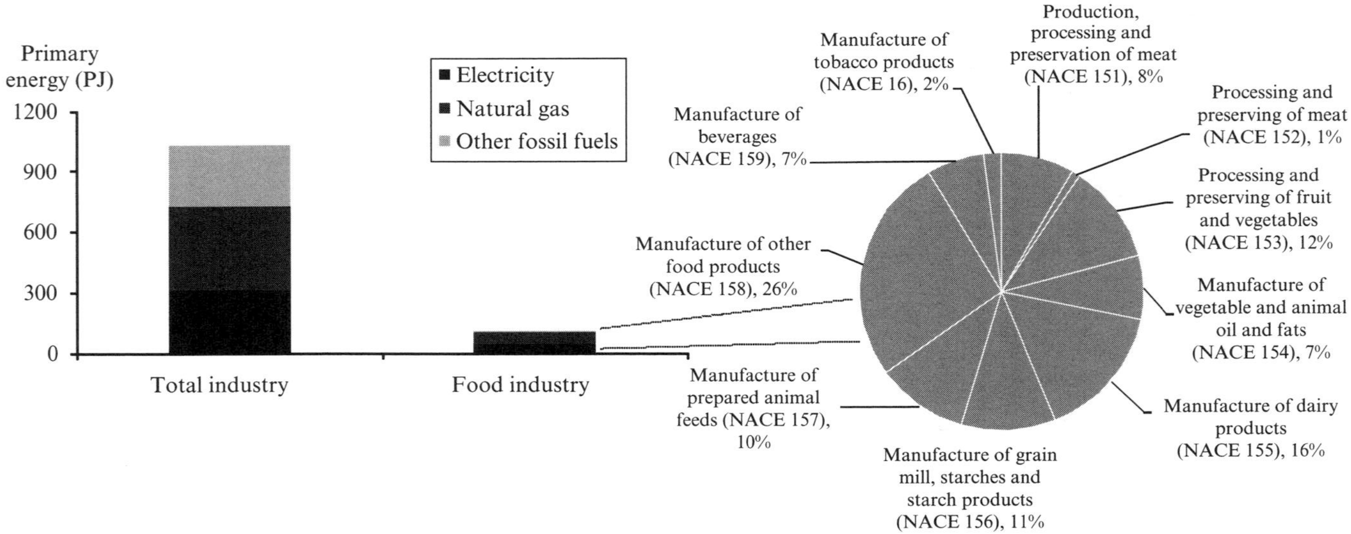

Figure 4.2 Primary energy use in the Dutch food and tobacco industry in PJ and percentages, 2001

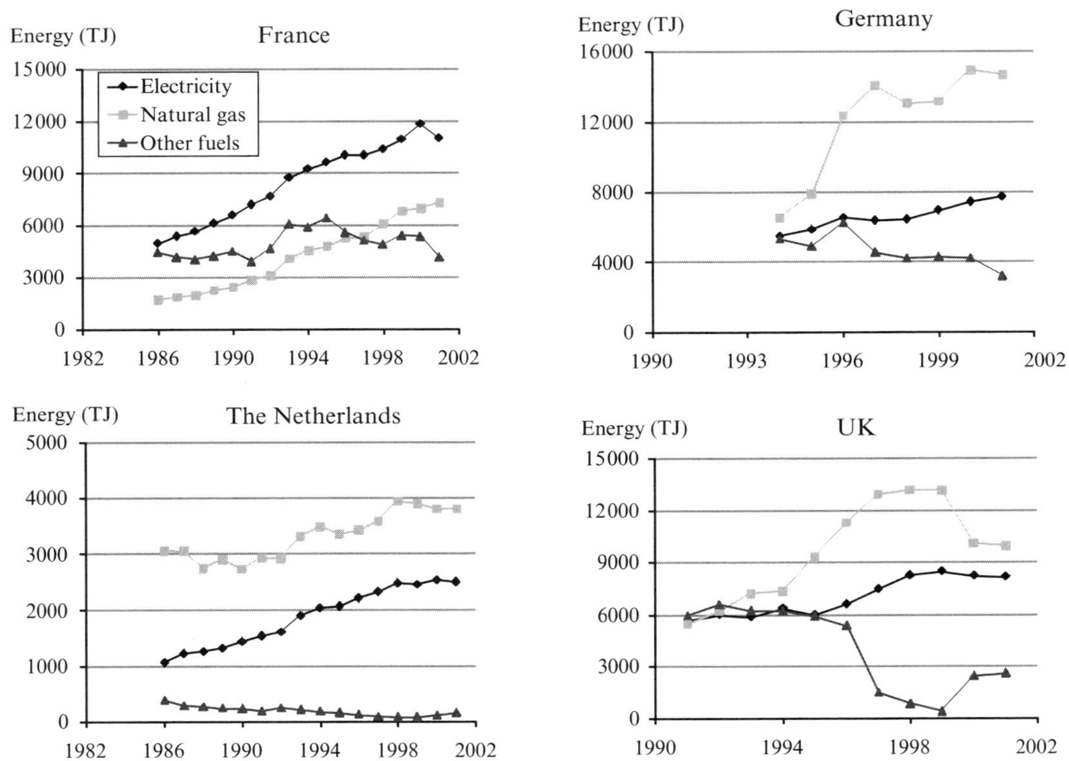

Figure 4.3 *Trends in final net energy consumption in the meat industry by type of fuel and country*

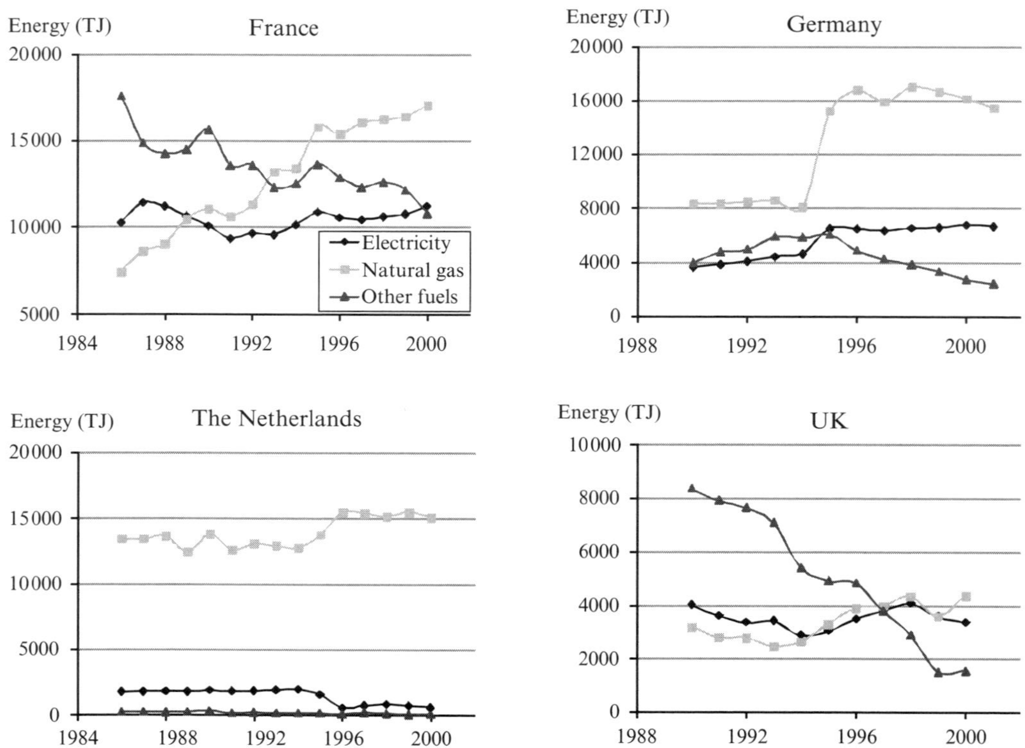

Figure 4.4 *Trends in final net energy consumption in the dairy industry by type of fuel and country*

the Netherlands has been characterized as 'the highest in the world, virtually every home, office and factory is connected to the gas grid' (IEA, 2000). The replacement of coal and petroleum products such as heavy fuel oil by natural gas can to some extent be associated with increasingly stringent environmental regulations, and increased competitiveness of natural gas and in some cases electricity.

5 DEVELOPMENTS IN ENERGY EFFICIENCY

Figures 4.5 to 4.8 show the development of the physical energy efficiency indicators of the meat and dairy sectors. We plot two different kinds of physical energy intensity indicators. The first indicator, PEI_1, is the ratio of energy use per unit of raw material processed (animal slaughtered or raw milk processed). The second indicator, PEI_2, is the comparison of realized energy use and a frozen energy efficiency development. The results in both figures are expressed in terms of primary energy. Figures 4.5 and 4.6 reveal a similar conclusion: the meat industry has increased its primary energy consumption per unit of weighted output during the last decade. The difference lies in the magnitude of the increase, with PEI_1 showing higher requirements of energy per unit of output. However, Figures 4.7 and 4.8 reveal quite different trends. The PEI_1-indicator shows that with exception of the British dairy industry, which reduced the use of energy

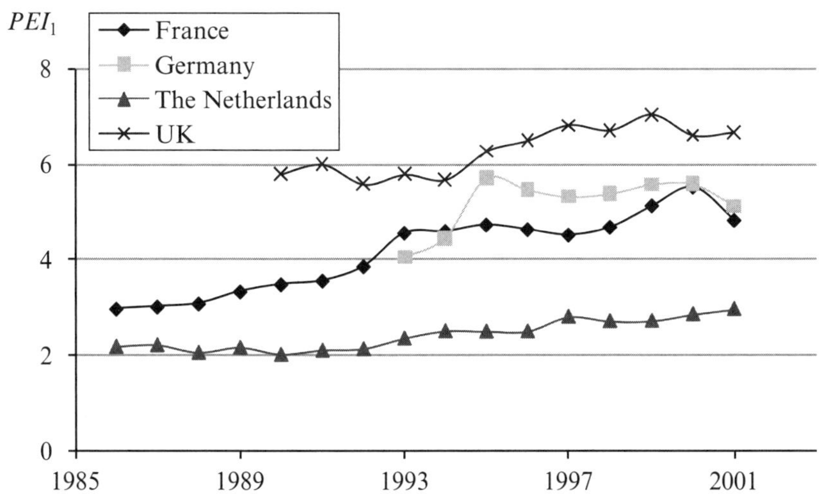

Figure 4.5 PEI$_1$ *in the meat industry by country, 1986–2001*

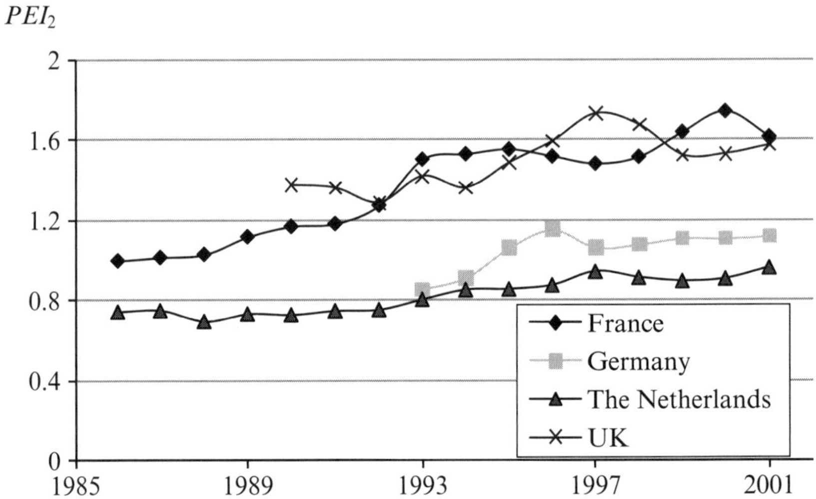

Figure 4.6 PEI_2 *in the meat industry by country, 1986–2001*

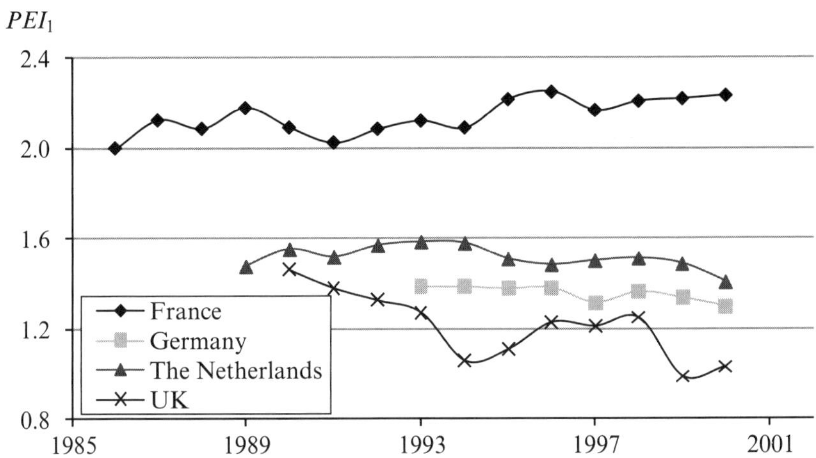

Figure 4.7 PEI_1 *in the dairy industry by country, various years–2000*

per tonne of milk processed at an average rate of 3.2 per cent per annum, improvements in the indicator have been either minor, as in the Dutch and German dairy industry (< 1 per cent per annum), or non-existent (the French industry shows increases of 0.7 per cent per annum). However, since PEI_1 does not take into account the differences in production mix

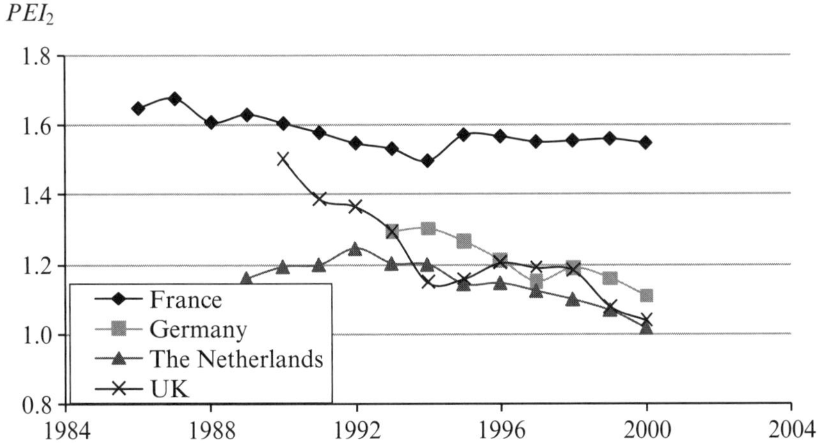

Figure 4.8 PEI_2 *in the dairy industry by country, various years–2000*

across countries and over time, the changes shown by this indicator may be caused by both changes in efficiency and changes in product mix, for instance, production of more energy-intensive products such as milk powders. The results for PEI_2, which accounts for structural changes, are quite different. PEI_2 indicates that the dairy industries in Germany, the Netherlands and the UK have reduced their PEI_2 values by more than 1 per cent annually, and that PEI_2 values for the French dairy industry have declined at a significantly lower rate (0.4 per cent per annum). The results also show that in absolute terms, the French dairy industry demands a higher amount of energy per unit of product.

6 ANALYSIS OF RESULTS

The comparison of the two indicators PEI_1 in Figures 4.5 and 4.7, and PEI_2 in Figures 4.6 and 4.8 allows us to explore the importance of changes in structure, such as changes in the product mix. For instance, when differences in product mix are not taken into account, the British dairy industry appears the least energy intensive, while accounting for product mix identifies the Dutch dairy sector as the industry using the least energy per weighted product. This is not an unexpected result. The UK stands out for processing over half of their raw milk for consumption as liquid milk and fresh milk products. The fact that these products demand the least energy during processing (see Table 4.1) explains why the British dairy industry shows a lower demand of energy per unit of processed raw milk.

Furthermore, the fact that average annual changes in both indicators are almost equal for the British dairy industry reflects that the industry has not witnessed significant product mix changes during the time period considered.

We found that although PEI_1 has several appealing characteristics (it is easy to calculate, requires only few data, can be understood by non-specialists and can be easily communicated), it has a main drawback: it does not reflect important differences in product mix among countries, nor does it account for changes in the product mix during the time period under consideration. The PEI_2-indicator is a much more complex indicator to calculate and the data burden is higher. However, it accounts for differences in structure among countries and changes in product mix. Since we found these factors to be important in the dairy and meat industry, PEI_2 should be preferred when developing indicators for energy efficiency. For this reason, the rest of the analysis in this chapter will be based on the trends revealed by the PEI_2 indicators.

Once changes in product mix have been taken into account our results show opposite developments in the energy efficiency indicators for the dairy and meat industry; see Figures 4.6 and 4.8. For the meat sector, we found an increase in primary energy consumption per unit of weighted output during the last decade (3.2 per cent per year in France, 3.4 per cent in Germany, 1.4 per cent in the UK and 1.6 per cent in the Netherlands).[9] For the dairy sector, we found an improvement in the physical energy intensity indicator. The values of the indicator PEI_2 have decreased in each country: in France by 0.4 per cent, in Germany 2.1 per cent, in the Netherlands 1.2 per cent, and in the UK 3.8 per cent per year. In the rest of this section we explore some of the drivers behind these trends.

6.1 The Meat Sector

The increasing energy demanded by the meat sector can be explained by two main factors. On one hand, increasing demand is caused by the change in consumer preferences from beef towards pig meat and poultry[10] since poultry and pork slaughtering are more energy intensive than cattle slaughtering, which is due mainly to the energy consumed in dehairing, defeathering and singeing operations.[11] On the other hand, the meat industry has progressed from the sales of carcasses, complete with fat and bone, to large bone-in or boneless cuts that are vacuum packaged, fully trimmed products and pre-cooked and ready-to-eat products, which increase the demand in electricity and fuel. To understand the deterioration in energy efficiency is, however, more complex. We therefore explore two main trends in the meat industry and assess their influence on the PEI_2-indicator. These

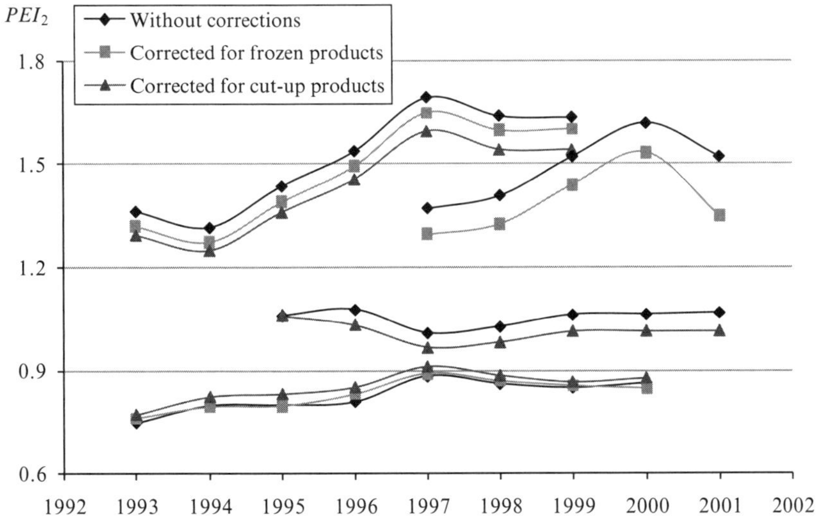

Figure 4.9 Influence of different corrections in PEI_2 *of the meat sector by country*

trends pertain to shifts in meat products towards fast food and cut-up and deboned products, and the increasing demand for food security.

6.1.1 Fast food and cut-up and deboned products

Two main factors that could explain changes in the demand of energy per unit of product in slaughterhouses are changes in the share of frozen products, and changes in the share of cut and deboned products. From an energy point of view the increase or decrease of frozen products is important because freezing increases electricity demand. Likewise, the trends towards cut-up and deboned products imply an increase in electricity and fuel demand. Higher electricity consumption is a consequence of the increasing use of automated cut-up equipment and the temperature conditions in which cutting and deboning can take place. Increased heat consumption is a consequence of an increasing demand for hot cleaning water. Figure 4.9 shows the effect of correcting PEI_2 by including frozen products or by cut-up products for the whole meat and for the slaughtering sector. We find that correcting for cutting and freezing decreases the absolute value of the indicator. It also tends to decrease the gap between countries and the correction can explain some of the fluctuations showed by the trends at a disaggregate level, although it does not explain the increase of the indicator for the whole meat sector.

Another possible reason for the increase in PEI_2 of the meat sector as shown in Figure 4.8 could be the use of more energy-intensive processes in the processed meat sector (NACE 1513). The central problem we face is that not enough information is available to distinguish among different types of processed meat products and the related energy consumption, thus the PEI_2 trends do not correct for changes in the production mix between further processed meat products. It is possible however to highlight some developments that potentially explain the increasing energy demand per product. One is the increasing number of new products, which could increase energy demand because, if processed in small amounts, production lines would be used for different products with a larger amount of energy used during cleaning and start-ups. A report from the Dutch Ministry of Economic Affairs (1999) on energy efficiency developments in the Dutch food sector indicates that the introduction of new products in the meat sector resulted in higher energy consumption, due to the introduction of extra process stages (including frying). Furthermore, the Dutch Long-Term Agreements for the meat sector stated that in the year 2000 about 8 per cent of the total energy used by the companies that signed the covenant was due to the introduction of new products. Other factors are structural changes in production, particularly towards more energy-intensive products, increasing packaging and automation, which increases the amount of electricity demand, and increasing food hygiene standards.

6.1.2 Increasing demand for food security

The food scandals associated with the food sector in general and the meat sector in particular, have generated a strong focus on food safety issues.[12] As a reaction to public concern, stronger hygiene requirements in food companies and pressure to meet the improved standards have been established. Temperature has been a fundamental factor in increasing food safety, because cold and hot treatments are effective ways of controlling microbiological growth and eliminating pathogens. Since the food scandals have mainly affected the meat sector, we explore whether increasing food security can explain the increasing trend in energy consumption per unit of weighted product in this sector.

An increase in the energy demand of the meat sector due to strict hygiene regulations has been recognized in several reports. For instance, a report made about the Dutch meat sector in the context of the Long-Term Agreements pointed out that 'the introduction of the Hazard Analysis Critical Control Points method for food health, safety and quality led to an increase in energy consumption which eliminated the effect of conservation measures' (Ministry of Economic Affairs, 1999, p. 52). The increased demand occurs for fossil fuel as well as for electricity. Part of the

Table 4.2 Scenarios electricity use as a consequence of more stringent EC hygiene regulations

Scenario	Share of the electricity used to comply with EC regulations, in per cent	
	Slaughterhouses	Meat processing facilities
1	10	10
2	25	10
3	25	25

increased amount of fuel originates from the higher demand for hot water, which is related to more stringent hygiene requirements (for example, hot water is used for the sterilization of tools and bacterial decontamination of carcasses). Furthermore, the increase in electricity is mainly due to stringent requirements for refrigeration. Some studies have already shown that slaughterhouses complying with the European Commission's temperature legislation use more electricity than those that do not (European Commission, 2003b).

There are not enough data to establish at which penetration rate slaughterhouses in the European Union are implementing EU legislation. In order to nevertheless estimate to which extent EU legislation has increased energy demand,[13] we have assumed that:

1. in 1990, slaughterhouses were using water for cleaning and sterilization at 60 °C, and in 2001 the temperature was 82 °C;
2. in 1990, poultry slaughterhouses were not using hot pre-wash before evisceration, while this was the case in 2001; and
3. in 1990, slaughterhouses did not chill the blood recollected while in 2001 all blood was chilled to 5 °C.[14]

On the basis of the above assumptions we estimate an increase in energy demand of approximately 340 TJ in France, 320 TJ in Germany, 140 TJ in the Netherlands and 370 TJ in the UK between 1990 and 2001. Because there is not enough information to calculate the additional demands on electricity for cooling as a consequence of stringent food security in a bottom-up way, we have proceeded as follows. Since Germany, France and the UK publish detailed energy figures for slaughterhouses and meat-processing factories and the shares of electricity used for cooling are known, we conduct a sensitivity analysis using assumptions for the share of additional electricity used to comply with EU regulations (see Table 4.2). The results are shown in Figure 4.10, and they depict what would have been the physical energy intensity indicator PEI_2 of the meat sector

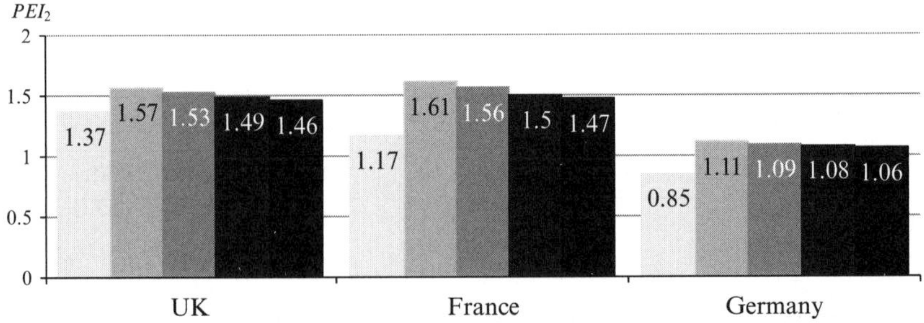

Note: The respective columns show: 1990 without corrections, 2001 without corrections, 2001 electricity share used to accomplish food hygiene standards of 10 per cent in slaughterhouses and 10 per cent in meat processing factories; ditto with 25 per cent in slaughterhouses and 10 per cent in meat processing factories; ditto with 25 per cent in slaughterhouses and 25 per cent in meat processing factories.

Figure 4.10 Influence of hygiene regulations for physical energy intensity of the meat sector

in the year 2001 if the regulations had not been implemented. The influence of increasing electricity use in the indicator is higher than the influence of increasing fuel demand. Increased use of electricity could explain between 3 and 8 per cent of the increase in PEI_2 for the meat sector in the UK, between 3 and 9 per cent in France, and between 2 and 4 per cent in Germany, during the period 1990–2001.

Another consequence of the implementation of the new EU measures has been felt in the rendering sector (treatment of meat by-products such as bones, blood, feather and offal). In the late 1980s the trend in the rendering industry was to move from a batch process towards a continuous process (Coopers & Lybrands, 1996). The continuous process is considered more reliable, because the process generates less water and air pollution, it involves less manual work and does not use organic solvents. Furthermore, it consumes approximately 30–40 per cent less heat per tonne of product as compared to the batch process (Krenk, 1991; Prosper de Mulder Ltd and Croda International plc, 1991).[15] However, as a consequence of European Council Directives 90/667[16] and 96/449[17] processing conditions have changed. In the UK for instance, specified bovine material (SBM)[18] has to be rendered in separate plants or dedicated lines within the plants, generally in *batch* units; the same development has occurred in France (Coopers & Lybrands, 1996). In the Netherlands, the introduction of the new legislation and the opinion of the steering committee of the EC[19] caused a reintroduction of batch sterilization in *all* Dutch rendering plants in 1996. Hence, the EC directives have contributed to a rise in energy demand in the rendering industry, with the exception of Germany, where Directive 90/667/EEC was essentially German law already and had been applied since the beginning of the 1980s (Fodgen, 1991). In order to estimate the increase in energy demand we assume that in 2001, SBM,[20] meat-and-bone meal for animal feed and fallen stock was pressure-treated in batch vats. The specific energy consumption of such a process is reported to be 1960 MJ of primary energy/tonne raw material (Schreurs, 2004). Under these assumptions, we estimate an approximate increase of 480, 310 and 360 TJ in energy demand of the rendering process in France, the UK and the Netherlands, respectively, in 2001.

In sum, stringent hygiene regulations can explain a significant share of the increasing physical energy intensity apparent in the meat sector. For instance, depending on the impact on electricity demand, stringent hygiene regulations explain between 42 and 78 per cent of the 32 per cent increase in PEI_2 in the UK, between 1990 and 2001. Comparable figures for France and Germany are 18–37 per cent and 14–32 per cent, respectively. Because of the aggregate nature of the Dutch data we could not estimate the influence of hygiene on electricity demand. However, if the effect were 3 per cent (that is, the minimum in France and the UK, and the average in Germany), hygiene regulations could explain already one-third of the 35 per cent increase in PEI_2 in the Dutch meat sector.

6.2 The Dairy Sector

Our results show that France, Germany, the Netherlands and the UK reduce their PEI_2 levels by 0.4, 2.1, 1.2 and 3.8 per cent per year, respectively.[21] The results also show that, in absolute terms, France has not only the higher indicator values but it also shows the lowest rate of decrease. Two factors appear to be the main driving forces behind the PEI_2 decline in the British dairy industry: the concentration process and fuel switching (Stace, 2003). A study about the British dairy industry highlights the period 1993–97 as a significant period of restructuring and investment in production capacity (Milk Development Council, 2003). This study identifies three key drivers: the acquisition of brands, cost reduction through rationalization, and vertical integration by milk groups to develop processing capacity either through acquisition or greenfield site development. In the case of the Netherlands, the drivers for the decline that has mainly occurred between 1992 and 2000 are the voluntary agreements between the dairy industry and the Dutch government,[22] the rationalization process, and the increasing use of combined heat and power systems (CHP).[23] The first and the last driver are actually interconnected since the dairy industry considers CHP the main option to achieve their energy efficiency goals.

In order to understand the drivers behind the improvement in efficiency and the differences between the countries, especially between France and the UK, and Germany and the Netherlands, we look at a series of potential causes. First, we assess whether differences in system boundaries among countries can explain the differences. Second, we identify whether important changes in product mix between France and the other countries exist, which are not captured by PEI_2 but could nevertheless explain its higher use of energy per unit of product. Third, we investigate whether the gap in PEI_2-values between France and the other countries occurs across all dairy branches. Finally, we assess the extent to which differences in average size, capacity of utilization or differences in technologies can explain the gap.

6.2.1 System boundaries

As far as we could check, the time series used in this chapter are consistent in terms of system boundaries and product definition. There is however a point that needs some further exploration. Energy data for the dairy industry are only available in terms of net available energy. Net available energy implies that electricity produced by CHP is not taken into account as electricity, while the extra fuel needed is accounted for. The difference between net available and final energy for the dairy industry is not important in all countries. According to personal communication with the British Dairy Industry Association and with the Statistical Office of the UK, CHP played a negligible role in the British dairy industry during the last decade: in the year 2000, only one medium-size site in the dairy sector had CHP (Gardiner, 2003; Stace, 2003). While there are no data available on the amount of electricity or heat produced by CHP, French energy statistics reported that in 2000 only 4 per cent of the electricity was auto-produced (Agreste, several years). Compared to France and the UK, the penetration rate of CHP in the Netherlands is much higher at 49 per cent. Comparable data for Germany are lacking. The relatively high penetration rate in the Netherlands can explain the relatively low use of energy per amount of product of the Dutch dairy industry (without CHP, the value of the physical energy intensity indicator would increase from 1.02 to 1.12, in 2000). The increasing penetration rate of CHP can also explain that in terms of net available energy, the Dutch dairy industry increases fuel consumption, while concurrently the amount of electricity bought is substantially reduced. Thus, while the indicators for the French and British dairy industry are essentially depicting physical energy intensity developments due to changes in processes, for the Dutch dairy industry they depict both changes in processes and changes in industrial power generation (that is, more application of CHP units and a steep decrease in the PEI_2 indicator

Table 4.3 Average annual changes in PEI_2 *by fuel and country*[a]

PEI	France 1986–2000	Germany 1993–2000	Netherlands 1989–2000	UK 1990–2000
Fuels	0.5	−2.6	0.4	−6.5
Electricity	−0.6	−1.7	−14.0	−2.1
Primary	−0.4	−2.1	−1.2	−3.8

Note: [a] In percentage per year.

for electricity; see Table 4.3, and note that the last row corresponds to the results plotted in Figure 4.8).

6.2.2 Production mix

Exploring the importance of production mix differences across countries requires a more detailed comparison of product shares. Table 4.4 shows that the production mixes in France and Germany are quite similar, despite differences in their PEI_2 values of about 33 per cent.[24] The main differences in product mix between both countries are the shares of UHT milk, sterilized milk, sweetened condensed milk, and in cheese production, the share of soft and hard cheese.

We check whether further detailing the product mix would lead to different results. To correct the PEI_2 indicator for these differences we need *SEC* reference values for the individual products, and hence several assumptions need to be made. First, the energy demand of processing UHT milk depends greatly on the type of process employed.[25] If indirect UHT is applied, the *SEC* reference value is almost the same as the *SEC* for pasteurization (782 and 784 MJ/tonne for pasteurization and indirect UHT, respectively),[26] and therefore the difference in shares between pasteurized, UHT and sterilized milk does not explain the differences in PEI_2. We therefore assume that UHT milk is processed by direct application of steam, leading to a *SEC* value of 1116 MJ/tonne. For sterilized milk we assume that the corresponding *SEC* is 1153 MJ/tonne. Second, since the production of sweetened condensed milk is less energy intensive than that of unsweetened condensed milk (the latter requires higher temperatures and a sterilization step that increases the demand for energy) and as we could not find *SEC* reference values that differentiate between both products, we assume that the sterilization step accounts for 15 per cent of the primary energy used per tonne of product (Arcadis, 2000; Hvid, 1992). These assumptions lead to *SEC* reference values of 2125 and 2500 MJ/tonne for sweetened and unsweetened milk, respectively. Combining

Table 4.4 *Detailed distribution of dairy products by type and percentage, in 2000*[a]

Product	Germany	France	Netherlands	UK
Milk and milk products				
Milk	63	64	54	93
Pasteurized	37	3	96	90
UHT	63	90	2	NA
Sterilized	~0	7	2	NA
Yogurt & other fresh milk products	32	31	44	3
Cream	5	5	2	4
Cheese				
Rennet cheese	47	59	96	83
Hard & semi-hard[b]	82	50	100	97
Soft[c]	18	50	–	3
Processed cheese	8	8	3	8
Fresh and quark	45	33	2	9
Condensed milk				
Sweetened	27	48	24	25
Unsweetened	73	52	76	75
Milk powders				
NFMP	52	53	33	44
WMP	48	47	67	56

Notes:

[a] The meaning of the symbols is: – not produced, ~0 less than the unit, and NA for not available.

[b] For instance, Edam and Gouda.

[c] For instance, Camembert and Brie.

Source: Own calculations based on data published by La Maison du Lait.

these *SEC* values with the product distribution shown in Table 4.4, we found that including specifications for UHT, sterilized milk and sweetened condensed milk in the French dairy industry lower the gap in PEI_2 values between France and Germany by only 1 per cent (see Figure 4.8).

We subsequently turn our attention to the difference found in cheese production when comparing Germany and France. Although there is not enough information to estimate *SEC* reference values by type of cheese, after comparing temperature requirements during cheese making, we found that soft cheese and fresh cheese seem to demand less severe conditions than hard cheese processing. We therefore expect that the share distribution in France would imply a lower demand of energy per tonne of

Table 4.5 Physical energy intensity (PEI_2) in primary energy at the sub-sectoral level for the French and Dutch dairy industry in 1998

Branch	PEI_2		Relative difference[a]
	France	Netherlands	
Milk and fermented products	2.16	1.06	71
Butter	4.78	2.15	79
Cheese	10.90	3.23	121
Other milk products[b]	1.06	0.83	24

Notes:

[a] The relative difference is calculated using log percentage ($L\%$) change, which is equal to the relative difference with respect to the logarithmic mean; so, for two measurements X_1 and X_2 the relative difference is $L\% = \ln(X_2/X_1) \times 100$. The metric $L\%$ has all the desirable properties for measuring relative changes, because it is symmetric (it is independent of which point is taken as point of comparison), additive (successive relative changes can be added up), and normed.

[b] Specifically, milk powder, whey products, condensed milk, caseines and lactose.

product instead of a higher share, as would be the case if this could explain the difference in PEI_2 values. In conclusion, the results show that differences in production mix are not large enough to explain the higher PEI_2 values for France. The results also confirm that the selection of products (see Table 4.4) does cover the main differences in production mix across countries.

6.2.3 Branch differences

In order to explore the gap in PEI_2 values between France and the other countries in more detail, we compare physical energy intensity indicators at the sub-sectoral level (for example, fluid milk plants, cheese-makers, and so on). Net available energy consumption data at this level of aggregation are only available for France (published by Agreste for the 1993–2000 period) and for the Netherlands (only for 1998, using data reported in Arcadis, 2000). The results based on the use of equation (4.2) for different sub-sectors are shown in Table 4.5. The French dairy industry shows higher PEI_2 values for every case. This gives us confidence that the higher values in the indicator for France are caused by real differences in energy efficiency.

6.2.4 Firm size, capacity utilization and technology

Differences in energy efficiency can be explained by several other factors not discussed so far, for instance differences in the size of the companies, the capacity utilization of plants, and differences in technologies,

Table 4.6 Development of the structure of dairy companies in four European countries

Country	Number of dairies	Average milk input per dairy (1000t)	Number of dairies	Average per dairy (1000t)	Number of dairies	Average per dairy (1000t)
	1985		1991		2000	
France	1332	19.5	966	24.6	710	31.9
Germany	515	60.0	379	90.9	250	134.1
Netherlands	38	321.9	22	478.9	15	555.6
UK	336	45.4	340	41.5	102	105

Source: Own calculations based on data published by statistical offices.

including the penetration of CHP mentioned above. We discuss these aspects below.

An important aspect of the dairy industry with obvious implications for average firm size is the trend towards increased concentration of the dairy industry during the past two decades. An indication of the present degree of concentration is the high share of processing by the four largest companies, which amounts to 51 per cent in France, 40 per cent in Germany, 96 per cent in the Netherlands, and 50 per cent in the UK. Table 4.6 shows that there has been a reduction in the number of companies and that among those still in business a shift has taken place in favour of larger companies.

It is important to take into account that it is not just the number of companies that has decreased while the output has increased. We also need to investigate whether the pace of 'rationalization' in France differed from other countries, and whether French individual processors have a smaller scale. We found that while the dairy French industry has indeed concentrated at a slower rate than the British and Dutch industry, the concentration rates of the French and German industries are quite similar. More importantly, the average size of the French dairy industry is much smaller. This is very likely to have important technology implications: reaping the benefits from scale economies requires significant capital investment, which includes investing in modern technologies that could increase production levels. These new technologies are generally able to produce larger quantities of product with improved energy efficiency and at lower cost. A study about the British dairy industry shows that in 2000, actual production costs for smaller plants of about 100 million annual litre capacity are about £0.05 per litre. Production costs decrease to about £0.03 litre for plants with an annual capacity of approximately 200 million litres, and to about £0.02 per litre for plants with an annual capacity of over 300 million litres (MDC, 2003).

One should also note that the ranking shown in Figure 4.8 for the year 2000 is quite similar to the size ranking showing the Netherlands below Germany, and the UK below France. In addition, the gap in the cumulative size distributions of dairy companies between France and the other countries is largest for cheese and smallest for other milk products (see Figure 4.11). This is in agreement with the differences in the indicator found at the sectoral level (see Table 4.6). The large differences in size for cheese and the small differences for other milk products can be explained by the fact that there are certain product types where scale of production is not the critical factor to increase price margins, but rather product innovation and specialization. This is particularly valid for the production of specialized cheeses, which is significant in France. For commodity type products such as milk powder and condensed milk, there is a trend towards larger-scale plants to achieve economies of scale. Even though there is not much information available, it is plausible that larger dairies have a better energy efficiency than smaller establishments.[27] This may be due to larger firms: using technologies with cost and efficiency advantages related to scale, utilizing continuous process lines rather than less efficient batch lines, spending more resources on energy management, and reaping the benefits of scale advantages such as fewer boundary losses.

Another factor that can affect the amount of energy used per unit of output is capacity utilization. For example, Cox and Miller (1986) showed that fuel consumption of a milk powder plant increased by 7 per cent when production throughput decreased by 12 per cent. Capacity utilization provides an indication of how efficient the plant and equipment are utilized. It is difficult to quantify capacity utilization accurately. Since there is a lack of official information, we have used monthly data to calculate rough estimates provided in Table 4.7.[28] Germany and the Netherlands show almost identical patterns, whereas the UK and France seem to be slightly falling behind. However, with the exception of NFMP and condensed milk, the differences are not significant. As a result, there is not enough evidence to support the fact that a lower capacity utilization explains the higher PEI_2 value for France.

Finally, we assess whether significant differences in production technologies exist. Specifically, France could be using more energy intensive processes. For example, a difference in technology used for concentration alone (for example, combination of a six-stage film evaporator with thermal vapour recompression (TVR) versus a three-stage evaporator with TVR)[29] could account for a 30 per cent increase in fuel demand, and

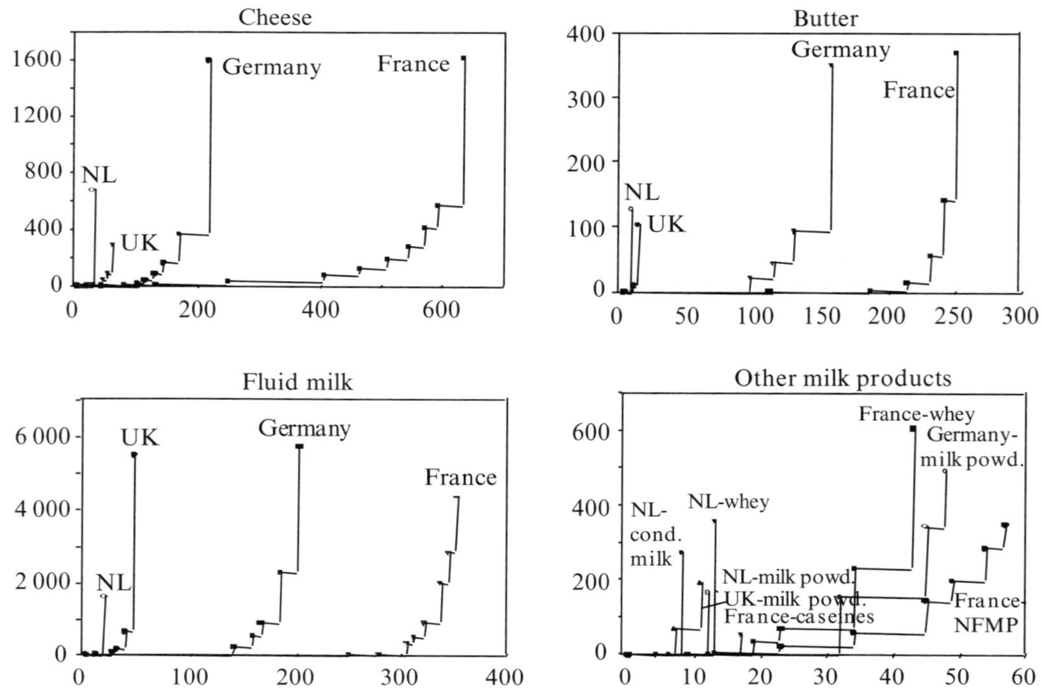

Note: Size is on the vertical axis, measured in thousands of tonnes, and the number of companies is on the horizontal axis.

Figure 4.11 Cumulative size distribution of dairy companies by country in 2000

Table 4.7 Approximated capacity utilization by country and product for the year 2000[a]

Product	France	Germany	Netherlands	UK
Liquid milk	90	92	95	96
Butter	84	86	86	89
Cheese	91	93	97	81
Condensed milk	71	95	95	84
NFMP	67	85	81 (NFMP and WMP combined)	78
WMP	82	84		64
Whey powder	NA	90	88	NA

Note: [a] In per cent; NA refers to not available.

Source: Own calculations based on monthly production data published by industrial associations.

explain 83 per cent of the gap in the indicator for other milk products[30] shown in Table 4.5. Unfortunately, we did not find sufficient substantive information to explore the differences in technologies used in dairy industry processes across countries.

In spite of the lack of detailed data, the available information strongly suggests that the lower energy efficiency of the French dairy industry can be related to its small scale. The French dairy industry is highly fragmented, and it has shown a relatively low pace of concentration. In addition, the French dairy industry could be using different technologies, especially more energy-intensive technologies.

These factors may also contribute to the low decrease in the PEI_2 indicator for France. A better understanding of why France shows a slower rate of decrease in physical energy intensity can be gained from making comparisons in terms of each sub-sector's contribution the overall change in energy intensity. Figure 4.12 indicates that the cheese and butter sectors established substantial decreases in physical energy intensity, but these decreases are offset by an increase in the other milk products sub-sectors, particularly between 1993 and 1996. As a result, the overall physical energy intensity of the French dairy sector appears constant.

7 CONCLUSIONS

In this chapter we presented an analysis of the meat and dairy industry in terms of energy consumption and energy efficiency in four European countries. Changes in energy efficiency were measured using two different

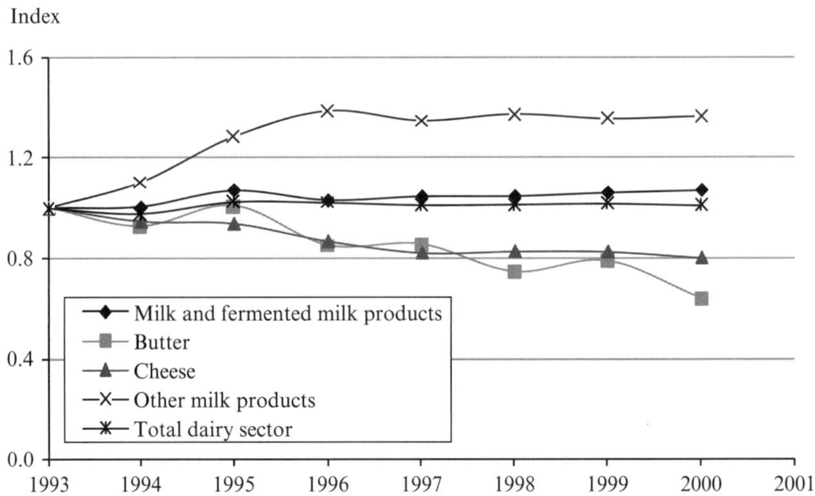

Figure 4.12 *Index for* PEI_2 *(1993 = 1) in the French dairy industry by branch, 1993–2000*

indicators based on physical data. The first indicator, PEI_1, has several appealing features: it is easy to calculate, requires only a small amount of data, and the indicator can be easily communicated and understood by non-specialists. The main drawback of the PEI_1 indicator is that it is not sensitive enough to changes in product mix. The second indicator, PEI_2, is a much more complex indicator, and therefore obviously associated with a correspondingly higher data burden. However, the PEI_2 indicator accounts for differences in structure across countries as well as for changes in product mix, and it is therefore to be preferred when comparing levels of energy efficiency across countries.

Our results show that the dairy industry has achieved considerable improvements in energy efficiency, contrary to the developments that took place in meat industry. In the latter, we found significant increases in primary energy used per weighted tonne of product, amounting to 3.2, 3.4, 1.4 and 1.6 per cent per year in France, Germany, the Netherlands and the UK, respectively. We found a trend in all countries towards higher electricity use due to increasing demands for refrigeration and motor drive power generation. In fact, correcting the PEI_2 indicator for changes in the shares of frozen, cut-up and deboned meat products would decrease the absolute values of the indicators as well as the gap between countries, and therefore help explaining some of the fluctuations showed by the trends at a disaggregate level. We also found that stringent hygiene regulations play a significant role and can explain an important share of the increasing amount of energy per unit of weighted product identified in the meat industry.

In the case of the dairy industry, we found that while in Germany, the Netherlands and the UK the dairy industry has reduced its PEI_2 values by more than 1 per cent per year, PEI_2 values for the French dairy industry have declined at a significantly lower rate of 0.4 per cent per year. In absolute values the French dairy industry shows a higher energy demand per unit of product as well. The higher PEI_2 values calculated for the French dairy industry can be related to the fact that this industry is highly fragmented with relatively many small-scale establishments and a relatively slow pace of concentration. Although there is a lack of detailed information, our results suggest that some sub-sectors of the French dairy industry use technologies that are rather energy intensive. Specifically, although the cheese and butter-making sub-sectors have substantially decreased their PEI_2 values, this decrease has been offset by an increase of the PEI_2 value in the sector producing milk powders and condensate products. As a result, PEI_2 values for the French dairy sector as a whole appear to remain nearly constant. Finally, the results also show that the UK, Germany and Dutch dairy industry have converged towards similar (lower) values in their energy efficiency indicator and that the French dairy industry can potentially save approximately 30 per cent primary energy if it were to converge to the same efficiency level.

In a more general sense, this chapter has shown the feasibility of monitoring changes in energy efficiency based on physical production data in heterogeneous, energy-extensive sectors, eventually correcting for structural differences. Provided that detailed and reliable data are available and that detailed studies of energy efficiency at the industrial level for a base year exist, historical changes in energy efficiency based on physical production data can be monitored without the need for the implementation of task forces that have to rely exclusively on confidential reporting at the firm level.

Data availability and data quality are of pivotal importance. A sizeable amount of research time was spent on obtaining reliable time series with a sufficient level of detail to allow the drivers behind energy efficiency changes to be analysed. Data impediments can therefore constitute a major deterrent to performing this type of analysis for different sectors and/or other different geographical areas. However, understanding and addressing the issue of energy efficiency requires the analysis of production processes in both physical and economic terms. Ultimately, it is difficult to see how private actors and policy makers can promote and achieve sustainability in the absence of information regarding energy and material flows in important sub-sectors of the economy.

NOTES

1. Strictly speaking, the energy efficiency indicator and energy efficiency are inversely related. The first concept measures energy use per unit of activity, while the latter refers to the output or activity delivered per unit of energy.
2. The Long Term Agreements (LTAs) are voluntary agreements between different sectors of the economy and the Dutch national government. The main goal of the LTAs is to increase the energy efficiency of the sector by a given percentage in a predetermined period of time. The information provided by the companies is confidential. For more information see: http://www.senternovem.nl/lta/lta/index.asp.
3. The difference between gross and net calorific value is the latent heat of vaporization of the water produced during the combustion of the fuel.
4. These conversion efficiencies reflect losses in transport and fuel conversion, which are considered to be relatively small (Phylipsen et al., 1998). The official Dutch Bureau of Statistics (CBS) has reported energy losses in transformation processes of approximately 7 to 10 per cent for the oil sector and 1 to 5 per cent for the gas sector.
5. We keep the electricity generation efficiency constant over time in order to isolate the effect of efficiency improvement on the energy demand side. Alternatively, the effect of efficiency improvements in power generation (that is, the energy supply side) could be included as well.
6. We allocate fuel inputs of CHP plants on the basis of the exergy content of the products as described in Phylipsen et al. (1998). We use an exergy factor for steam of 0.36 (Blok, 1991). In 1995, steam is valued at 83 per cent of its energy content and electric efficiency is set to 43 per cent.
7. The temperature correction method of Eurostat is based on the share of fuels used for heating purposes. Hence the temperature corrected energy (E_{nt}) is given by $E_{nt} = E_{ht}/d_t + E_{pt}$, where E_{ht} is the energy used for heating purposes, E_{pt} the energy used for non-heating purposes, and $d_t = D_t/D$, where D_t and D correspond to the actual and the long term degree days, respectively (Brook, 2001).
8. The dairy sector covers activities related to the treatment of milk for alimentary use, milk-derived products and by-products. The meat sector comprises the production and preservation of meat plus the further processing of all types of meat. Production refers to slaughtering, rendering and meat processing, while husbandry and retail are not included.
9. The time periods are slightly different: 1986–2001 for France and the Netherlands, 1993–2001 for Germany, and 1990–2001 for the UK.
10. There are several reasons for this shift: the damage of public confidence in beef due to food scandals, a decrease in cattle available for slaughtering as a side effect of the European milk quota system, low producer prices for poultry and higher animal efficiency (to produce 1 kg of broiler meat it takes 3.1 kg of dry matter, while non-dairy cattle needs 24 kg).
11. Dehairing/defeathering and singeing operations account for about 75 per cent of the fuels consumed in pork and poultry slaughtering.
12. For instance, E. coli in beef trimmings, the use of hormones in cattle, the bovine spongiform Encephalophaty (BSE) crisis, the Creutzfeld-Jacob disease (vCjd) and the foot-and-mouth disease.
13. We found partial information for EC approved slaughterhouses in the UK. The report states that in 1992 only 12 per cent of the slaughterhouses were approved while the percentage increased to 62 per cent in 1996 (Shaoul, 1996). These figures support our assumptions.
14. The values were calculated assuming that water use (at 82°C) is about 30 litres per pig, 105 per cattle, and 4.3 per chicken (Pontoppidan and Hansen, 2000, 2001); the amount of water use for hot washing before chilling in poultry, according to EU legislation, is 1.5–3.5 litres per carcass (European Commission, 1996); the average amount of blood that can be collected per animal is 3.6 kg per pig, 13.6 per cattle, 2.5 per lamb and 0.06

kg per chicken (Gracey, 1998); and the requirement for blood refrigeration is 30.5 kWh/tonne of blood (European Commission, 2003b).

15. The reason for this is that in the continuous process there is reutilization of heat.
16. It establishes that high-risk material must be heated to a core temperature of at least 133 °C for 20 minutes at a pressure of 3 bar.
17. It establishes the approval of alternative heat treatment systems for processing animal waste (European Commission, 1996).
18. Specified meat offal comprises the skull, including the brain and the eyes, tonsils, intestines, thymus and the spinal cord of bovine, ovine and caprine animals.
19. When the batch system was changed to continuous it was overlooked that the heat treatment in some of the systems were less than in conventional batch systems. Continuous systems were indeed able to get rid of salmonella and vesicular disease, however, their effectiveness to eliminate BSE is doubtful.
20. In the UK, sbm is calculated to be 46.5 kg per cattle and 2 kg per sheep and goat, while corresponding figures for France and the Netherlands are 17 kg per cattle and 2 kg per sheep and goat (MLC, 1998). The higher amount per cattle in the UK reflects the fact that British law (Specified risk material order, 1997) considers as sbm the head and spleen of the animals, while in France and the Netherlands they are not.
21. If PEI_1 were used, we would obtain a quite different picture. With the exception of the British industry, who decline the use of energy per tonne of milk processed at an average rate of 3.2 per cent per year, improvements in this indicator have been either minor, as in the Dutch and German industry ($<$ 1 per cent per year) or non-existing (the French dairy industry shows 0.7 per cent increases per year). The main reasons for the divergence in the results between PEI_1 and PEI_2 are discussed in section 6. The decrease in the British dairy industry is a consequence of their product mix, specifically a higher share of milk and fermented products, which are the least energy-intensive products of the milk sector. The increase for the other three countries is mainly a consequence of a structural shift from fluid milk products to cheese and milk powders, which require more energy per unit of raw milk.
22. In 1992, several sectors and the Dutch Ministry of Economic Affairs signed a covenant aiming to improve energy efficiency by a specific percentage within an agreed period (the so-called LTAs; see note 2). The dairy industry (excluding ice cream) signed this agreement in 1994 and the goal was to increase energy efficiency by 20 per cent in the year 2000 in comparison to 1989.
23. Combined heat and power (CHP) systems combine the production of electrical (or mechanical) and useful thermal energy from the same primary energy source.
24. Categories such as caseins and lactose were not further decomposed due to lack of data.
25. Ultrahigh temperature (UHT) milk can be processed using direct or indirect UHT treatments. In direct UHT, milk is heated from 75 °C to 135 °C by steam injection. Indirect UHT is basically the same process as pasteurization, but the operating temperature is higher. Indirect UHT is more energy efficient since internal heat recovery is possible: part of the heat can be recovered from the direct UHT treatment by the use of vapour recompression, but this is seldom done in practice (European Commission, 2003b).
26. We have used data published by Hvid (1992) to replace the pasteurization step used in Arcadis (2000), which is the source used to calculate PEI_2 in section 6.
27. A Canadian study on fluid milk plants shows that dairies producing 20–50 million litres of milk per year consume 30 per cent more energy (in final terms) than dairies producing 50–100 million litres (NDC, 2001). Similarly, a Norwegian study on industrial milk production shows that small diaries consumed about 55 per cent more energy (in final terms) than large dairies (Høgaas, 2002).
28. Utilized capacity is estimated as the ratio between the annual production of a product and its peak production multiplied by 12.
29. TVR uses pressure steam to increase the pressure of part of the vapour generated in an evaporator effect so it can be used again to drive the evaporation process.

30. Note that the sub-sector other milk products is composed of dairy products which are essentially the result of concentration and drying processes.

REFERENCES

Agreste, annual publications, *Energy Consumption in the Agro-food Industry*, Paris: Ministry of Agriculture, Agreste.

Arcadis IMD (2000), 'Reference method for dairy: assessment of available energy saving potentials', report IMD/MA00/2211/30185, Arnhem: Arcadis IMD.

Blok, K. (1991), 'On the reduction of carbon dioxide emissions', PhD thesis, Department of Science, Technology and Society, Faculty of Chemistry, Utrecht University, The Netherlands.

Brook, K. (2001), *Temperature Correction Procedures to Monitor Changes over Time of Energy Consumption Data*, London: Department of Trade and Industry, http://www.dti.gov.uk/energy/inform/energy_trends/tempmethod1.htm.

Coopers & Lybrands (1996), *The UK Beef Industry: Report on the Rendering Sector*, London: Coopers & Lybrands.

Cox, G.C. and E.J. Miller (1986), 'Comparative energy efficiencies of dairy manufacturing and processing industry: Australia and New Zealand', *Engineering Costs and Production Economics*, **10**, 293–303.

European Commission (1996), 'Commission Decision 96/449/EC of 18 July 1996', *Official Journal of the European Communities*, L184, 43.

European Commission (2003a), 'Proposal for a Directive of the European Parliament and of the Council on Energy End-use Efficiency and Energy Services (presented by the Commission), COM 2003, 739 final report', Brussels: European Commission.

European Commission (2003b), 'IPCC draft reference document on best available technologies in the food, drink and milk industry', Sevilla: European Commission, http://eippcb.jrc.es/pages/BActivities.cfm.

Farla, J. (2000), 'Physical indicators of energy efficiency', PhD thesis, Department of Science, Technology and Society, Faculty of Chemistry, Utrecht University, The Netherlands.

Farla J., K. Blok and L. Schipper (1997), 'Energy efficiency developments in the pulp and paper industry: a cross country comparison using physical production data', *Energy Policy*, **25**, 745–58.

Fodgen, M. (1991), 'By-product of animal origin and animal waste: 1991 investigation into EC collection and rendering practices', London: Meat and Livestock Commission.

Freeman, S.L., M.J. Niefer and J.M. Roop (1997), 'Measuring industrial energy intensity: practical issues and problems', *Energy Policy*, **25**, 703–14.

Gardiner, A. (2003), Personal communication on future energy solutions, 24 September.

Golove, W.H. and L.J Schipper (1997), 'Restraining carbon emissions: measuring energy use and efficiency in the USA', *Energy Policy*, **25**, 803–12.

Gracey, J.F. (1998), *Meat Plant Operations*, Shedfield: Chalcombe Publications.

Haas, R. (1997), 'Energy efficiency indicators in the residential sector: what do we know and what has been ensured?', *Energy Policy*, **25**, 789–802.

He, K., H. Huo, Q. Zhang, D. He, F. An, M. Wang and M.P. Walsh (2005), 'Oil consumption and CO_2 emissions in China's road transport: current status, future trends and policy implications', *Energy Policy*, **33**, 1499–507.

Høgaas, E. (2002), 'Life cycle assessment of industrial milk production', *The International Journal of Life Cycle Assessment*, **17**, 115–26.

Hvid, J. (1992), 'Review of energy efficient technologies in the dairy industry sector', Taastrup: Energy Centre Denmark.

International Energy Agency (IEA) (2000), 'Energy policies of IEA countries: France 2000 Review', Paris: Organisation for Economic Co-operation and Development, International Energy Agency.

Krenk, P. (1991), 'An overview of rendering structure and procedures in the European Community', in R. Bradley R., M. Savey and B. Marchant (eds), *Sub-acute Spongiform Encephalopathies*, Brussels: European Community.

Meat and Livestock Commission (MLC) (1998), 'A study of the competitive position of the red meat industry in Great Britain in relation to the collection, processing and disposal of animal by-products both in Great Britain and other EU member states', London: Meat and Livestock Commission, Planning and Forecasting Group.

Milk Development Council (MDC) (2003), 'Prices and Profitability in the British Dairy chain', London: Milk Development Council.

Ministry of Economic Affairs (1999), 'Long-term agreements on energy efficiency: progress in 1998', The Hague: Ministry of Economic Affairs.

Nanduri, M., J. Nyboer and M. Jaccard (2002), 'Aggregating physical intensity indicators: results of applying the composite indicator approach to the Canadian industrial sector', *Energy Policy*, **30**, 151–63.

National Dairy Council of Canada (NDC) (2001), 'Energy performance indicator report: fluid milk plants', Toronto: National Dairy Council of Canada.

Nielsen, L. (2004), Personal communication, 20 April.

Nyboer, J., J. Olivier and A. Paper (1996), 'Aggregating physical intensity indicators for the Canadian industry 1990–1995', Burnaby: Canadian Industry Energy End-Use Data and Analysis Centre.

Ozawa, L., C. Sheinbaum, N. Martin, E. Worrell and L. Price (2002), 'Energy use and CO2 emissions in Mexico's iron and steel industry', *Energy*, **27**, 225–39.

Phylipsen, G.J.M., K. Blok and E. Worrell (1998), *Handbook on International Comparisons of Energy Efficiency in the Manufacturing Industry*, Utrecht: Department of Science, Technology and Society, Utrecht University.

Pontoppidan, O. and P. Hansen (2000), 'Implementation of cleaner technology in the poultry abattoir branch', Project Report Environmental Project No. 573, Copenhagen: Ministry of the Environment (Miljøstyrelsen), http://www.mst.dk/udgiv/publikationer/2000/87-7944-306-0/html/.

Pontoppidan, O. and P. Hansen (2001), 'Cleaner technology in the poultry and cattle slaughtering branch', Environmental Report No. 9, Copenhagen: Ministry of the Environment (Miljøstyrelsen), http://www.mst.dk/udgiv/publikationer/2000/87-7944-306-0/html/.

Prosper de Mulder Ltd and Croda International plc (1991), 'A report on the merger situation', presented to Parliament by the Secretary of State for Trade and Industry, Cm 1611, London: Monopolies and Merger Commission, http://www.competition-commission.org.uk/rep_pub/reports/1991/307properdemulder.htm.

Rafiquil, I., C. Weber, B. Lehmann and A. Voss (2005), 'Energy efficiency improvements in ammonia production – perspectives and uncertainties', *Energy*, **30**, 2487–502.

Ramírez, A. (2005), 'Monitoring energy efficiency in the food industry', PhD thesis, Utrecht University, The Netherlands.

Rijkinstituut voor Volksgezonheid en Milieuhygiëne (RIVM) (1995), 'Samenwerkingproject Procesbeschrijving Industrie Nederland', Bilthoven: National Institute for Public Health and the Environment (RIVM).

Ross, M. and R. Hwang (1992), 'A model for long-term industrial energy forecasting', Washington, DC: US Department of Energy.

Schreurs, M. (2004), Personal communication, 29 April.

Shaoul, J. (1996), 'BSE: for services rendered? The drive for profits in the meat industry', Manchester: Department of Accounting and Finance, University of Manchester.

Sinton, J.E. and M.D. Levine (1994), 'Changing energy intensity in the Chinese industry', *Energy Policy*, **22**, 239–52.

Stace, G. (2003), Personal communication on dairy industry association limited, 14 October.

Suijkerbuijk, M., C. van Oosterhout and A. Hoogenkamp (1995), 'Slaughtering and the meat industry', Report 773006173, Samenwerkingproject Procesbeschrijving Industrie Nederland (SPIN), Bilthoven: National Institute for Public Health and the Environment (RIVM).

Williams, R.H., E.D. Larson and M.H. Ross (1987), 'Materials, affluence and industrial energy use', *Annual Review of Energy*, **12**, 99–144.

Worrell, E. (1994), 'Potentials for improved use of industrial energy and materials', PhD thesis, Department of Science, Technology and Society, Faculty of Chemistry, Utrecht University, The Netherlands.

5. Monitoring energy use and energy efficiency in the Dutch service sector

Andrea Ramírez, Martin K. Patel and Kornelis Blok

1 INTRODUCTION

Energy efficiency is considered a key policy strategy for the abatement of greenhouse gases (EC, 2005; IEA, 2006; IPCC, 2007). It is an attractive policy option for various reasons. The most important reasons are: energy efficiency would allow a (partial) decoupling of economic growth from environmental pressures, it enhances the reliability of energy supplies by reducing system loads and stresses, and it often has short pay-back periods. Additional benefits can include noise reduction, labour and time savings,[1] improved process control, and water and waste minimization (Worrell et al., 2003).

The potential for larger savings through energy efficiency has long been recognized. In 1993 for instance, the EC Directive 93/76/EEC had as its main objective limiting carbon dioxide emissions by improving energy efficiency. In April 2000, the European Commission presented an action plan to improve energy efficiency in the European Community (EC, 2000). In this action plan an estimated economic potential for energy efficiency improvement of more than 18 per cent was reported. In 2005, the European Commission's Green Paper on Energy Efficiency pointed to the fact that the European Union (EU) could save at least 20 per cent of primary energy by implementing energy efficiency measures (EC, 2005). In turn, the 2006 EU Directive on the promotion of end-use efficiency and energy services stated overall national indicative energy savings targets of 9 per cent by 2016 to be reached by each member state (EC, 2006a). Furthermore, member states were required to submit Energy Efficiency Action Plans (EEAPs) with indicative efficiency targets per sector for the year 2010 and 2016. Table 5.1 shows as an example the EEAP submitted by the Netherlands in 2007.

Up to now, most in-depth research on energy efficiency potentials

Table 5.1 Indicative annual energy savings targets according to the 2007 Netherlands Energy Efficiency Action Plan

National targets				
National indicative annual energy savings target 2016 (GWh)				51 190
National indicative annual energy savings target 2010 (GWh)				11 376
Package of measures to improve energy efficiency planned for achieving the target	Annual energy savings expected by end of 2010 (GWh)		Annual energy savings expected by end of 2016 (GWh)	
	Low	High	Low	High
Residential sector	5527	5527	23 576	34 257
Tertiary sector	1515	1515	9112	15 499
Industry (non-ETS)[a]	249	874	636	1778
Transport sector	3807	5293	17 613	26 939
Agriculture	845	1158	2917	5528
Total ESD energy savings expected[b]	11 943	14 366	53 854	84 001

Notes:
[a] ETS = EU Emissions Trading Scheme.
[b] ESD = EU Energy Service Directive.

Source: The Netherlands Energy Efficiency Action Plan 2007, The Hague, 12 September.

at the sector level has targeted the industrial and residential sectors, although the transport sector has recently drawn attention as well. Despite being the major economic sector in industrialized countries in terms of the share of value-added and employment,[2] the service sector (also called tertiary sector) has remained at the bottom of the priority list.

Figure 5.1 shows the trends in the share of total value-added of the service sector for the European Union and the Netherlands. Note that in the second half of the 1990s, the Dutch service sector went through a period of particularly strong growth, amounting to 3 per cent per year between 1982 and 2003. A study published by the Netherlands Bureau for Economic Policy Analysis (abbreviated as CPB in Dutch) identified the service sector as the largest and fastest growing sector of the Dutch economy (Suijker et al., 2002). It is expected that the share of the service sector in the Dutch economy will further expand over the next two decades (CPB, 1998). A similar trend is forecasted for other developed countries. In terms of energy consumption about 13 per cent

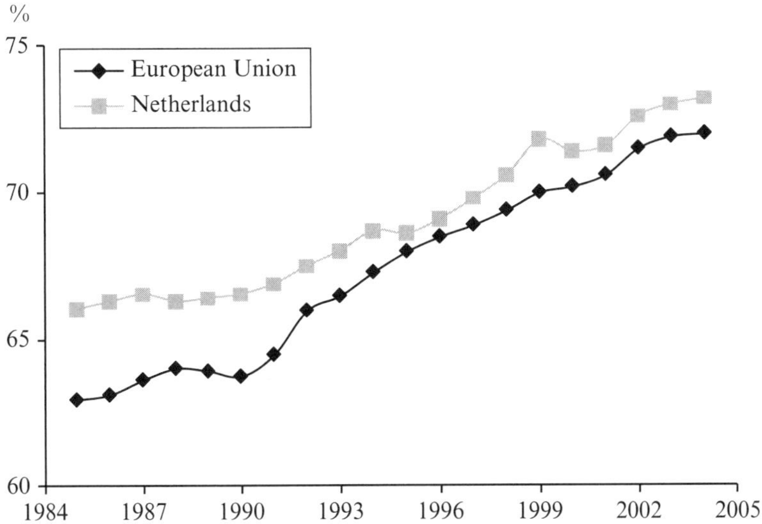

Figure 5.1 Share of value added by the service sector in the total economy, in the Netherlands and the European Union

of the total final energy consumption and 30 per cent of the total electricity consumption in Organisation for Economic Co-operation and Development (OECD) countries can be allocated to the service sector (see Figure 5.2). In a recent report of the Intergovernmental Panel on Climate Change (IPCC, 2007) a baseline scenario with global projections of primary energy use for the service sector is estimated to be doubling between 1990 and 2020. The baseline scenario shows the most rapid growth with a projected average of 2.6 per cent per year. These trends point to the need to develop and implement energy efficiency measures that would offset increasing energy demand driven by rapid growth. The 2006 Communication on the EU Action Plan for Energy Efficiency has already stated that goal. Partly because of its large share of total consumption, the largest cost-effective energy savings potential is expected in the residential (households) and commercial buildings sector (tertiary sector) where the full potential is estimated to be around 27 per cent and 30 per cent of total primary energy use, respectively (EC, 2006b).[3]

Monitoring the effects of energy efficiency policies is necessary for the evaluation and steering of these policies. However, the lack of interest in the service sector due to its comparatively low energy use, the low quality and incompleteness of currently available data for

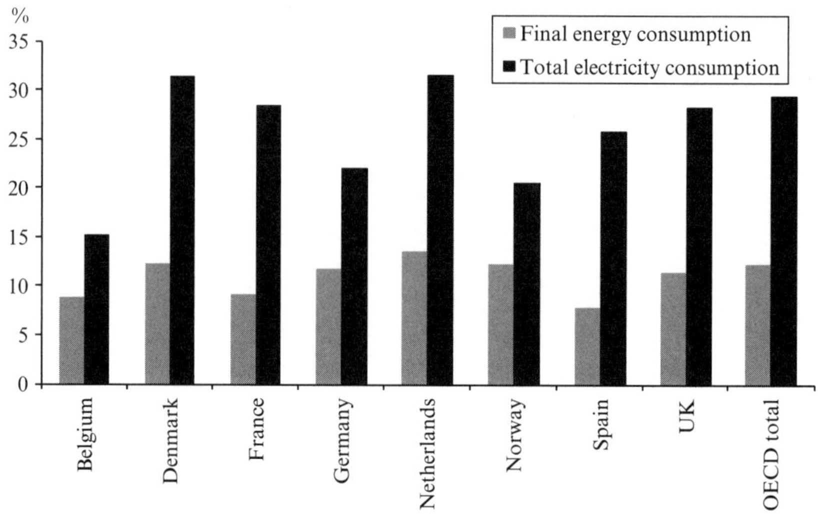

Figure 5.2 Share of the service sector in total final energy and total electricity consumption in several European countries in 2002

this sector (for example, few data are publicly available, high levels of aggregation, poor statistical definitions), and the heterogeneity of its constituents, make it difficult for researchers to provide policy-makers with *reliable* information on the changes and drivers of energy efficiency developments. In this chapter we exemplify this problem by presenting an analysis of the availability and quality of energy efficiency indicators for the Dutch service sector. The overall aim of this chapter is to explore whether currently available data allow for the development of reliable energy efficiency indicators for the service sector in the Netherlands.

The layout of this chapter is as follows. We start by defining the service sector, describing data sources and the general methodological background for the analysis of energy intensity indicators in section 2. In section 3, historical developments in energy consumption for the Dutch service sector are presented, while section 4 shows the trends in energy intensity and the results of the decomposition methodology. Section 5 deals with the developments in energy consumption and energy intensity in three sub-service sectors. Conclusions are drawn in section 6.

2 SECTOR DEFINITION, DATA SOURCES AND DECOMPOSITION METHODOLOGY

2.1 Sector Definition

The service sector, also known as the commercial or tertiary sector, involves activities that take place in buildings used outside of manufacturing, agriculture and the household sectors (Krackeler et al., 1998). The Dutch Central Bureau of Statistics (abbreviated as CBS in Dutch) splits the service sector according to its main economic activity. Other agencies, such as the United States Energy Information Administration, use as main parameter to classify the service sector the activity occupying most of the floor space in the building (generally 75 per cent of the buildings' square footage). This is based on the premise that buildings with the same principal building activity have similar energy consumption patterns, for instance, hospitals, offices and schools (DOE/EIA, 1995). For the purpose of this chapter, we use an economic classification (the Dutch Standard Industrial Classification SBI 3) that is equivalent to the NACE (Nomenclature générale des Activitiés économiques dans les Communautés Européenes) classification, because this is the approach used by CBS. The service sector is thus composed of activities falling into 23 sub-sectors, namely wholesale and retail trade (SBI 50–51 and 55), financial intermediation (SBI 60–67), real state renting and business activities (SBI 70–74), education (SBI 80), health care (SBI 85), recreation and others (SBI 90–93).

2.2 Data Sources

A main problem in developing historical trends is that data availability is not consistent. Since this problem was also encountered in this study, several data sources were used. For the period 1984–2001, energy consumption data for the whole service sector were published by CBS in 'De Nederlandse Energiehuishouding Jaarcijfers'. After 1993, gas and electricity consumption data are also published through the Statline service of the CBS. At the sub-sector level, until 1999 energy consumption data was reported in dedicated studies such as 'Energieverbruik van Bank- en Verzekerinswezen en Zakelijke Dienstverlening' (1994, 1999), 'Energieverbruik van Scholen' (1993) and 'Energieverbruik in de Intramurale Gezondheidzorg' (1982–88, 1990, 1992, 1994, 1998). Since 2000 there are no publicly available data for these sub-sectors. It should be noted that not all published values at the sector or sub-sector level are the result of surveys. For years between surveys, which in some cases constitute the majority of observations in a time series, CBS reports estimated values based on extrapolation procedures, which are mainly driven by the number of employees per sector. The quality of the data resulting from such extrapolation will be discussed later in this chapter.

An important indicator generally used to follow energy efficiency developments in the residential and service sector is floor space. Unfortunately, floor space data in the Netherlands are not available for most sectors and years. Due to the importance of this parameter, several agencies have developed their own estimation methods. For instance, the Netherlands Bureau for Economic Policy Analysis (CPB) assumes that floor space is proportional to employment size. In other words, the amount of m^2 office floor space needed per working person is fixed and it is assumed that floor space coefficients are constant over time (Schuur, 1999). In this chapter a different approach is used since we would like to develop energy efficiency indicators which are simultaneously based on employment and floor space and it is therefore desirable that the two values are independent.

Floor space data have been estimated based on building permits, progress on building projects reported by the CBS in the 'Maandstatistiek Bouwnijverheid' and the floor space data reported for the year 1995 by ECN, the Energy Research Center for the Netherlands (van Arkel et al., 1999). Reports compiled by the CBS show building permits issued by type of work, type of building and by size. It also reports buildings completed by type of building and economic activity as well as demolition figures. The first data set has been used to find a correlation between building costs and floor space. This relation is important because figures for new square meters are only published in monetary units. Thus, by using these two data sets the number of new square metres has been estimated for the time period 1981–99. To obtain the total values of floor space, we combined the data reported by ECN for 1995 and the data on new square meters. Due to the preliminary nature of these estimates they are only used for descriptive analysis.

Economic data such as value added, deflators, output, and labour are published by the CBS in the 'National Accounts of the Netherlands'. In this chapter, value-added data have been deflated to 1995 values and they are reported in euros. Additional data on employment and the number of firms were obtained from the 'Statistiek Werkzame Personen' and 'Bedrijven in Nederland', respectively, published by the CBS. At the international level, data on energy consumption was taken from the energy statistics of countries belonging to the OECD, as published by the International Energy Agency.

Table 5.2 Fuel consumption share for households

Year	1980	1985	1990	1991	1992	1993	1994	1995
%	88	87	80	80	78	77	77	76
Year	1996	1997	1998	1999	2000	2001	2002	
%	80	79	78	76	76	77	77	

Source: EnergieNed, reported in Goldewijk et al. (2004).

2.3 Methodological Aspects

2.3.1 Energy

In this study, and unless otherwise specified, when referring to energy we mean *primary energy*. We have assumed an efficiency of 40 per cent for electricity, which is representative for the Netherlands with its relatively high share of natural gas in the power sector. For every MJ of oil products delivered, we assume a primary energy input of 1.05 MJ, and for natural gas a primary energy equivalent of 1.01 MJ. In addition, energy consumption data have been adjusted by taking into account the influence of weather. The correction can be expressed as (Farla, 2000):

$$E_{i,corr} = f_{sh} \times E_i \times \frac{N_{yr,ref}}{N_{yr}} + (1 - f_{sh}) \times E_i \tag{5.1}$$

where $E_{i,corr}$ is the temperature-corrected natural gas consumption, f_{sh} is the fraction of gas consumption assumed to be dependent on outdoor temperatures, and $N_{yr,ref}/N_{yr}$ is the ratio of the number of degree days for a specific year and the 30-year average number of degree days. The fraction of gas used for heating can change over time due to an increasing number of buildings with insulation and/or, specifically in the case of households, owing to increasing or decreasing amounts of fuel used for other applications than space heating (cooking, hot water supply for showers). In the residential sector the fraction of fuel use for space heating has been observed to decrease from 88 per cent in 1980 to 76 per cent in 2000. When correcting fuel consumption data in the household sector we use the shares shown in Table 5.2. For the service sector this kind of information is not available and we therefore use a fixed share (82.5 per cent) as provided by the Dutch Ministry of Economic Affairs.

2.3.2 Energy efficiency and energy intensity

Energy efficiency refers to using less energy for producing the same amount of services or useful output. The measurement of energy efficiency at the lowest level of aggregation, for instance a machine, is simple and straightforward. However, policy-makers are generally interested in understanding developments at higher levels of aggregation, such as energy efficiency of an industrial sector or a country. In this case, energy efficiency cannot be directly measured and it therefore has to be analysed by the use of appropriately defined indicators.[4]

To provide meaningful information to policy-makers and researchers, an energy efficiency indicator must be easily observable with little or no time lag, it should be closely associated with the policy actions in the sense that is easily affected by the policy undertaken, and it should be related to the target and goal variables. There is general consensus that an energy efficiency indicator should relate the amount of energy used to the useful output or activity, by means of a ratio. Hence an energy efficiency indicator equals the amount of used energy divided by the amount of activity. Activity can be measured in economic (for example, gross domestic product, value added, value of production) or physical terms (for example, tonnes, kilometres, square metres). To differentiate between both approaches, the term 'economic energy intensity' is used for the ratio of energy use to unit of economic activity, and we use 'physical energy intensity' for the ratio of energy used to the unit of physical activity.

The first step when designing indicators for energy use is to understand the factors that can affect energy use in a given process or sector. Energy use in the service sector is affected by factors such as the physical characteristics of the buildings, the age of the buildings, the efficiency of the equipment, the occupants' energy-related behaviour, and by location and structural effects. Activity data that could reflect the level of energy use of the service sector are, for instance, value added, floor space, operating hours, number of employees, number of beds and number of firms.

2.3.3 Decomposing the changes in energy consumption

Several factors may lead to changes in the energy consumption of a sector, for instance, changes in activity (increases in total production), structure (sub-sectoral growth differences), intensity (energy per unit of output) and labour (labour intensity of production). Although it can be argued that a more representative measure of the amount of activity in equation (5.2) would be floor space, we decided to use the number of employees for the decomposition since no official data exist for floor space and the number of employees is often considered the second best choice (for example, Kerssemeeckers, 2001; Krackeler et al., 1998; Murtishaw and Schipper, 2001; Schipper et al., 1986). In order to quantify the contribution of these factors to the total change in energy consumption in the Dutch service sector, statistical decomposition methodologies can be used. In this chapter we use a multiplicative log-mean divisia method. For a detailed overview and discussion of different decomposition methodologies we refer to Ang and Zhang (2000).

For decomposition purposes we describe energy consumption of the service sector as:

$$E_t = \sum_i I_{i,t} \cdot LB_{i,t} \cdot S_{i,t} \cdot A_t \tag{5.2}$$

where i refers to sub-sectors of the service sector and t to time, E is primary energy consumption, $I = E/Emp$ is energy intensity, $LB = Emp/A$ is labour intensity, $S = A/\Sigma_i A_i$ is the share of value added, with A is value added and Emp the number of employees. For a given final point in time T, the total effect on primary consumption D_{tot} can then be decomposed into an activity (D_{act}), a structural (D_{str}), an intensity (D_{int}) and a labour effect (D_{lab}), as shown in the following equations:

$$D_{tot} = \frac{E_T}{E_0} = D_{act} \cdot D_{str} \cdot D_{int} \cdot D_{lab} \tag{5.3}$$

$$D_{act} = \exp\left\{ \sum_i \left(\frac{L(\omega_{i,T}, \omega_{i,0})}{\sum_i L(\omega_{i,T}, \omega_{i,0})} \ln\left(\frac{A_{i,t}}{A_{i,0}}\right)\right)\right\} \tag{5.4}$$

$$D_{str} = \exp\left\{ \sum_i \left(\frac{L(\omega_{i,T}, \omega_{i,0})}{\sum_i L(\omega_{i,T}, \omega_{i,0})} \ln\left(\frac{A_T}{A_0}\right)\right)\right\} \tag{5.5}$$

$$D_{int} = \exp\left\{ \sum_i \left(\frac{L(\omega_{i,T}, \omega_{i,0})}{\sum_i L(\omega_{i,T}, \omega_{i,0})} \ln\left(\frac{I_{i,t}}{I_{i,0}}\right)\right)\right\} \tag{5.6}$$

$$D_{lab} = \exp\left\{ \sum_i \left(\frac{L(\omega_{i,T}, \omega_{i,0})}{\sum_i L(\omega_{i,T}, \omega_{i,0})} \ln\left(\frac{LB_{i,t}}{LB_{i,0}}\right)\right)\right\} \tag{5.7}$$

where $L(x, y) = (y - x)/\ln(y/x)$ and $\omega_i = E_{i,t}/E_t$.

To restrict the effect of anomalies in the data due to the extrapolation procedures performed by the CBS, the decomposition analysis has been conducted for the years 1994 and 1999, which correspond to the years in which surveys were done for most of the 23 sub-sectors.

3 ENERGY CONSUMPTION IN THE DUTCH SERVICE SECTOR

Primary energy consumption in the Dutch service sector increased by about 63 per cent during the period 1985–2002 (from 336 PJ to 546 PJ), effectively making the service sector the fastest growing segment in total energy demand (Figure 5.3) and reflecting the ongoing restructuring of the economy towards services.[5] The drop in primary energy use shown in the figure for the years 1993–94 and 1998–99 was caused by a reported decrease of gas consumption. However, the level of data available does not make it possible to find the reasons behind these changes and in fact, they may only reflect errors in the values reported to the CBS (Tinbergen, 2002).

In Figure 5.4 we plot the growth rates of primary energy consumption, value-added, total number of employees, number of enterprises and floor space for the Dutch service sector. On average, primary energy consumption increased at an average rate of 3.7 per cent per year over the period 1985–2002. Value-added showed two periods with different trends, the first period from 1982–88 characterized by slow growth (less than 1 per cent per year) and a second period from 1989–2002 with an annual growth rate of about 3 per cent. These values are in line with the observed trend in

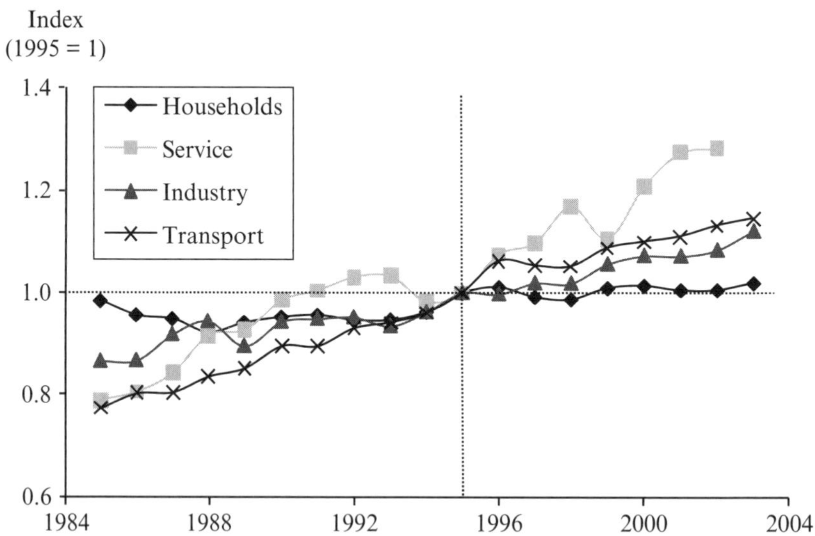

Figure 5.3 Energy consumption growth of the different sectors of the Dutch economy, normalized by the year 1995

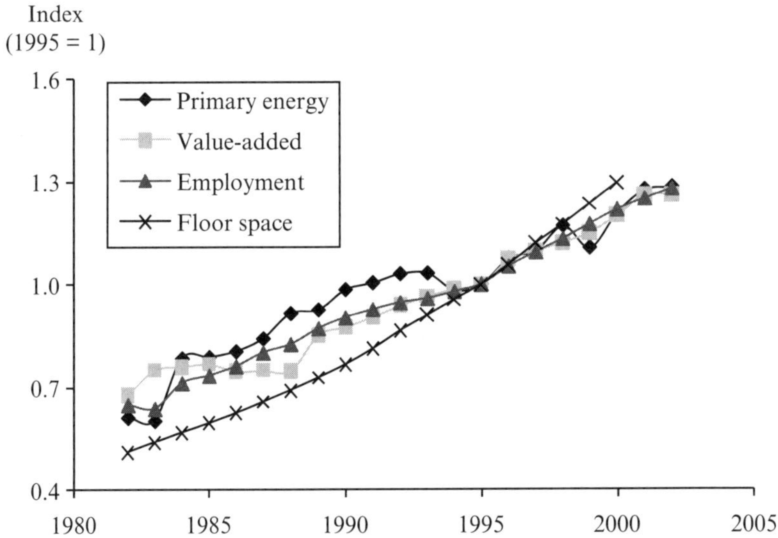

Figure 5.4 Activity indicators for the Dutch service sector, normalized by the year 1995

the number of firms, which amounts to approximately 0.6 per cent growth per year between 1983 and 1988, and 3.1 per cent between 1989 and 2001. Contrary to the fluctuations shown by value-added and number of firms, the number of employees increased at a steady rate of 3.6 per cent per year between 1984 and 2002. Preliminary figures for floor space show an increase of about 5 per cent per year.

Table 5.3 shows the share of gas and electricity according to the type of energy services. Developments in gas and electricity consumption are plotted in Figure 5.5. Electricity consumption has grown at a faster pace than final energy consumption and hence also faster than fossil fuel use. It is expected that, due to the rapidly expanding use of information technology and air conditioners in businesses, total electricity consumption will be between 30 and 100 per cent higher in 2020 as compared to 1995 (Krackeler and Schipper, 1998). The increasing electricity demand can be mainly attributed to the penetration of computers and electronic equipment as well as the adoption of office automation and climate control (Lee et al., 2001; Krackeler et al., 1998; Bern, 1998; van Dril et al., 1998).

Table 5.3 Average percentage of energy demand by fuel and energy service in the Dutch service sector

Share of gas use, %		Share of electricity use, %	
Heating	86	Lighting	42
Warm water	8	Appliances	37
Cooling, humidifiers, etc.	6	Ventilation	10
		Cooling	6
		Pumping, warm water and heat	5

Note: The role of climate control reported by van Arkel et al. (1999) seems much lower than in Cuelenare and Blok (1993). The latter concluded that in the service sector the major determinant of the growth of electricity consumption during the period 1980–90 was the larger penetration of climate control equipment as compared to other factors such as the number of appliances, lighting and other applications.

Source: Van Arkel et al. (1999).

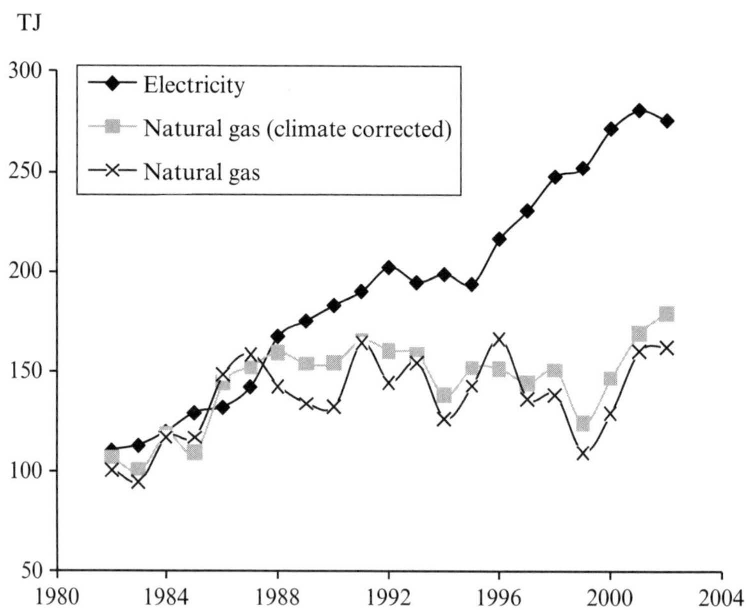

Note: The difference between the gas and climate corrected gas data reflect the influence of outdoor temperature.

Figure 5.5 Gas and electricity consumption in the Dutch service sector, in terajoules

4 ENERGY INTENSITY IN THE DUTCH SERVICE SECTOR

Changes in energy efficiency can be examined by looking at the ratio of energy consumption and activity values. Figure 5.6 shows the results of four different indicators: energy per unit of value added, energy per employee, energy per firm and energy per square meter. Two trends are apparent from these data. First, since 1993, developments in energy per employee, energy per firm and energy per unit of value added have followed a similar pattern. Second, for the entire period changes in energy intensity have been rather minor at less than 0.6 per cent per year, with the exception of the indicator energy per square metre which is based on preliminary values.

In order to better understand the increasing energy demand of the whole sector and the impact of different drivers we applied the decomposition technique described in section 2. Figure 5.7 depicts the results of the period-wise analysis for 1994–99. As mentioned previously, these years have been selected because energy data reported by the CBS are based

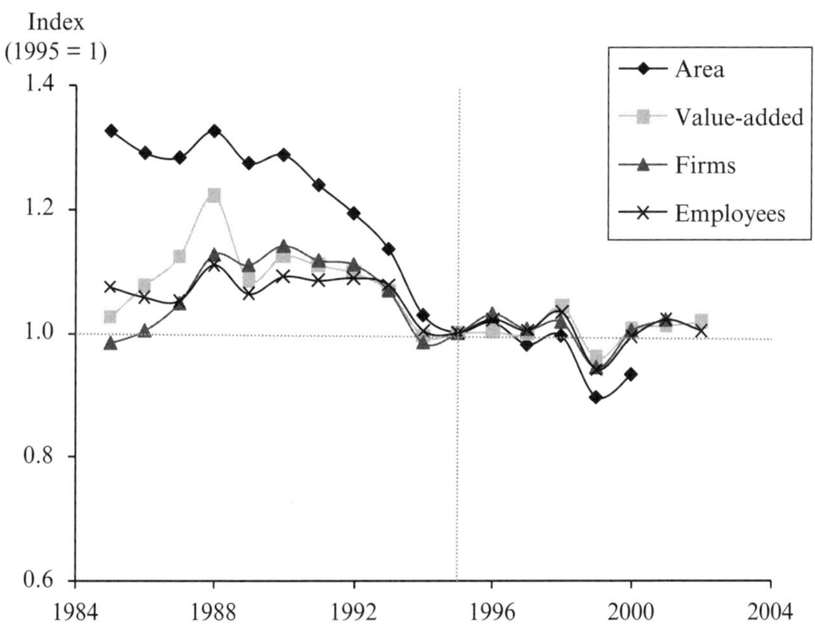

Figure 5.6 Energy intensity indicators for the Dutch service sector, normalized by the year 1995

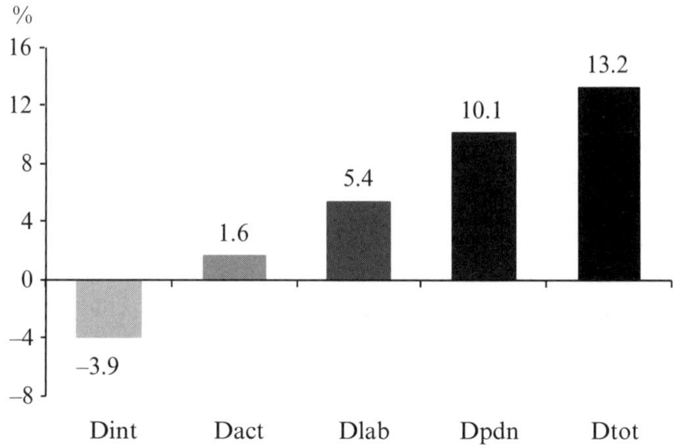

Figure 5.7 Decomposition of energy consumption in the Dutch service sector, 1994–99

on surveys rather than extrapolations, and as a result this diminishes the level of uncertainty of the results. The analysis shows that, as expected, the strongest driving force behind the increasing energy consumption of the service sector has been the activity effect, specifically the sectoral development of value-added and the associated labour intensity. In fact, the strong role of value-added is merely a reflection of the increased importance of the service sector in the national economy. Labour intensity has boosted energy use because growth in employment has outpaced growth in value-added.[6] The analysis also shows that the effect of deviating growth rates of sub-sectors within the service sector is rather small. Finally, energy intensity measured as the amount of primary energy per employee seems to have decreased, which points to improvements in energy efficiency in the sector. Nevertheless, these improvements have not been large enough to offset the increased energy demands associated with a growing output.

5 RESULTS FOR THREE SPECIFIC SUB-SECTORS

In this section we take a closer look at the developments in energy consumption and energy efficiency of three sub-sectors of the service sector: (1) trade, accommodation and food services, (2) financial intermediation, and (3) intramural health care. Together these sectors account for approximately 60 per cent of the primary energy consumption in the Dutch service sector.

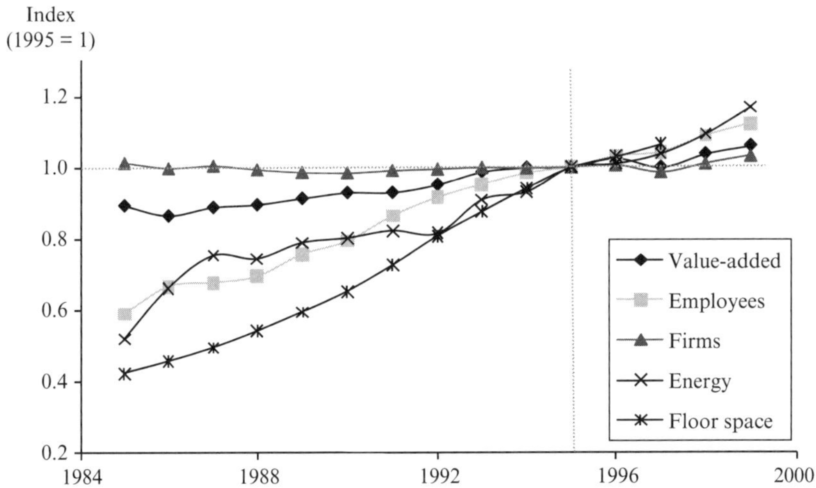

Figure 5.8 *Activity indicators for the subsector trade, accommodation and food services, normalized by 1995*

5.1 Trade, Accommodation and Food Services

We refer to the sub-sector 'trade, accommodation and food services' as a catch-all term for wholesale and retail trade, repair of motor vehicles, motorcycles and personal and household goods, and hotels and restaurants. Following the official Dutch classification of sectors, these are the sub-sectors SBI 50–55.

In 2000 this sector accounted for 13 per cent of total output, 15 per cent of the total value-added, and 20.5 per cent of the total personnel employed in the entire Dutch economy. Figure 5.8 plots the development of the main activity indicators. Floor space showed the highest growth with a total relative increase of 92 per cent, followed by energy consumption (62 per cent), employment (60 per cent), value-added (16 per cent) and number of firms (6 per cent). About one-third of the total change in energy consumption is reported for the period 1994–99. The energy consumption values reported for this period are the result of CBS extrapolations, which are based on the number of employees. One should also note that employment grew at a higher rate than energy during the period 1985–94; employment growth outpaced the growth of value-added during the period 1993–99, and employment rates also outpaced our estimates for floor space. As a result, it is likely that the use of employment as the base for projecting energy use values during the period 1994–99 overestimates the actual energy consumption of this sector.

Developments in electricity consumption and gas consumption at the sectoral and sub-sectoral level are plotted in Figure 5.9. The resulting trends exemplify several of the problems that are encountered when trying to analyse energy consumption in the service sector. For instance, after 1993 CBS used the same growth rate to extrapolate the consumption of gas and electricity, and therefore both energy carriers show the same increase in consumption of 20 per cent between 1994 and 1999. An additional problem is that the years used for the extrapolations (1993 and 1994) diverge from the historical trend (1984–92). This deviation could be caused by the significantly lower number of companies that answered the CBS survey within this sector (Tinbergen, 2002). Therefore, consumption figures may not be representative of the behaviour in the entire sub-sector. As an example, data reported for the period 1992–94 for the sub-sector wholesale trade (SBI 51) show a decrease in electricity consumption of about 25 per cent and an increase in gas consumption of about 46 per cent. The data also reveal that during these years the number of employees only decreased by 6 per cent while the value-added and number of firms increased by 2 per cent. Furthermore, there is no evidence in this period of significant changes in the number of heating degree-days that could partially explain the increasing gas consumption, neither was evidence found for special policy measures that could have affected electricity demand. Consequently, no explanation is available for the observed changes and they may therefore merely be statistical artefacts related to data collection and estimation.

These problems affect every analysis of energy efficiency in the sector. Figure 5.10 shows the ratios of energy use to different activity values. The figure illustrates that, using this data, the identified magnitude and direction of the changes in energy intensity crucially depends on the indicator that is examined. The trend in the indicator energy per value added shows a strong increase in energy intensity throughout the entire period, which could be interpreted as a decrease in energy efficiency. The largest improvements in energy efficiency are shown by the indicator energy per square metre. However, strong conclusions cannot be drawn from this indicator alone since the data for floor space are of a preliminary nature (see section 2) and there is no data available for any year that allows us to judge the accuracy of the estimated values. As discussed earlier, in the absence of reliable figures of floor space, number of employees is considered a second-best activity measure to be used in the development of energy intensity indicators. However, since energy values after 1994 have been mainly based on the changes in employment, the resulting ratio between energy and employment is approximately constant. The indicator energy per employee therefore only reflects the methodology used and

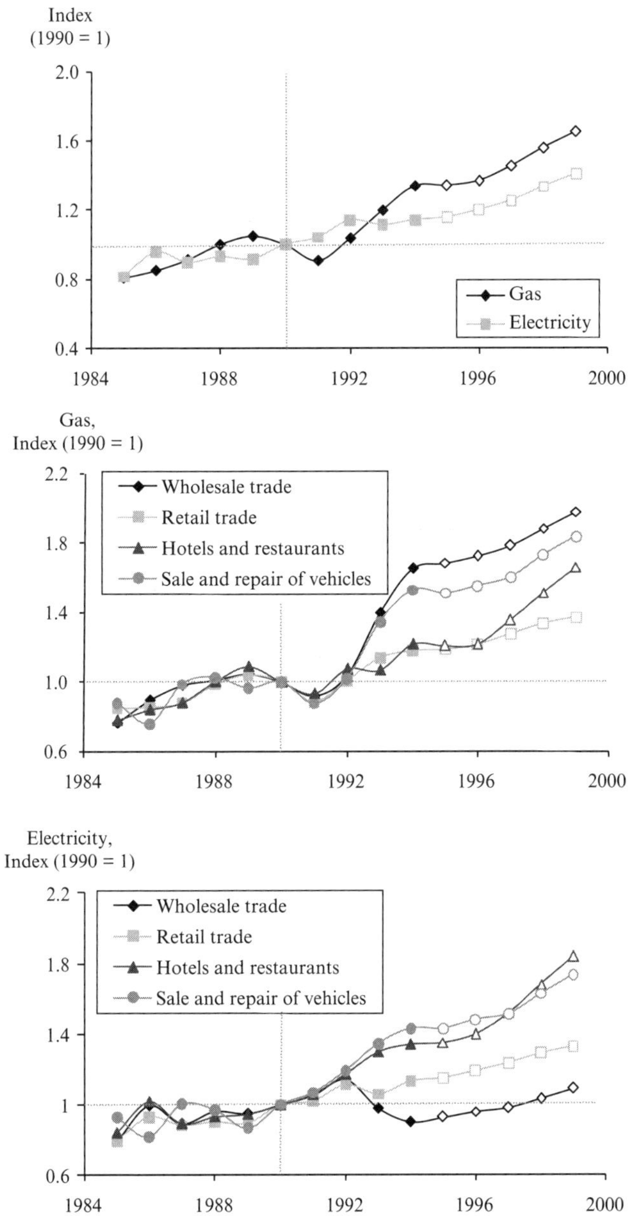

Figure 5.9 *Gas and electricity consumption for the sector trade and hotels and restaurants (top) and by sub-sector (middle and bottom), normalized by 1990*

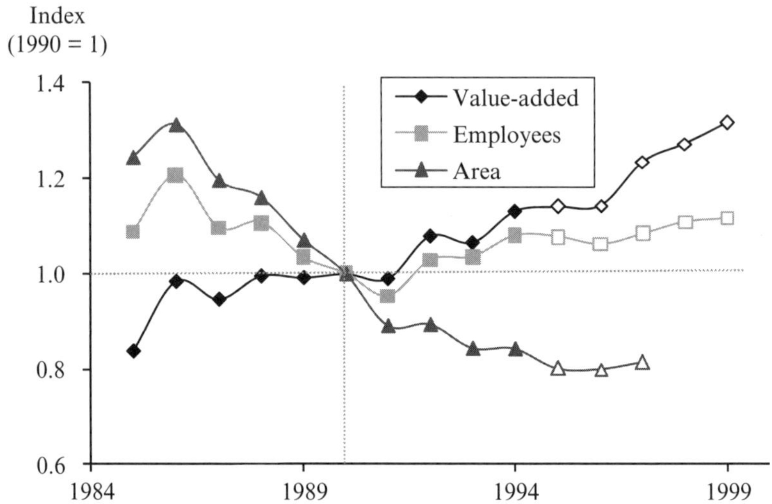

Figure 5.10 Energy intensity indicators for the sub-sector trade, accommodation and food services, normalized by the year 1990

does not provide any information on energy efficiency developments. In sum, with the available data it is not possible to provide reliable estimates of the historical changes in energy consumption and energy efficiency for the trade, accommodation and food services sector.

5.2 Financial Intermediation

The financial sector comprises financial intermediation[7] (SBI 65), insurance and pension funding (SBI 66) and activities auxiliary to financial intermediation[8] (SBI 67). In 2000, the financial sector accounted for 6 per cent of total value-added, 4 per cent of employment and 4 per cent of the primary energy consumed by the Dutch service sector. Energy data for the financial sector are only available since 1993. Data for 1994 and 1999 are the result of CBS surveys, while published data for 1993 and the period 1995–98 are the result of extrapolations. There are no data available for the period after 1999. Contrary to the energy figures of the sub-sector trade, accommodation and food services the fact that there are survey-based values for 1999 for the financial sector allow us to evaluate the accuracy of the extrapolations made by the CBS. It is also helpful that the surveys reported floor space data for the years 1994 and 1999 for the financial intermediation sector.

Among all activity indicators, energy showed the highest growth in the seven-year period amounting to approximately 22 per cent. Firms, employment and value-added grew by 18 per cent, 17 per cent and 12 per cent, respectively. Floor space grew by 19 per cent in the five-year period. Figure 5.11 shows the trends in electricity and gas consumption at the sub-sectoral level. The decrease in energy consumption shown for the sub-sector insurance and pension funding (SBI 66) is the result of a period of economic decline in the sector (see Table 5.4).

Figure 5.12 depicts the trends for different energy intensity indicators. The results highlight the high degree of uncertainty of the extrapolated values reported by CBS. A comparison of the relative changes shown by the indicators for the period 1994–99 points out that the results for the indicators energy used per employee and energy used per square metre are similar, whereas the indicator energy per unit of value-added shows an increase that is three times larger. Since energy per value-added is considered a less accurate indicator of energy developments in the service sector, we conclude that primary energy intensity in the financial intermediation sector increased by about 3 per cent between 1994 and 1999. In terms of final fuel use, the sector showed an increase in gas and electricity intensity of 2 per cent and 3.5 per cent, respectively.

5.3 Intramural Health Care

The intramural health care sector consists of primary care provided by hospitals, psychiatric hospitals, nursing homes and institutions for mentally handicapped people. Figure 5.13 depicts the main indicators for this sector: energy consumption, number of employees, number of beds, number of institutions and floor space. Data are only available for the period 1985–98. Note that, contrary to the two sub-sectors examined previously, all data published by the CBS for this sector are the result of surveys. The highest growth rate is shown by the number of employees and the number of square metres with average values of about 30 per cent for the period 1985–96, and 36 per cent for the 1985–97 period, respectively. Energy consumption shows a fairly increasing trend between 1985 and 1998 of 16 per cent growth. The trend for the number of beds indicates a slight increase of about 2.5 per cent during the 1985–1998 period, which was caused by restrictions in the number of beds enforced on the basis of national health policies in the Netherlands. Finally, the number of institutions is the only indicator that shows a declining trend of 12 per cent during the period 1985–98.

In terms of final energy consumption, electricity demand increased by 26 per cent while gas consumption increased by 6 per cent between

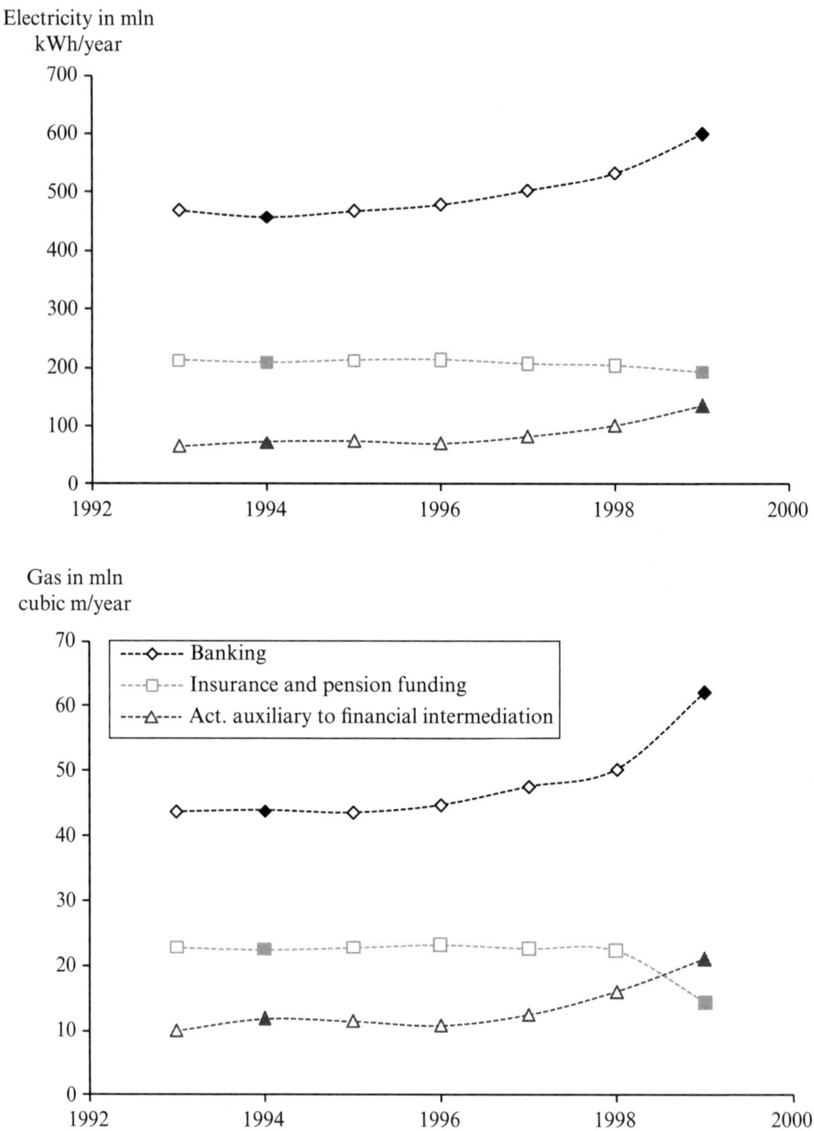

Note: Unfilled dots represent values obtained from extrapolation estimation by the CBS.

Figure 5.11 Electricity and gas consumption at the sub-sectoral level

Table 5.4 Comparison of percentage growth rates for the sector financial installations during the 1993–99 period

Sub-sector	SBI	Growth in primary energy usage[a]	Growth in number of employees	Growth in square metres
Financial intermediation	65	+26	+16	+30
Insurance and pension funding	66	−16	+1	−12
Activities auxiliary to financial intermediation	67	+76	+46	+35

Note: [a] Refers to 1994–99.

1984 and 1998. Kerssemeeckers (2001) pointed out that 51 per cent of all electricity is used for lighting, which makes it the highest demand for this purpose of all service sectors. The high consumption of electricity may be due to the long hours of operation and the relatively large concentration of workers/patients. Figure 5.14 plots the development in energy intensity indicators. Two contradictory trends are visible: energy per employee and energy per square metre exhibit a declining trend, while energy per bed shows a slightly increasing trend. Since the number of beds has been kept constant as a result of the previously mentioned policy measures, its value as a measure of activity is limited. The results of the other indicators point out energy efficiency improvements of about 1.8 per cent per year.

6 CONCLUSIONS

In this chapter we explored whether available data allow the development of *reliable* energy efficiency indicators for the service sector in the Netherlands. It turned out that the quality of reported energy data hinders a meaningful analysis of energy efficiency developments, particularly at the sub-sectoral level. Although the general opinion is that having some data is better that having no data at all, this situation does not allow for an accurate monitoring of the effectiveness of energy policies. We recommend that in the case where annual surveys at the sub-sector level are not a feasible way of data collection, periodic energy surveys (for instance, every three or four years) should be conducted in order to evaluate the accuracy of extrapolation procedures used by statistical offices.

The results of our research also show that conclusions concerning the energy efficiency behaviour of the service sector and its constituent sub-sectors strongly depend on the type of energy intensity indicator that is

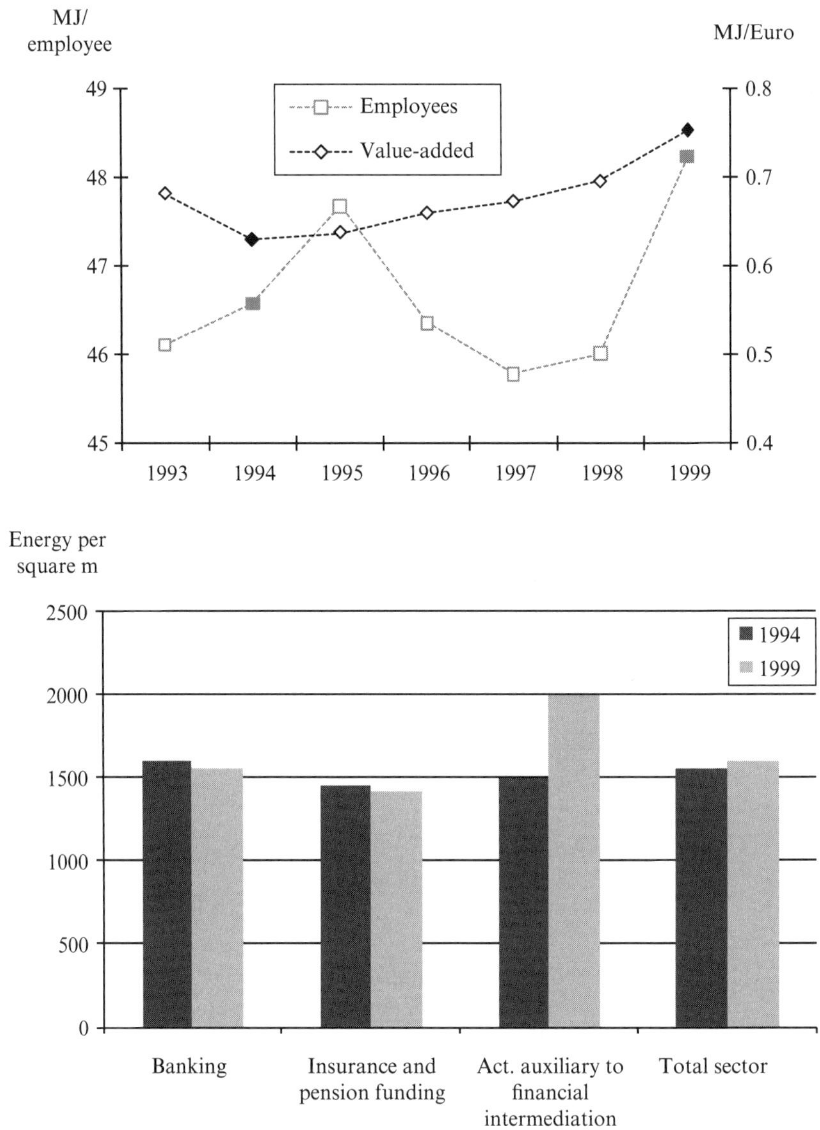

Note: Unfilled dots indicate published values based on extrapolations.

Figure 5.12 Energy per employee, value-added (top) and per square metre (bottom) for the financial sector

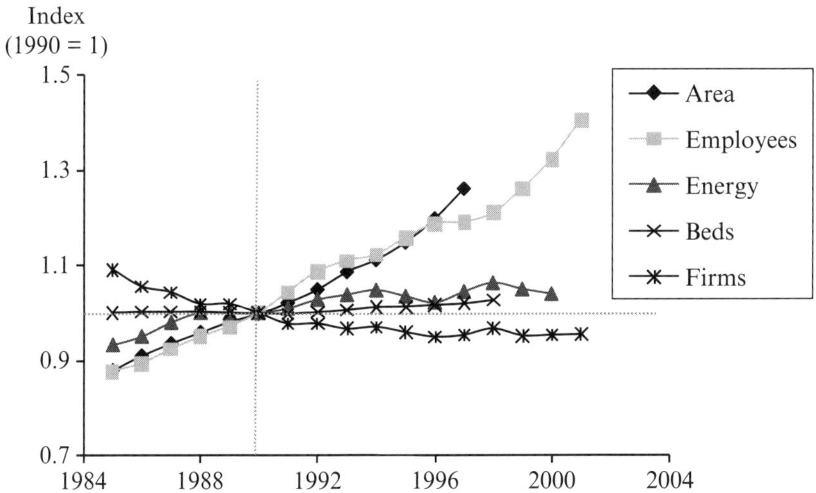

Figure 5.13 *Activity indicators for the Dutch health sector, normalized by the year 1990*

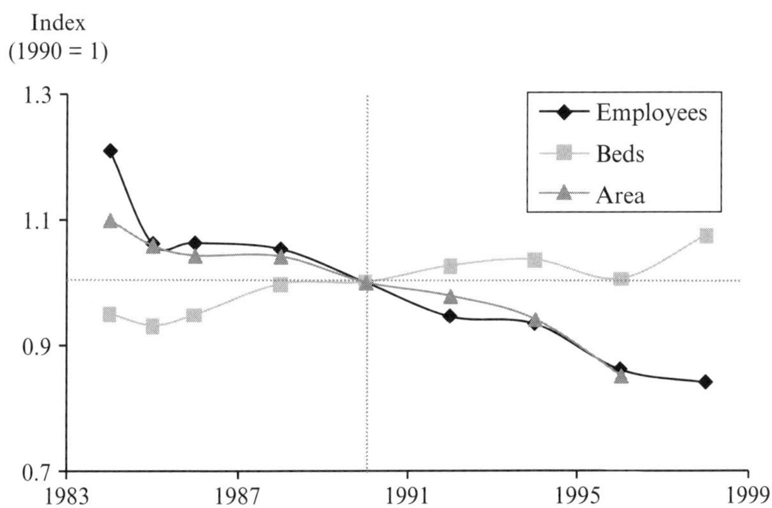

Figure 5.14 *Energy intensity indicators for the Dutch health care sector, normalized by the year 1990*

selected. We also found that when value-added was used as measure of activity, changes in energy intensity tended to be amplified.

At the aggregate level, our results show a strong increase in energy consumption, especially in electricity use. The results of the decomposition analysis showed that the activity effect (value-added) has been the main driving force behind the increase of energy consumption of the service sector, an increase in labour intensity has also boosted energy consumption, and energy intensity (in terms of energy per employee) has not been large enough to offset energy demands associated with growing output.

These trends in conjunction with the current targets for energy efficiency savings for the service sector (specifically, 15.5 GWh by 2016 in the Netherlands) demand a better understanding of energy use in the service sector. The demand for more and more thorough analysis is amplified by the fact that the available data cannot fully explain developments within the sector and as a result, significant uncertainties remain in the actual explanations of apparent trends in energy usage. This lack of knowledge will make it difficult to pursue an evidence-based energy policy. In sum, a considerable research effort is still required to obtain insight into the forces driving energy consumption in the service sector. In particular, there is a need for an in-depth assessment of the developments in specific sub-sectors and the way in which energy-related agendas are formulated and managed.

NOTES

1. For instance, according to estimates of the German Council for Sustainable Development, more than 2000 full-time jobs could be created for each million tonnes of oil equivalent that will be saved as a result of measures and/or investments specifically taken to improve energy efficiency as compared to investing in energy production. For more information see: http://www.nachhaltigkeitsrat.de/service/download/publikationen/broschueren/Broschuere_Kohleempfehlung.pdf.
2. The World Bank estimated an annual growth rate in Western Europe for the service sector of 1.9 per cent during the period 1990–99. This value is high if compared with the 0.8 per cent annual growth in the manufacturing sector (World Bank, 2001).
3. Interestingly, this figure is in accordance with the global potential savings for the residential and commercial sector (30 per cent cost-effective by 2020) as reported by IPCC (2007).
4. Indicators have played an important role in scientific analysis and policy-making. A comprehensive definition has been given by the United States Department of Health, Education and Welfare: 'An indicator is a statistics of direct normative interests which facilitates concise, comprehensive and balanced judgements about the condition of major aspects of society. It is in all cases a direct measure of welfare and is subject to the interpretation that, if it changes in the "right" direction, while other things remain equal, things have got better, or people better off' (Godin, 2003, p. 681).
5. In the same time period, the manufacturing sector increased its consumption by 26 per cent, while the agricultural and transport sector increased demand with 39 per cent and 46 per cent, respectively.

6. The increase in labour supply was the dominant factor affecting the Dutch labour market in the 1990s. Between 1990 and 1999 more than 110 000 persons entered the labour market each year. The fast growth in employment occurred especially during the period 1994–99. Among the main reasons for this growth are: an increase in participation, the business cycle through what has been called 'the discouraged worker' effect (the growth rate of labour supply dipped in the early nineties, when economic growth was slow, and took off afterwards, when the economy started to boom), and the government policy aimed at reducing the use of social benefits (Huzinga, 2001).
7. Specifically, this refers to banking, leasing and other credit granting.
8. This comprises administration of financial markets, security broking and fund management, and activities auxiliary to financial intermediation and pension funding.

REFERENCES

Ang, B.W. and F.Q. Zhang (2000), 'Survey of dex decomposition analysis in energy and environmental studies', *Energy*, **25**, 1149–76.

Bern, V. (1998), 'Novem activities in commercial buildings', opening address during the conference for improving electricity in commercial buildings, 21 September.

CPB (1998), 'The economy and its physical surroundings: policy challenges and possible solutions 1995–2020', Extensive Summary, CPB working paper 102, The Hague: Netherlands Bureau for Economic Policy Analysis (CPB).

Cuelenare, R. and K. Blok (1993), 'Ten years of development in Dutch electricity consumption', NWS report 93050, Utrecht: Department of Science, Technology and Society, Utrecht University.

Department of Energy, Energy Information Adminstration (DOE/EIA) (1995), *Energy Consumption Series, Measuring Efficiency in the United States Economy: A Beginning*, Washington, DC: Department of Energy.

European Commission (EC) (2000), *Action Plan to Improve Energy Efficiency in the European Community, Communication from the Commission to the Council and the European Parliament, the European Economic and Social Committee and the Committee of the Regions*, Com (2000)247 Final, Brussels: European Commission.

European Commission (EC) (2005), *Doing More with Less*, Green Paper on energy efficiency, Luxembourg: Office for Official Publications of the European Communities.

European Commission (EC) (2006a), Council Directive 2006/32/EC of the European Parliament and of the Council of 5 April 2006 on Energy End-use Efficiency and Energy Services, OJ L 114, 27.4.2006, 64–85.

European Commission (EC) (2006b), *Action Plan for Energy Efficiency: Realising the Potential*, Communication from the commission, COM (2006) 545 final, Brussels: European Commission.

Farla, J. (2000), 'Physical indicators of energy efficiency', PhD thesis, Department of Science, Technology and Society, Faculty of Chemistry, Utrecht University, The Netherlands.

Godin, B. (2003), 'The emergence of S&T indicators: why did governments supplement statistics with indicators', *Research Policy*, **32**, 679–91.

Goldewijk, K., J. Olivier, J. Peters, P. Coenen and H. Vreuls (2004), 'Greenhouse gas emissions in the Netherlands 1990–2002', National Inventory Report, Bilthoven: National Institute for Public Health and the Environment (RIVM).

Huzinga, F. (2001), 'Employment in the 1990s', CPB Report 2001/1, The Hague: Netherlands Bureau for Economic Policy Analysis.
Intergovernmental Panel on Climate Change (IPCC) (2007), 'Issues related to mitigation in the long-term context', contribution of working group III to the fourth assessment report of the Intergovernmental Panel on Climate Change, in B. Metz, O. Davidson, P. Bosch, R. Dave and L. Meyer (eds), *Climate Change 2007: Mitigation*, Cambridge: Cambridge University Press, pp. 169–250.
International Energy Agency (IEA) (2006), 'Energy technologies perspectives: scenarios and strategies to 2050', Paris: International Energy Agency.
Kerssemeeckers, M. (2001), 'Icarus 4: sector study for the service sector', report as part of the project EB/99044645, Ecofys and environment, Utrecht: Utrecht University.
Krackeler, T. and L. Schipper (1998), 'Carbon dioxide emissions in the OECD service sector: the critical role of energy use', *Energy Policy*, **26** (15), 1137–52.
Krackeler, T., L. Schipper and O. Sezgen (1998), 'Carbon dioxide emissions in OECD service sectors: the critical role of electricity use', *Energy Policy*, **26**, 1137–52.
Lee, W., F. Yik, P. Jones and J. Burnett (2001), 'Energy saving by realistic data design for commercial buildings in Hong Kong', *Applied Energy*, **70**, 59–75.
Murtishaw, S. and L. Schipper (2001), 'Disaggregate analysis of US energy consumption in the 1990s: evidence of the effect of internet and rapid economic growth', *Energy Policy*, **29**, 1135–356.
Schipper, L., S. Meyers and A. Ketoff (1986), 'Energy use in the service sector: an international perspective', *Energy Policy*, **14**, 201–18.
Schuur, J. (1999), 'Industrial land-use planning in the Netherlands', CPB Report, The Hague: Netherlands Bureau for Economic Policy Analysis.
Suijker, F., A. Kuypers, M. van Dijk, H. Kox and H. van der Wiel (2002), 'De Commerciële Dienstverlening: Een Heterogene Sector met Gunstige Groeiperspectieven', CPB document 17, The Hague: Netherlands Bureau for Economic Policy Analysis.
Tinbergen, W. (2002), Personal communication, CBS Statistics Netherlands.
van Arkel, W., H. Jeeninga, M. Menkveld and G.J. Ruig (1999), 'Energieverbruik van Gebouwgebonden Energiefuncties in Woningen en Utiliteitsgebouwen', report ECN-C-99-084, Petten: Energy Research Center of the Netherlands.
van Dril, A.W.N., W.G. van Arkel, P.G. Dougle, H. Jeeninga, P. Kroon and M.A. Uyterlinde (1998), 'Energy efficiency indicator in the Netherlands', Dutch contribution to the project 'Cross country comparison on energy efficiency', phase 4, ECN report ECN-C-99-038, Petten: Energy Research Center for the Netherlands.
World Bank (2001), *World Development Indicators 2001*, Washington, DC: World Bank.
Worrell, E., J.A. Laitner, M. Ruth and H. Finman (2003), 'Productivity benefits of industrial energy efficiency measures', *Energy*, **28**, 1081–98.

PART II

Technology and Investment Behaviour

6. Adoption criteria, diffusion and firm size: the role of learning in reconciling theories of endogenous technical change with empirical evidence

Greg W. Hunter

1 INTRODUCTION

The promise of progress through scientific inquiry gives rise to moments when empirical evidence disagrees with the predictions of theory. Typically, the pattern of developments following such moments begins with an explosion of new theories, each an iteration building on some prior theory, and each able to explain some part of the aberrant data. Eventually, one of these theories, through a novel approach, lays the groundwork for a new perspective on the relationships to be explained. In what follows, the literature on the economics of innovation will be presented in a way that alludes to this sort of 'creative destruction'.

The empirical literature on technical change has showcased two anomalies, adoption lag and slow diffusion. These studies have masked a third anomaly that is important to understanding not only how technical change influences decisions of the firm, but also the entire structure of an industry, which is firm size. In what follows, we will explain the connection between the three anomalies using a specific mixture of two rarely blended strands of economic theory. Drawing upon the economic literature that views information as a public good and from the theory of cartels, we show that a new technology can spur cooperation among firms. In particular, we show that in response to a new technology, a subset of firms in an industry may rationally decide to merge, or form a joint venture, to experiment with the technology, thereby sharing the costs of providing information about the suitability of the innovation for firms in the industry.

Our analysis begins by reviewing the empirical literature on endogenous technical change with a special emphasis on exposing the role of firm size in explaining other aspects of the data in section 2. Next, in section 3, a review of the theoretical literature is presented. Section 4 exposes a simple model that marries the issues of adoption, diffusion and firm size. In Section 5 a solution is provided. Conclusions and directions for further investigation complete the chapter.

2 THE EMPIRICAL LITERATURE

The empirical literature on technical change is marked by two frequently mentioned characterizations. The first is that there is a tendency for decision-makers to 'irrationally' forestall the adoption of new technology. The second is that innovations tend to slowly diffuse into use. In our discussion of the empirical literature, we use this division as an organizing principle. Before moving forward, it should be said that the empirical evidence for both of these characterizations is scant. Despite the dearth of evidence, both ideas have received a considerable amount of attention from theorists.

Empirical research on technology adoption timing has noted that both firms and consumers appear to postpone adoption of a new, positive net present value technology. For firms, the strongest presentation of this point is made by DeCanio (1998), who presents an econometric analysis of data gathered from firms (and governmental entities) participating in a voluntary programme sponsored by the US Environmental Protection Agency (EPA). The programme, known as Green Lights, provides participants with information on energy-efficient lighting technologies and the participants, in exchange, provide information to the EPA on their decisions to upgrade lighting. Specifically, the firms provide information on the 'payback' used to decide whether to go forward with lighting projects, data on the cost and other economics factors related to each project, and data related to the organizational characteristics of the firm. DeCanio (1998, pp. 453–4) interprets his findings in the following way:

> While economic forces play a role, . . . [t]hese returns are higher than the cost of capital for projects of comparable risk, and they vary systematically according to a number of non-economic characteristics. In general, the data reinforce the view that there is a large potential for profitable energy-saving investments that is not being realized because of impediments that are internal to private and public sector organizations.

For DeCanio (1998, p. 453) the non-economic characteristics are what elsewhere is referred to as 'organizational and institutional factors'. The organizational variables include several measures that relate directly to or may correlate with firm size. For example, DeCanio (1998, p. 451) examines the influence of the area of the building that the firm occupies and finds that larger firms are 'more willing to accept a longer payback' on energy-efficient lighting projects. This effect is strongly significant, but DeCanio (1998, p. 451) downplays its importance in interpreting his findings, noting that the impact on the payback is small relative to the mean. As the quote above implies, DeCanio's (1998) main thrust is to emphasize that there are internal organizational failures that cause firms to miss profitable opportunities. The problem with having too much confidence in this conclusion is that there is not enough detailed analysis of the data to know if the organizational factors are multicollinear with firm size. One might argue that firm size should encourage the adoption of new technologies, *ceteris paribus*. By no means are we the first to suspect such a connection (Rose and Joskow, 1990).

The firm size effect has shown up in a more recent contribution to the literature on the energy-efficiency paradox. Diederen et al. (2003) examine the adoption behaviour of the Dutch horticulture industry with respect to two energy-saving technologies. The focus of the study is to argue that an options approach (Dixit and Pindyck, 1994) to modelling the adoption criterion of firms explains a large portion of the adoption lag perceived when one looks at the net present value of an investment. Diederen et al. (2003) have firm-level data from 433 firms that is detailed enough to allow for the calculation of the hurdle rate for adopting a new technology, assuming the option criterion is used, and examine whether it does a good job of predicting the firms' investment decisions. Apart from finding general support for the option criterion, they find that the size of the firm, as measured by the area under cultivation, is strongly significant in predicting the adoption decision of firms. Diederen et al. (2003, p. 390) conclude: 'The size of the area under cultivation contributes positively, and significantly, to the probability of adoption. Our findings therefore suggest that larger firms are more likely to invest in energy-saving installations.'

Diederen et al. (2003) and DeCanio (1998) echo the findings of earlier empirical literature on the adoption and diffusion of new technology. Without citing studies whose results are subject to 'considerable criticism' based on econometric issues (Rose and Joskow, 1990, p. 356), we will briefly mention three relatively recent studies that support the conclusion regarding the negative correlation between firm size and adoption time. Rose and Joskow (1990) analyse the adoption behaviour of firms in the electric utility generation industry with respect to two generations of technologies. The newer of the two technologies, 'super critical' high-pressure generators, were found to have a higher probability of being adopted early by larger firms. Levin et al. (1987) investigate the diffusion of optical scanning technology in the grocery store industry. Using a hazard rate model similar to the one used by Rose and Joskow (1990), Levin et al. (1987) find that a 1 per cent increase in firm size, as measured by total sales, increases the probability that a store quickly adopts the new scanning technology by over 100 per cent (Levin et al., 1987, p. 16). Lastly, Hanna and McDowell (1984) examine the adoption decision of banks with respect to the introduction of automated teller machines (ATMs). Their results show that 'larger banks are found to exhibit a higher conditional probability of adoption, all else being equal' (Hanna and McDowell, 1984, p. 332).

Evidence on the speed of diffusion has focused mainly on identifying the factors that influence the speed of adoption. Allen Blackman (1999) does a good job of succinctly describing contributions to the literature in this area. The main point of studies on diffusion/adoption timing is to acknowledge the correlation between an adoption lag on the one hand, and firm, industry and innovation characteristics on the other. For example, the literature finds that simple innovations tend to spread more quickly in an industry than more complex innovations. Blackman (1999) also notes that empirical evidence suggests that the speed of diffusion is influenced by firm size, market structure, factor prices, and regulation. For our part, we wish to emphasize that the empirical literature suggests that differences in firm size within an industry can generate technological diffusion. Typically, the pattern of diffusion observed begins with large firms adopting first and smaller firms adopting later.

In sum, the empirical literature tends to support three conclusions regarding firms' decisions to embrace technological advances. First, firms tend to delay the opportunity to invest in a new technology beyond what one would predict using the net present value criterion. Second, new technologies tend to be adopted in a more-or-less sequential manner or diffused into use. Third, larger firms tend to adopt new technologies before smaller firms. The modelling section of this chapter develops a model that generates all three of these empirical regularities in tandem.

3 THE THEORETICAL LITERATURE

The theoretical literature on the adoption and diffusion of new technology is expansive. The contributions are so many and of such a varied extent that it would not serve the interest of this chapter to attempt to summarize a broad section of them. Rather, we take this opportunity to refer the interested reader to two excellent surveys of the literature and then discuss the subset of the literature that is most closely related to our model. The most recent of these surveys is by Hoppe (2002). She breaks the literature into four categories based on whether each contribution fits either of two criteria. The first criterion asks if a particular contribution models the adoption behaviour of firms in a strategic setting. More precisely, models satisfying this criterion make the payoff to adopting the new technology depend on the behaviour of other firms in the industry. For a particularly simple example of this type of model, imagine a new technology that would lower the marginal production cost of an adopting firm in an industry whose output market equilibrium is determined by Cournot competition. The second criterion asks if a particular model captures uncertainty with regard to the either the value of an innovation or its date of being introduced. Reinganum (1989) also provides an excellent survey of the earlier literature.

Using Hoppe's (2002) classification scheme, we discuss elements of the model that we introduce in the following section and contributions in the literature that are closely related. The model to be introduced will have elements of uncertainty and strategic interaction among firms. So, it satisfies both of the criteria put forward in Hoppe (2002). However, the uncertainty in our model is relatively simple compared to the stochastic-dynamic analyses appearing in the literature (Doraszelski, 2001; Farzin et al., 1998; van Soest and Bulte, 2001). We consider only two states of nature and two dates. Our model is connected to the stochastic-dynamic models in that they also contemplate uncertainty in the value of adopting a new technology. However, the impact of the uncertainty on the firm's investment decisions, the manner in which the uncertainty is resolved, and the associated policy implications are completely different. In the continuous-time stochastic-dynamic, or 'real options' literature an element of irreversibility in the adoption decision ensures that the opportunity cost of investment is higher than the net present value criterion predicts. The reason for the increased opportunity cost has to do with the evolution of the stochastic process describing the value of the investment over time. Because the investment decision is irreversible, the firm cannot recover losses from adopting the new technology if its value declines. It is also typically assumed that by waiting, the firm can observe the value of the investment opportunity as it changes. By deciding to invest, the firm gives up the value of future information in making its decision. This makes the options investment criterion more conservative than the net present value rule. We do not dispute this finding. Yet on the basis of empirical evidence such as DeCanio (1998), it seems doubtful that the real options approach in its starkest form, would account for the magnitude of hurdle rates actually observed. In our model, firms learn about the new technology by observing the experiences of others. This public good component of learning is not unique to our model. For example, Hoppe (2000) introduces a model of firms that can learn from one another's experience with a new technology. However, our approach allows the exploration of issues of firm interaction that have not previously been addressed in the literature. In the following paragraph, we are more explicit. For the time being, we mention one more paper that is similar in spirit to ours. Jaffe and Stavins (1994) introduce a model of technological adoption where firms are able to learn from the cumulative industry experience with a technology, as well as from their own use of the technology.

The second component of the theoretical literature identified by Hoppe (2002) involves papers that deal with the influence that strategic interaction in the product market has on incentives to invest. We look at a similar issue, in particular, strategic interaction with regard to sharing the cost of learning. In our model, firms can learn from the collective experience with a new technology. Thus, in some instances some firms may be tempted to wait until they can see how other firms have fared with the new technology. Firms may also form coalitions to promote the sharing of information. In this respect our model is most similar to Yi and Shin (2000), who also examine the formation of research coalitions. Their emphasis, however, is on examining the formation of competing joint research ventures. Ours is to examine the impact on adoption decisions of existing technology. In addition, their model is concerned with the strategic interaction in the output market, whereas we assume the output market is competitive. Lastly, we explicitly model the cost of coordination, which grows as the size of a coalition grows. The explicit cost of coordination clearly marks our point that firm size will be partly determined by innovations, rather than strategic interaction in the output market.

4 A SIMPLE MODEL

Consider an industry with a competitive output market consisting of N identical, risk-neutral firms. Imagine the choice of a typical firm in the industry faced with a decision regarding the adoption of a new production technology. To capture the dynamic effects of the firm's choice, the model includes two dates, $t = 1, 2$, over which the firm must make decisions regarding its level and technique of production. At date $t = 1$, the firm has the opportunity to adopt a new technology, $c_n(q)$, or remain with the status quo, $c_0(q)$, where q is the firms output. However, the suitability of the new technology is uncertain and after being adopted the true cost of the new process will be revealed at date $t = 2$. At $t = 1$, the firm only knows the distribution of values related to the new technology's cost and the associated expected value. If the model is constrained to include only two states of nature, the information available to the firm for its decision at date $t = 1$ would, for example, be $E(c_n(q)) = \phi c_L(q) + (1 - \phi)c_0(q)$, where φ is the probability that the new technology reveals a lower cost, $c_L(q)$, than the cost of the status quo technology, $c_0(q)$. Throughout our exposition, we assume all production techniques exhibit diminishing returns. To save notation, we also assume that at worst the new technology has a marginal cost equal to the status quo technology. However, this could easily be changed without loss of generality.

Thus, the firm's choice is whether to adopt the technology in light of the information available at $t = 1$ or wait until $t = 2$ to make a decision. It is assumed that at $t = 2$ all firms in the industry are able to completely observe the mutual success or failure associated with implementing the technology at $t = 1$, even if the new technology is only used by a fraction of the firms at $t = 1$. In fact, it is this process that leads to the revelation regarding the true cost of the technology at date $t = 2$. Of course, we could end up with a waiting game of infinite duration. Our model allows for the waiting game outcome, but we do not emphasize it because the outcome has already received ample attention in the literature (Hoppe, 2002). In this context, we also want to mention that our model assumes no passive information arrival. We stress this point in order to ensure that there is no suspicion that our model bares any direct resemblance to a real option model.

Information about the suitability of the technology takes on the character of a public good in our setting. To more fully capture the range of incentives created by information spillovers, we assume that the expected payoffs associated with adoption of the new technology are such that without coordination to share the costs of experimentation, firms may unilaterally adopt the new technology. Our objective is to capture the trade-off that firms face in coordinating the adoption of the new technology at date $t = 1$.

There are two essential elements to this trade-off. The first is the cost of monitoring and enforcing coordination among firms participating in the coalition of adopters, $\gamma(m)$, where m is the number of firms in the coalition using the new technology with $\gamma'(m) > 0$ and $\gamma''(m) > 0$. We assume firms in the coalition are symmetric in all respects and thus pay an equal share of the coordination costs, $\gamma(m)/m$. Monitoring and enforcement is necessary to ensure that firms joining the coalition of adopters in the first period faithfully reveal the results of their experience with the technology, and actually carry out adoption of the new technology in a way that generates information regarding its cost-effectiveness. So, while information regarding the cost-effectiveness of the new technology is a public good in the sense that the information will predictably 'leak' to all firms by the second date, information regarding the cost of using the technology at the first date is essentially private.

Second, as members of the coalition of adopting firms at $t = 1$, firms are able to reduce the cost of experimenting with the technology, $\alpha(m)$. We assume that the process of gathering information about the new technology's cost-effectiveness is similar to a statistical sampling problem. Thus, to gain a certain level of confidence regarding the efficiency of the new technology requires a sample of a certain size. Each unit produced with the new technology is a member of the sample being collected by the coalition. This view of the information production process implies that the costs of the information are essentially independent of the size of the coalition. So, we treat $\alpha(m)$ as a constant, of which each member of the coalition pays an equal share $\alpha(m)/m$. The exact mechanism for reducing the cost of experimentation at the firm level is implicitly defined in the model presented. One might imagine that firms could reduce the burden of gathering information about the technology by coordinating their efforts in many ways. For example, by sharing information, each firm will need to take fewer production runs with the technology before knowing whether it is cost effective. Firms might also learn more quickly how to calibrate the equipment for optimal use when they act collectively.

Operationally, our model of firm behaviour in this setting takes the following form:

$$\Pi = \text{Max}(\Pi_0, \Pi_n, \Pi_m) \tag{6.1}$$

This nested functional of the firm's problem shows that the firm can choose the maximum of two production techniques and one strategic variable. In equation (6.1), Π_0 represents the firm's set of profit opportunities if it decides to produce with the status quo technique at date $t = 1$, which requires no coordination and represents the firm's reservation payoff. In the same equation, Π_n represents the opportunities if the firm unilaterally experiments with the new technique at $t = 1$. Finally, Π_m represents the payoff the firm receives from coordinating with other firms in the industry who are experimenting with the new technology at date $t = 1$. The strategic aspect of the firm's situation is the decision whether or not to join a research coalition or not. This sort of strategic interdependence is similar to the decision of whether to join a cartel and is well exposed in the literature. The novelty of our model is that we look specifically at the information created by experimenting with a new technology. It is typical in the cartel and public good literature to dwell on the non-cooperative aspects of the solution to problems like ours. In our setting, however, it is natural to assume that firms can legally share information regarding their experiences with a new technology, and form binding agreements to ensure that members of the coalition do not shirk. It is also true that, depending on the size of the market one might be concerned about how experimentation with the new technology affects market power. Throughout the analysis, we assume price-taking behaviour and therefore neglect this topic, which may be of interest for further research. Using the notation developed to this point, we extend upon the two-state example mentioned earlier:

$$\underset{q_1, q_L, q_H}{\text{Max}}\, \Pi_0 = pq_1 - c_0(q_1) + \phi(pq_L - c_L(q_L)) + (1 - \phi)(pq_H - c_0(q_H)) \tag{6.2}$$

$$\begin{aligned}\underset{q_1, q_L, q_H}{\text{Max}}\, \Pi_n = pq_1 - \mathrm{E}(c_n(q_1)) - \alpha(0) + \phi(pq_L - c_L(q_L)) \\ + (1 - \phi)(pq_H - c_H(q_H))\end{aligned} \tag{6.3}$$

$$\begin{aligned}\underset{q_1, q_L, q_H}{\text{Max}}\, \Pi_n = pq_1 - \mathrm{E}(c_n(q_1)) - \frac{\alpha(0) + \gamma(m)}{m} + \phi(pq_L - c_L(q_L)) \\ + (1 - \phi)(pq_H - c_H(q_H))\end{aligned} \tag{6.4}$$

We mention a few initial observations regarding this problem before describing the intriguing aspects of its solution. Note that the expression capturing the firm's payoff if it decides to unilaterally proceed with implementing the technology, Π_n, differs from the expression describing the payoff to coordinated experimentation, Π_m, because the former expression assumes that when the firm experiments unilaterally it must bear the full cost of experimenting (terms involving α), and has no coordination costs (terms involving γ). Intuitively, coordination costs must satisfy the restriction $\gamma(1) = 0$. Because we are interested in the properties of interior solutions to the firm's problem, explicit modelling of the inequality constraint $m \geq 1$ is neglected. However, for coalition formation to be optimal, we require the average cost of formation to be U-shaped in m. Formally, we require $(\alpha(0) + \gamma(m))/m = \gamma'(m)$ for some $m > 1$.

It should also be noted that even though the problem involves the resolution of uncertainty over time, there is no option value to waiting. The lack of an option value (Dixit and Pindyck, 1994) is due to reversibility in the firm's decision at date $t = 2$. A casual inspection of the equations (6.2), (6.3) and (6.4) indicates that the firm has the same technology options available at the second date regardless of the state of nature revealed and regardless of participation in the experimental use of the new technology at $t = 1$. This simplification ensures that our analysis focuses on the public good component of the information and not on issues linked to irreversibility of the firm's decision. Moreover, the assumption re-enforces the idea that use of the new technology is initially just an experiment to give the firm firsthand knowledge of how well the technology will serve the firm's specific needs. Thus, *ex post* with new information in hand, the firm can choose to revert back to its old production technique, $c_0(q)$, if the old technology is superior to the new one. Also, observe that uncertainty regarding the cost effectiveness of the new technology is embodied at $t = 1$ in the expectation of the random variable $c_n(q)$. Since the firm is risk-neutral, this representation of the uncertainty is sufficient.

5 A SOLUTION

In our discussion of the solution to the firm's problem, we will proceed by emphasizing those aspects of the solution that relate to firm size. Specifically, we are interested in examining the circumstances that give rise to coordination among firms in producing information about the suitability of the new technology. Ultimately, when coordination does arise in our model it is due to economies of scale owing to the non-rival nature of information. However, our model also makes clear that firms must respect the costs associated with coordinating activities within a large coalition. The two competing concerns give rise to an optimal coalition size. These information coalitions require binding commitments to avoid the free-rider problem associated with the production of information, a pernicious issue in the literature dealing with public goods provision in the absence of commitment mechanisms (Barrett, 1994).

Looking at the solution to (6.1), we state without proof that any of the three components of the maximand, Π_0, Π_n, or Π_m, is a candidate for the maximum. However, we can learn more from the structure of the model by examining the relationship between the components as the coalition size, m, is allowed to vary. To begin, we note that each of the components is additively separable at each date/state, which means that the firm's state contingent production plans for each date/state can be formulated independently. If the firm chooses to produce as part of a coalition, the firm will, by signing a binding contract, submit to the authority of the coalition in choosing the number of members of the coalition and the firm's level of output. It turns out that because of the price-taking assumption – if the coalition's objective is to maximize the joint profit of its membership, as we assume – there is no difference between the firm's unilateral output choice and the level of output solving the joint profit maximization problem. The membership, M, chosen by the coalition, on the other hand, is potentially quite different from the membership level that would be chosen by the firms if they were not acting in concert. In the framework of binding contracts, the offer to join a coalition is a take-it-or-leave-it offer in which the potential member will look at the payoff available from the coalition against outside alternatives, in this case Π_0 and Π_n. Our aim is to point out that the desirability of such an offer depends on the proposed size of the coalition, M, in ways that are meaningful for the empirical patterns of firm behaviour noted previously.

This argument requires two steps. The first step is to formulate a means for comparing the firm's options with offers made by the coalition. The second is to show that there are circumstances under which we can expect the coalition to make a desirable offer for firms to join. We begin with the first step by showing that there are cases in which the firm might view joining a research coalition as the most profitable opportunity. Consider a comparison of the two alternatives available to the firm if the firm decides to experiment with the new technology, Π_n and Π_m, on the domain of feasible research coalitions, $m \geq 1$. Since there are no costs of coordination when $m = 1$, the firm is indifferent between the two alternatives at this point in the domain. At other points in the domain, the firm will rank Π_n and Π_m differently depending on the level of M planned by the coalition. In these cases, it is helpful look at the firm's view of different levels of m. To this end consider the sub-problem solved by the firm at the first date/state if it had complete autonomy over the size of the coalition:

$$\underset{q_1, m}{\text{Max}} \tilde{\Pi}_m = pq_1 - \text{E}(c_n(q_1)) - \frac{\alpha(0)}{m} - \frac{\gamma(m)}{m} \tag{6.5}$$

with first order conditions:

$$\frac{\partial \tilde{\Pi}_m}{\partial q_1} = p - \text{E}(c_n'(q_1)) = 0 \tag{6.6}$$

$$\frac{\partial \tilde{\Pi}_m}{\partial m} = \frac{\alpha(0) + \gamma(m) - m\gamma'(m)}{m^2} = 0 \tag{6.7}$$

The Hessian matrix gives the second order conditions:

$$H = \begin{pmatrix} \partial^2 \tilde{\Pi}_m / \partial q_1^2 & 0 \\ 0 & \partial^2 \tilde{\Pi}_m / \partial m^2 \end{pmatrix}$$

$$= \begin{pmatrix} -\text{E}(c_n''(q_1)) & 0 \\ 0 & \dfrac{-m^2(\gamma''(m)) - 2(\alpha(0) + \gamma(m) - m\gamma'(m))}{m^3} \end{pmatrix} \tag{6.8}$$

If the second order conditions to this problem are satisfied on the domain of feasible coalitions, then the firm will have an optimal coalition size, m^*, that is greater than unity. We have ensured this result by assuming that the average costs associated with experimenting are U-shaped. Thus, compared to experimenting alone, Π_n, there is always some coordinated experimentation arrangement that is better. However, this conclusion requires that we allow for the possibility of fractional membership within a coalition, which from this point forward we will rule out.

To fully compare the options available to the firm, we must also consider the profits associated with not experimenting, Π_0. Since our previous results show that coordinated experimentation dominates experimenting alone on the domain of feasible coalitions, we begin looking at the circumstance where no experimentation might dominate coordinated experimentation, $\Pi_0 > \Pi_m$. Given the flatness of Π_0 and the concavity of Π_m, we can define coalition sizes that indicate when a firm is indifferent between producing with the status quo technology and participating in a research coalition. We refer to a coalition size of $m^\vee$ as the *smallest viable coalition* (SVC) when $\Pi_0 > \Pi_m$ for all $m < m^\vee$. We will refer to a coalition of size $m^\wedge$ as the *largest viable coalition* (LVC) when $\Pi_0 > \Pi_m$ for all $m > m^\wedge$. It is easily shown that under general conditions, an SVC and LVC exist for the firm. In cases where they do not exist, the implication is that the firm should not experiment with the new technology. Our next proposition elucidates this point.

Proposition 1: Let $\Delta = \Pi_m - \Pi_0 + \alpha(0)/m + \gamma(m)/m$. When $\Delta < \alpha(0)$, an SVC and LVC exist for some $\gamma(m)$ satisfying $\gamma'(m) > 0$ and $\gamma''(m) > 0$.

Proof: This proposition relies on Brower's fixed point theorem. Because of the continuity of Π_m and Π_0 on the domain of viable coalitions, our task is to show the existence of $\Pi_m - \Pi_0 = 0$, which can be written as $\Delta - \alpha(0)/m - \gamma(m)/m = 0$. The fixed point version of this 'zeros problem' is $m = (\alpha(0) - \gamma(m))/\Delta$. Based on our previous assumption regarding the derivatives of $\gamma(m)$, we need only show that at $m = 1$ the expression $(\alpha(0) - \gamma(m))/\Delta > 1$, which is satisfied when $\Delta < \alpha(0)$. Q.E.D.

Now, we turn to the second step in our argument. When can we rely on a coalition that is being managed to maximize the joint profit of its membership to make a desirable offer of membership to firms outside the coalition? This question is easy to answer by looking at the coalition's problem:

$$\underset{q_1,m}{\text{Max}}\Sigma\tilde{\Pi}_m = m[pq_1 - \text{E}(c_n(q_1))] - \alpha(0) - \gamma(m) \tag{6.9}$$

with first order conditions:

$$\frac{\partial\Sigma\tilde{\Pi}_m}{\partial q_1} = m[p - \text{E}(c_n'(q_1))] = 0 \tag{6.10}$$

$$\frac{\partial \Sigma \widetilde{\Pi}_m}{\partial m} = pq_1 - \mathrm{E}(c_n(q_1)) - \gamma'(m) = 0 \tag{6.11}$$

The Hessian matrix gives the second order conditions as:

$$H = \begin{pmatrix} \partial^2 \Sigma \widetilde{\Pi}_m / \partial q_1^2 & \partial^2 \Sigma \widetilde{\Pi}_m / \partial q_1 \partial m \\ \partial^2 \Sigma \widetilde{\Pi}_m / \partial q_1 \partial m & \partial^2 \Sigma \widetilde{\Pi}_m / \partial m^2 \end{pmatrix}$$

$$= \begin{pmatrix} -m\mathrm{E}(c_n''(q_1)) & p - \mathrm{E}(c_n'(q_1)) \\ p - \mathrm{E}(c_n'(q_1)) & -\gamma''(m) \end{pmatrix} \tag{6.12}$$

Within the bounds of our previous assumptions, it is easy to establish cases that satisfy the second order condition for a maximum M^* to the coalition planner's problem given in (6.9). Note that if profits for the firm's sub-problem (6.5) are (weakly) positive, the implication is that $pq_1 - \mathrm{E}(c_n(q_1)) \geq (\alpha(0) - \gamma(m))/m$, which provides a means for comparing the firm's preferred level of membership, m^*, with the coalition's preferred level, M^*. This comparison is valuable for determining when a coalition should arise and is summarized in Proposition 2.

Proposition 2: When the costs of coordination in a coalition increase monotonically, $\gamma'(m) > 0$ on the domain of feasible coalitions, and the profits to joining a coalition are positive at the firm level, $\widetilde{\Pi}_m > 0$, then the coalition's preferred level of membership will exceed the level preferred by an individual firm, $M^* > m^*$.

Proof: This follows directly from the monotonicity of $\gamma'(m)$ and the first order conditions (6.7) and (6.11).

5.1 Coalition Formation

Now, we can illustrate the circumstances in which we would expect a coalition for sharing information about the efficiency of a new technology to arise and characterize the size of the coalition. First notice that if $\Pi_m > \Pi_0$ for some $m \in [1,N)$, then we would only expect the firm to join if the proposed size of the coalition, M, is between the firm's SVC and LVC, $m^\wedge \geq M \geq m^\vee$. We also know from our examination of the coalition's problem (6.9), that the coalition's profits, Π_m, are monotonic over the range $M^* \geq m \geq 1$. This means that the coalition will make a rational offer to non-members to join the coalition that is as close as possible to the coalition's optimum M^* without exceeding the LVC, $m^\wedge$. Proposition 2 ensures that when it exists, this coalition M will always be at least as large as the firm's most preferred coalition size, m^*.

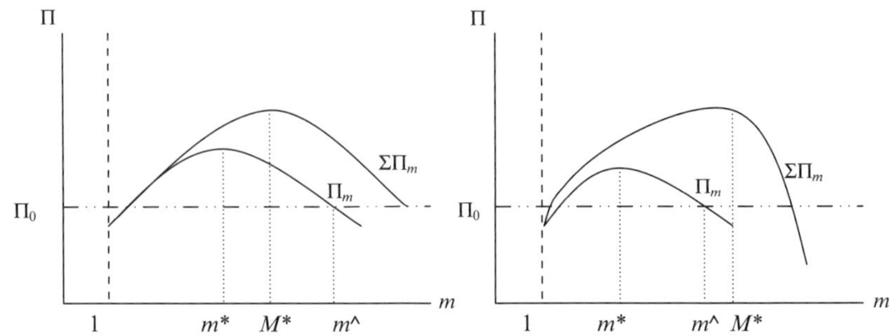

Note: Interior solutions where the coalition's proposed level of membership M is less than the LVC (left), and where M is equal to the LVC (right).

Figure 6.1 Coalition formation

Figure 6.1 depicts the two situations implied by the foregoing discussion. In the left panel, the firm receives an offer to join a coalition that is smaller than the LVC, $M < m^\wedge$. In the right panel, the firm is asked to join a coalition equal in size to the LVC, $M = m^\wedge$. In both cases, the firm's best response to such an offer is to join the proposed coalition.

The cases in Figure 6.1 demonstrate the existence of solutions to our innovation adoption problem that involve a rational choice on the part of some firms, who are otherwise identical, to behave differently with respect to a single technological opportunity. The introduction of the new technology generates a circumstance where it is beneficial for some subset of the existing production capacity in the industry to coordinate activities, while others find that they were better off remaining independent. The idea is that if the optimal feasible coalition is M for a particular technology, then at most N/M coalitions can form from the existing firms. Since fractional membership is precluded by the atomistic nature of firms, then at least some firms will be left outside a complete coalition in most cases. For the remaining firms, the question is whether forming a fractional coalition is the best way to cope with the new technology. These remaining firms may choose not to adopt the technology or to form a coalition that is smaller than M. Which choice is best depends entirely on whether the proposed smaller coalition is an element of the SVC/LVC interval.

It is important to point out that our conception of the Nash equilibrium in this game between the coalition planner and the firms, who are potential members of the coalition, relies on the exclusive availability of a certain type of 'contract'. By using the word 'contract' here we need to be clear that we mean the arrangement of how the costs of innovation are being shared among the members of the coalition and the objective of the coalition planner. Had we allowed for other arrangements in sharing the costs between the members of the coalition, our equilibrium would have been a bit different. However, our basic conclusion would still be the same. Since coordinated experimentation is optimal, conditional on the size of the experimenting coalition, firms capable of finding a suitable coalition will experiment in the first period. If firms cannot find a suitable coalition, they will continue using the status quo technology and make an adoption decision in period two. What then is the substance on which such decisions rely? Our model points to innovation characteristics, namely the 'size' of the innovation, $c_0(q) - \mathrm{E}(c_n(q))$, and the cost of experimenting $\alpha(0)$. Information leakage regarding the cost-effectiveness of the innovation is also very important, as are the costs of coordinating to realize the gains from information sharing. We now turn to the implications of our analysis for industry structure.

5.2 Industry Structure

There is an interesting interpretation to the results of this type of model. To see this interpretation vividly, think of the firms described above as independent, atomistic production units prior to the introduction of the new technology. One fitting description of this type of coordination is what might be commonly referred to as a 'joint research venture'. It might be equally well argued that this type of coordination could be called a 'merger'. There is not sufficient detail in the model to make a distinction between these two labels, but it is worthwhile to note that the coordination that takes place is profitable. Thus, if the managers of the firms resist the opportunity to experiment in coordination with other firms, it is reasonable to expect that the firm will be acquired, in the absence of other impedimenta. Even if the output market is competitive, as we have assumed throughout, the merged firms/coalitions will have output proportional to the number of merged production units, M. This means that the merged firms will be large relative to those who remain independent. Another way of looking at these results is to say that larger firms have a greater potential for spreading the fixed costs associated with innovation, which implies that they will adopt new technologies more readily.

6 CONCLUSIONS

We have shown that the introduction of new technology is a *casus origine* of industry structure, and may easily be implicated by extension to differences in adoption timing between firms and a generally perceived adoption lag. A small extension of our model to multiple time periods and more controlled leakage of the out of coalition information spillover could easily result in firms of different sizes being formed as the technology diffuses into general use – we state this point without proof. It is also easy to extrapolate from the results presented that a distribution of innovations that differs according to the size and novelty of the innovation will influence the extent of concentration in an industry over time. We could iterate forward our analysis starting with the heterogeneous industry structure that we ended with in our example and construct another example in which some collection of the production units reorganize in response to another innovation. Thus, characteristics of the innovation distribution could be identified as a determinant of industry structure.

We could add another layer of detail to our model showing that contractual commitments of this nature are often easiest to include in the bundle of contracts called a 'firm'. However, we find nothing novel in such an undertaking and prefer to merely point out that the formation of new firms is one the possible outcomes of the exposed model.

It is reassuring that our main conclusion of firm size being driven by technological opportunities, which is sometimes referred to as the Schumpeterian hypothesis, is supported by the weight of empirical evidence on adoption timing and diffusion. Our analysis lays the groundwork for future empirical study and it also points the way toward a more accurate interpretation of firm size effects in existing work. Still, there are many unanswered questions, which may benefit from further theoretical investigation. Chief among these are the questions of a comparative static nature. For example, given information regarding the costs of experimenting with a new technology and the costs of coordination, how much of a reduction in marginal cost, as measured against the status quo, is needed to induce a merger or joint venture? If we see an industry whose path of technological development can be characterized as one in which many 'large' innovations have occurred, should we expect to see a highly concentrated industrial structure emerge, *ceteris paribus*? These questions represent the kinds of falsifiable hypotheses that would enable further development of the proposed theory.

Finally, we wish to acknowledge that our theory is but one of many explanations of the adoption lag and slow diffusion paradox. We are slightly enamoured with our own interpretation, especially because it says something meaningful about the role of firm size in the appearance of the other anomalies. Yet, we would not argue that other explanations, such as anticipation of future improvements in the technology, the option value of waiting and strategic interaction more generally, are any less important in understanding the whole of what has been observed.

REFERENCES

Barrett, S. (1994), 'Self-enforcing international environmental agreements', *Oxford Economic Papers*, **46**, 878–94.

Blackman, A. (1999), 'The economics of technological diffusion: implications for climate policy in developing countries', Discussion Paper 99–42, Washington, DC: Resources for the Future.

DeCanio, S. (1998), 'The efficiency paradox: bureaucratic and organizational barriers to profitable energy-saving investments', *Energy Policy*, **26**, 441–54.

Diederen, P., F. van Tongeren and H. van der Veen (2003), 'Returns on investments in energy-saving technologies under energy price uncertainty in Dutch greenhouse horticulture', *Environmental and Resource Economics*, **24**, 379–94.

Dixit, A., and R. Pindyck (1994), *Investment Under Uncertainty*, Princeton, NJ: Princeton University Press.

Doraszelski, U. (2001), 'The net present value method versus the option value of waiting: a note on Farzin, Huisman and Kort (1998)', *Journal of Economic Dynamics and Control*, **25**, 1109–15.

Farzin, Y., K. Huisman and P. Kort (1998), 'Optimal timing of technology adoption', *Journal of Economic Dynamics and Control*, **34**, 292–318.

Hanna, T. and J. McDowell (1984), 'The determinants of technology adoption: the case of the banking firm', *RAND Journal of Economics*, **15**, 328–35.

Hoppe, H. (2000), 'Second-mover advantages in the strategic adoption of new technology under uncertainty', *International Journal of Industrial Organization*, **18**, 315–38.

Hoppe, H. (2002), 'The timing of new technology adoption: theoretical models and empirical evidence', *The Manchester School*, **70**, 56–76.

Jaffe, A. and R. Stavins (1994), 'The energy paradox and the diffusion of conservation technology', *Resource and Energy Economics*, **16**, 91–122.

Levin, S.G., S.L. Levin and J.B. Meisel (1987), 'A dynamic analysis of the adoption of a new technology: the case of optical scanners', *Review of Economics and Statistics*, **69**, 12–17.

Reinganum, J. (1989), 'The timing of innovation: research, development, and diffusion', in R. Schmalensee and R.D. Willig (eds), *Handbook of Industrial Organization*, Vol. 1, Amsterdam: Elsevier Science.

Rose, N. and P. Joskow (1990), 'The diffusion of new technologies: evidence from the electric utility industry', *RAND Journal of Economics*, **21**, 354–73.

van Soest, D. and E. Bulte (2001), 'Does the energy-efficiency paradox exist? Technological progress and uncertainty', *Environmental and Resource Economics*, **18**, 101–12.

Yi, S. and H. Shin (2000), 'Endogenous formation of research coalitions with spillovers', *International Journal of Industrial Organization*, **18**, 229–56.

7. A meta-regression analysis of the investment–uncertainty relationship

Mark J. Koetse, Henri L.F. de Groot and Raymond J.G.M. Florax

1 INTRODUCTION

Over the past three decades major doubts have been raised about the environmental sustainability of prevailing economic processes and behaviour, the most relevant starting point being the report to the Club of Rome (see Meadows et al., 1972). Limits to the capacity of our natural resource stock to replenish the fuels that we extract as well as coping with the pollution we cause are at the heart of the associated economic and political discussion. This discussion is still relevant and thriving. An important option to steer economic production into a more sustainable direction is to invest in energy-saving technologies. The adoption of these technologies is of political and social interest because it provides promising potential for reducing energy use and thereby the emission of environmentally harmful gasses. However, adoption is of interest from an economic perspective as well, because it may provide a means to producing more efficiently.

The standard framework for analysing the profitability of adopting a technology is the net present value (NPV) framework. In this framework the current and future costs and benefits of the investment are discounted using a critical discount rate to produce a net present value. A standard textbook rule is to invest when the net present value is non-negative. Using the NPV framework, it is widely recognized that many energy-saving technologies are cost-effective at current prices (Schipper and Meyers, 1992). Still, apart from large energy-intensive companies, the adoption of these technologies has been largely unsuccessful, especially in small and energy-extensive firms. Empirical studies on adoption rates and diffusion patterns show that very high implicit discount rates are needed to explain the actual diffusion of energy-saving technologies (Fawkes and Jacques, 1987). The contrast between the alleged economic profitability and low adoption rates of energy-saving technologies is generally referred to as the energy efficiency paradox.

Many possible explanations for this paradox, such as information gaps and capital constraints, have been brought forward. A potential explanation that has received relatively little attention in the literature so far is uncertainty. The theoretical literature on the investment–uncertainty relationship is extensive. One of its more salient features is that the literature is ambiguous about the direction of the relationship. Although many attempts have been made at resolving this issue empirically, these attempts have only added to the existing ambiguity. In this light, a thorough synthesis of the empirical investment–uncertainty literature is warranted. Carruth et al. (2000) provide a good overview of the investment–uncertainty literature. Although useful and important in its own right, a limitation of the Carruth et al. (2000) review is that their survey is qualitative in nature, and does not attempt to quantify the importance of the various study characteristics in explaining the variation in study outcomes. For these reasons we apply meta-analysis as our research method. A general setback is that, although the available studies on the impact of uncertainty on investment in energy-saving capital confirm that uncertainty is a relevant factor (de Groot et al., 2001; Koetse et al., 2006), empirical evidence on this specific topic is too scarce to warrant a separate analysis. In our meta-analysis we therefore focus on studies that analyse the relationship between aggregate investment and uncertainty.

Meta-analysis is a form of research synthesis in which previously documented empirical results are combined or re-analysed in order to increase the power of statistical hypothesis testing. It has therefore also been referred to as the 'analysis of analyses' (Hunter and Schmidt, 1990, p. 478). Some proponents maintain that meta-analysis can be viewed as quantitative literature review (Stanley, 2001), while others assert that meta-analysis can be used to pinpoint aspects critical to the future development of theory (Goldfarb, 1995; Rosenthal and DiMatteo, 2001). Although meta-analysis was originally developed in experimental medicine, it soon extended to fields such as biomedicine and experimental behavioural sciences. The advantage of applying meta-analysis in these academic fields is that underlying studies are generally homogenous in their setup, implying that empirical results can readily be compared across studies. Meta-analysis is currently also gaining ground in economics (Bateman and Jones, 2003; Card and Krueger, 1995; Görg and Strobl, 2001; Koetse et al., 2008, 2009a; Smith and Huang, 1995; Weichselbaumer and Winter-Ebmer, 2005). In contrast to the situation in medicine and psychology, the setup of empirical research in economics is largely heterogeneous. This induces omitted variable bias in at least a subset of the suite of empirical studies on a certain topic. Moreover, in a strict sense, effect sizes estimated with different sets of control variables cannot be assumed to represent identical population effect sizes. Differences in model specifications are, however, generally observable across studies. Therefore, as long as these differences evoke systematic variation in study outcomes, their impact can be controlled for in a meta-analysis.

The remainder of this chapter is organized as follows. In the following section we address some standard investment models. Section 3 discusses the extensive theoretical literature on the investment–uncertainty relationship, and in section 4 we address important moderators of the relationship. Sections 5 and 6 constitute the empirical part of this contribution. Section 5 addresses the sampling procedure and provides an extensive discussion of the differences in study characteristics that may systematically impact the outcome of a study. In section 6 we provide descriptive statistics of the sample of estimates and we present the results from a meta-regression analysis. Section 7 concludes.

2 INVESTMENT MODELS

Although models of investment have a long history of development, Dixit and Pindyck (1994) note that standard econometric investment models that do not explicitly account for uncertainty, have not been very successful in explaining or predicting investment spending. Not only do they explain only a small portion of the variation in investment, variables that should have strong explanatory power are in practice not very powerful (Dixit and Pindyck, 1994, p. 419). We present some of the seminal contributions to the investment literature below. We do not aim to provide an extensive, let alone exhaustive, overview of these models; excellent surveys already exist (Dixit and Pindyck, 1994; Jorgenson, 1971). Instead, we would like to give the reader a taste of the pivotal contributions, and concurrently provide a contextual framework for the analysis of the impact of uncertainty on investment spending.

An early model of investment behaviour is the accelerator model developed by Clark (1917). In this model investment is explained by changes in the desired capital stock, which in turn is determined by long-run business considerations. In a flexible version of this model, originating from Chenery (1952) and Koyck (1954), changes in the desired capital stock are not instantaneously transformed into actual investment, but rather through a geometric distributed lag function. A relevant discussion within this framework has been concerned with the determinants of the desired capital stock, and suggestions ranged from output (Clark, 1917) to profits (Tinbergen, 1938) and the cost of external capital (Meyer and Kuh, 1957). Since all three variables are theoretically acceptable as quantities explaining investment, empirical equivalents of these models usually include a combination of output, profit and the financial position of the firm as drivers of investment spending. A major drawback of these models, at least in the light of our analysis, is that they do not in any way include uncertainty as a potential explanatory variable of investment spending.

A second highly influential model is due to Jorgenson (1963). In this neoclassical model, changes in investment are determined by changes in the user cost of capital. Specifically, a firm's desired capital stock is determined by equating the marginal product of capital to its user cost. The actual stock is assumed to adjust to the desired level by a lag process, or in response to an explicitly defined adjustment cost function. Over the years many empirical variations on this model have been developed. A good example of an empirical neoclassical investment model is presented in Jorgenson and Stephenson (1967). In their model, the desired capital stock is determined by a weighted average of the ratio of value added to the price of capital services in previous periods. The desired capital stock is subsequently transferred to actual investment by a geometric lag process. In mathematical terms, the model looks as (Jorgenson, 1971, p. 1127):

$$I_t = \beta_0 + \sum_{j=1}^{2}\beta_j(I - \delta K)_{t-j} + \beta_3 K_t + \sum_{j=4}^{7}\beta_j\Delta(pQ/c)_{t-j} + \varepsilon_t \qquad (7.1)$$

where I_t and K_t are investment and capital stock at time t, respectively, pQ is gross value added in current prices, δ is the rate of depreciation, βs are parameters to be estimated and ε is an error term. The price of capital services, c, is defined as:

$$c = q\left(\frac{1 - uv}{1 - u} + \frac{1 - uw}{1 - u}r\right) \qquad (7.2)$$

where q is the price of investment goods, r the cost of capital, u the tax rate, and v and w the tax deductible proportions of the depreciation of capital and the cost of capital, respectively. In the model, a weighted average of the ratio of gross value added to the cost of capital, which is basically a measure of the rate of return to capital, in periods $t - 4$ to $t - 7$ is transferred to investment in $t - 2$ and $t - 1$ and current investment at time t. Current investment furthermore depends on the current capital stock, K_t, the reason being that a larger capital stock implies more replacement investment. Specifically, replacement investment is proportional to the net capital stock. In comparing different empirical investment models, Jorgenson (1971) claims that the performance of the empirical Jorgenson and Stephenson model is superior to the accelerator model and alternative neoclassical model specifications. Despite the model's attractive characteristics, the author himself already notes the necessity of including uncertainty as an additional explanatory variable for investment spending.

A final model to be discussed in this section is generally referred to as the *Q*-model of investment. It is basically a neoclassical model in which Tobin's *Q* instead of, for instance, the ratio of gross value added to the cost of capital, is used to measure capital profitability (Cuthbertson and Gasparro, 1995; Tobin, 1969). In this model, investment takes place if the marginal *Q*, which is the ratio of the marginal value of capital and the market price of capital, is larger than one. Since Tobin's *Q* is a marginal quantity, and hence difficult if not impossible to measure, most empirical studies use an average for *Q* such as the ratio of the market value of a firm to the replacement costs of its assets. The main difference between the *Q*-model of investment and the standard neoclassical model is that the *Q*-model incorporates expected profits in the future into current investment decisions. Since future expected profits are reflected in stock prices, and therefore in *Q*, this model has an additional feature above and beyond the standard neoclassical investment model. However, the issue of interest for our purposes is that, as *Q* represents the market value of capital usually measured by stock values, it should incorporate uncertainty about the profitability of capital as well. Therefore, the remaining question is whether explicitly accounting for uncertainty has power in explaining investment above and beyond *Q*.[1] This issue is explicitly addressed in our empirical investigations.

3 INVESTMENT UNDER UNCERTAINTY: THEORETICAL BACKGROUND

The theoretical literature on the investment-uncertainty relationship is extensive and incorporates various analytical frameworks. Hartman (1972) developed one of the first models. Using a neoclassical model without capital stock adjustment costs, his study focuses on the relationship between capital productivity and the uncertainty variable. Under convexity of this relationship, by Jensen's inequality, the incentive to produce and invest increases when uncertainty on that variable increases, implying a positive relationship between investment and uncertainty. Hartman's (1972) results show that both uncertainty about output prices and uncertainty about wage rates have a non-negative effect on investment, whereas investment is invariant to uncertainty about future investment costs. A weakness of this model is that it is restricted to markets with perfect competition. Moreover, it relies on the assumptions of constant returns to scale production technology and substitutability of capital with other input factors, since these two characteristics ensure that capital productivity is convex in the uncertainty variable. Finally, adjustment costs are assumed to be symmetric, which is equivalent to assuming that capital investments are reversible. This assumption is obviously violated for most capital investments.

Pindyck (1982) uses a neoclassical model that allows for asymmetric adjustment costs. He argues that the effect of uncertainty on investment spending depends on the characteristics of the adjustment cost function: uncertainty may increase (decrease) investment when adjustment costs are a convex (concave) function of investment. In contrast, using a variant of the Pindyck (1982) model, Abel (1983) suggests that increased uncertainty leads to increased investment spending regardless of the characteristics of the adjustment cost function. This would confirm the results of Hartman (1972). However, Abel also shows that adjustment costs *do* matter for the relationship between investment and Tobin's Q (the marginal value of capital expressed as a ratio of the market value of capital), based for instance on stock data, and the market price of capital. In the Q-model of investment, the growth rate of investment under certainty is equal to the growth rate of Q multiplied by the elasticity of investment to Q, whereas under uncertainty the growth rate of capital is smaller (greater) than this quantity if adjustment costs are a concave (convex) function of investment (Bo, 2001). Uncertainty in the Abel (1983) model therefore has a direct effect on investment, but also an indirect effect through Q. It is clear that in this model uncertainty has a positive effect on investment spending whenever adjustment costs are convex. However, the net effect of uncertainty on investment when adjustment costs are concave is uncertain, and thereby largely an empirical issue.

One of the most important contributions to the investment–uncertainty literature, judging by its dominant position in current thinking about investment and uncertainty, is the concept of capital investment irreversibility or real options theory. This concept goes back to the seminal theoretical work of Dixit and Pindyck (1994). In a neoclassical model with asymmetric capital adjustment costs (that is, a certain degree of irreversibility of capital investment), Dixit and Pindyck show that an increase in uncertainty around future values of relevant economic variables creates an option value of waiting for information on these variables to arrive in the future.[2] We illustrate their central point. A firm can invest in a project at investment equal to I at a risk-free interest rate equal to r. The return of the project at time $t = 0$ is equal to L_0 with certainty, but the future returns of the project are uncertain. The return of the project from time $t = 1$ onwards is $L_0(1 + a)$ with probability q, and $L_0(1 + b)$

with probability $(1 - q)$. We assume that the project is not profitable if the return from time $t = 1$ onwards is lower than L_0, that is, we assume that $I > L_0(1 - b)(1 + r)/r$. The net present value of this investment opportunity is:

$$NPV^I = -I + L_0 + q\sum_{t=1}^{\infty}\frac{L_0(1 + a)}{(1 + r)^t} + (1 - q)\sum_{t=1}^{\infty}\frac{L_0(1 - b)}{(1 + r)^t} \quad (7.3)$$

Waiting one period and investing at time $t = 1$ will only be profitable if the return has gone up, because the project was assumed not to be profitable in case the return decreases. The net present value at time $t = 0$ of waiting one period and investing only when profits have increased is given by:

$$NPV^W = -\frac{qI}{1 + r} + q\sum_{t=1}^{\infty}\frac{L_0(1 + a)}{(1 + r)^t} \quad (7.4)$$

When $NPV^W > NPV^I$ waiting is optimal for the firm. The costs of waiting consist of foregone revenues at time $t = 0$, or, more general, during the period of delay, while the benefit of waiting is that new information at time $t = 1$ enables a firm to avoid investing in a project which looks profitable *ex ante* but turns out to be unprofitable *ex post*.[3] In addition, the present discounted cost of the investment is lower when investment is delayed. The difference in the return on the two investment strategies can be called the 'option value of waiting', and is equal to:

$$NPV^W - NPV^I = I\left(\frac{1 + r - q}{1 + r}\right) - L_0\left(\frac{r + (1 - q)(1 - b)}{r}\right) \quad (7.5)$$

When analysing the effects of uncertainty on the option value of the investment, one should observe that the expectation and variance of the return on investment at time $t = 1$ are equal to:

$$\mathrm{E}(L_1) = L_0(q(1 + a) + (1 - q)(1 - b)) \quad (7.6)$$

and

$$\mathrm{var}(L_1) = \mathrm{E}(L_1^2) - [\mathrm{E}(L_1)]^2 = q(1 - q)(a + b)^2 \quad (7.7)$$

respectively. Now consider the special case with $q = b/(a + b)$, which represents an equi-proportional increase in both a and b. In this case, the expected returns of the investment as well as the probability of an upward swing in the returns are both unaffected, while the variance of the returns will increase. Note that an increase in the size of the downward jump of the returns will result in firms being more likely to delay the investment. Therefore, an increase in uncertainty, caused by an equi-proportional increase in a and b, leaves the expected returns on the project as well as the probabilities of an upward or downward jump unaffected, but it will *ceteris paribus* result in more projects being delayed. This is the basic result of the irreversibility or real options literature.

It is useful to state this insight in an alternative manner. If uncertainty around a certain economic variable increases, then the value of this variable at which investment will take place (the trigger value of investment) will increase. Note that under these circumstances there is an adverse effect as well. Increased volatility of the variable under consideration increases the chance that the (increased) trigger value will be reached sooner than under less volatile circumstances. The general conviction, however, is that the increase in the trigger value dominates the increase in the probability of hitting the trigger value, implying that the net effect is postponement of investment.

The central point of the real options or irreversibility literature is that an increase in uncertainty leads to an increase in the investment threshold level above the standard NPV level. Note, however, that although this argument has major implications for the timing of investment, it has no implications for the level of investment in the long run.[4] We can therefore distinguish between two general branches of research: a first branch in which uncertainty is related to the *timing* of investment, and a second branch that analyses the impact of uncertainty on investment *levels*.

An interesting difference between the two branches is the importance of irreversibility. Although irreversibility is certainly relevant in the second branch (Pindyck, 1982, 1988), it is not as dominant as in the first branch. For instance, where both Pindyck (1982, 1988) and Abel (1983) claim that the characteristics of the adjustment cost function are the main explanation for the differing insights on the investment–uncertainty relationship, Caballero (1991) shows that asymmetric adjustment cost are not sufficient to explain a negative relationship. In contrast, he focuses on the above-mentioned assumptions of perfect competition and constant returns to scale technology. Caballero (1991) shows that under decreasing marginal returns to capital, due to either imperfect competition or decreasing returns to scale, the results attained by Hartman (1972) and Abel (1983) also hold in the case of asymmetric adjustment costs. Moreover, in a highly competitive environment, the asymmetry of adjustment costs is virtually irrelevant for the direction of the relationship.

More recent theoretical endeavours focus on other arguments than irreversibility to show that the investment–uncertainty relationship can be negative. For instance, Nakamura (2002) shows that, under decreasing returns to scale, increased uncertainty about the future leads to a decrease in investment spending if the lifetime of capital is shorter than the firm's horizon. In another study by the same author uncertainty reduces investment spending when a firm is risk-averse (Nakamura, 1999). This result holds even under competitive conditions.[5] In a revision of the framework used in the Nakamura study, Saltari and Ticchi (2005) confirm the results, but based on a different line of reasoning. In addition, their analysis contains more detail and differentiation. Their central result is that the effect of uncertainty on investment consists of two separate effects. The first effect is coined the 'flexibility effect', and is basically the standard effect brought about by the convexity of marginal capital productivity in prices. The second effect is the 'risk aversion effect' brought about by the entrepreneur's position towards risk. Since the two effects work in opposite directions, the net effect in their model is ambiguous and depends on the relative magnitude of the two separate effects.[6]

Finally, some studies have tried to unify the timing and level effects of uncertainty by making a distinction between the short run and the long-run impact of uncertainty. Bar-Ilan and Strange (1999) argue that the effects of uncertainty on the timing and the level of investment are related when a clear distinction is made between incremental and lumpy investments, and between short-run and long-run investment behaviour. Since incremental investment can take place at any rate and at any point in time, firms only have to decide on the magnitude of the investment. Qualitatively, the same results apply as under the standard investment model: uncertainty depresses investment. However, for lumpy investments they show that, although investment at time t is postponed when uncertainty increases, the level of investment may be higher when investment eventually occurs, say at time $t + j$. Stated differently, it is likely that under uncertainty the price of output at time $t + j$ is higher than the certainty equivalent price at time t, thereby stimulating investment.

Similarly, Abel and Eberly (1999) focus on the long-run development of the capital stock under irreversibility and uncertainty. They argue that the standard user-cost effect on capital under irreversibility is counteracted by a so-called 'hangover effect'. The hangover effect results from the fact that when the marginal revenue product of capital is low, a firm would like to sell some of its already installed capital, but it cannot because investment is irreversible. In this case, the desired capital stock is lower than the actual capital stock. Since under reversible investment the actual capital stock can be instantaneously adjusted to the desired capital stock, the actual capital stock is higher under irreversibility than under reversibility when capital productivity is low. Under uncertainty this effect is reinforced.

Although Abel and Eberly's (1999) results mainly address the behaviour of capital stock under irreversibility and uncertainty, their inferences also point to the importance of making a distinction between gross and net investment in the investment–uncertainty debate.

4 POTENTIAL SOURCES OF VARIATION

Given the ambiguity of the theoretical literature it seems impossible to determine the sign of the relationship between investment and uncertainty a priori, let alone that we can make theoretical inferences regarding the magnitude of the effect and its economic relevance. Various explanations for the theoretical ambiguity have been brought forward, some of which have already been touched upon in the previous section. One of the most obvious sources of heterogeneity is the degree of irreversibility of investment itself: the smaller the possibilities to disinvest, the greater the impact of uncertainty on investment. A similar argument holds for risk aversion. Obviously, the higher the risk aversion of an investment decision maker, the smaller the willingness to invest under uncertain circumstances. Without trying to be exhaustive, other potential sources of heterogeneity are underlying market structure, the differential impact of industry-wide and firm-specific uncertainty and financial conditions of the firm. Below we briefly address these issues.

With respect to market structure Hartman (1972) and Abel (1983) provide evidence for a positive relationship in markets with perfect competition, while Caballero (1991) argues that the relationship is negative in markets with imperfect competition. The argument is that the marginal productivity of capital is different in different market regimes, that is, marginal productivity is convex under perfect competition and concave under imperfect competition. To complicate matters, Kulatilaka and Perotti (1998) show that strategic impulses in markets with imperfect competition may outweigh the effect of irreversibility on investment spending. For instance, an oligopolistic firm facing increased industry-wide uncertainty may want to delay investment because uncertainty has increased the option value of waiting. However, if the firm expects rivals to delay investment spending as well, it may want to invest now because of strategic reasons. This argument is partly related to the size of the opportunity costs of postponing an investment. Because profits are above zero in markets with imperfect competition, opportunity costs are likely higher than in markets with perfect competition, where profits are zero. Therefore, although the degree of competition in the underlying sales market is argued to be an important source of heterogeneity in the investment–uncertainty relationship, theory does not agree on the direction of its impact. This ambiguity is reflected in empirical research as well. Using firm–seller concentration as an indicator of the degree of competition, Ghosal and Loungani (1996) find a negative and statistically significant relationship for firms in markets with relatively high degrees of competition, while the effect in low competition markets is smaller and statistically insignificant. In contrast, using unit price profit margins as an alternative measure of market competition, Guiso and Parigi (1999) find a more substantial negative relationship in markets with lower degrees of competition.[7]

A general distinction can be made between industry-wide and firm-specific uncertainty. The models used by Abel (1983) and Caballero (1991) imply that there is a positive relationship between investment and 'idiosyncratic' uncertainty for firms with constant returns to scale that operate in a competitive environment. In contrast, Pindyck (1993) notes that if uncertainty is identical for all firms in an industry, it will be more difficult to disinvest than if only a single firm experiences increased uncertainty. He subsequently shows that industry-wide uncertainty has a negative impact on investment spending, even under perfect competition and constant returns to scale. However, under alternative assumptions, idiosyncratic uncertainty may be just as important for investment decisions. In fact, in an empirical investigation using firm-level data, Bo (2002) suggests that idiosyncratic uncertainty has a negative impact on investment and is more important than aggregate uncertainty. Since in empirical research it is difficult to distinguish between the two sources of uncertainty due to data constraints, there are hardly any other empirical studies that distinguish explicitly between industry-wide and firm-specific uncertainty. However, the effect may show up when different levels of data aggregation (country, industry or firm) are used in empirical studies. In an operational sense it is, however, problematic that differences between studies using different levels of data aggregation may incorporate other sources of data-related heterogeneity as well.

Some of the accelerator and neoclassical models discussed in Jorgenson (1971) include financial restrictions or debt as potential factors explaining investment. This is interesting since financial conditions of a firm potentially influence the relationship between investment and uncertainty as well. In a study with Belgian and Spanish firms, Peeters (2001) finds evidence for this claim by showing that uncertainty has a negative impact on investment spending in low leverage firms. This finding is interpreted as showing that firms that rely heavily on internal finance are more affected by uncertainty. Similarly, using firm size as a proxy for access to external capital markets, Ghosal and Loungani (2000) show that uncertainty has a larger impact on investment spending in small, firm-dominated industries.

They argue that, since firm size is closely related to the degree to which firms can communicate information to external financial institutions, the investment–uncertainty relationship is influenced by capital market imperfections; small-firm access to external financing under uncertain economic circumstances is more restricted than large-firm access in the same situation. However, there may be other reasons why the impact of uncertainty on investment spending differs between small and large firms. For instance, it is reasonable to assume that large firms have more financial expertise and know-how, largely because they are more specialized in certain (financial) areas, and they have greater access to external financial resources. For the same reasons they are likely more able to attain more and better information. These differences may give large firms an advantage over small firms in dealing with uncertainty. Another argument is that large firms can use the opportunity to hedge against risk and uncertainty, whereas small firms cannot, at least not to the same extent. Moreover, the probability of default is higher for small firms than for large firms, especially under uncertain economic circumstances. However, large firms generally engage in regional, national and international markets, and small firms operate in more localized economies. The absolute amount of uncertainty that large firms face might therefore be larger in absolute terms. The net effect is ambiguous in that case. In sum, there is more than one reason to assume that the investment-uncertainty relationship is different for firms of different sizes.

5 EFFECT SIZE AND STUDY CHARACTERISTICS

5.1 Sampling Procedure and Effect Size

In this section we describe the sampling procedure used to select study results, typically referred to as effect sizes. We also discuss several of the main differences between the theoretical investment models used in the primary studies as well as other salient characteristics of the empirical studies that may contribute to variation in the outcomes of the studies. A first sampling restriction is that we exclude studies or estimates for developing countries, mainly because investment behaviour and investment reactions to uncertainty in these countries likely differ substantially from those in developed countries. Online search engines and databases such as Econlit, Picarta and RePEc, were used to search the literature with keywords such as 'investment', 'uncertainty' and 'volatility'. We subsequently looked for papers and articles in the reference lists of studies that were collected. In general, our analysis focuses on studies that simply relate some sort of investment measure to the level of uncertainty. In general, the model that is estimated in these studies looks as:

$$I = \alpha + \lambda U + \sum_i \beta_i x_i + \varepsilon \tag{7.8}$$

where I is investment, U uncertainty with its associated parameter λ, β_i are parameters for additional explanatory variables x_i, α is a constant, and ε an error term. In the empirical literature, two distinct models are accelerator models and *Q*-models of investment. In the former, one of the explanatory variables is sales or GDP, depending on the aggregation level of the data. In *Q*-models of investment, which may in practice overlap with accelerator models, one of the explanatory variables is Tobin's average *Q*. Most of the models used in empirical research include a wide variety of additional explanatory variables. Among these are relative factor prices, capacity utilization, human capital, trade flows, governmental expenditures and the financial position of the firm. Although there are good arguments for including each of these variables as potential drivers of investment spending, none of the variables are mutually exclusive. As a result, investment models in empirical research differ widely.

We are specifically interested in the direction and magnitude of the investment–uncertainty relationship, which implies that the estimates provided by the studies have to be defined in a common metric. In our case the relevant metric is an elasticity, implying that the estimate is obtained directly from a double-log specification. Unfortunately, only three studies estimate a double-log model. Most studies use linear or log-linear models, implying that the estimates need to be transformed to elasticities. Specifically, the coefficients from these specifications have to be evaluated around a certain data point in order to obtain an elasticity estimate. In empirical research the general approach is to take the mean of the relevant data series as the point of evaluation. Specifically, for a linear model the point elasticity η^{pe} evaluated around the mean of the relevant variables is given by:

$$\eta^{pe} = \frac{\partial I}{\partial U}\frac{\overline{U}}{\overline{I}} = \hat{\lambda}\frac{\overline{U}}{\overline{I}} \tag{7.9}$$

where $\hat{\lambda}$ is an estimate of λ in equation (7.8), which includes investment and uncertainty in a linear form, and $\overline{U}$ and $\overline{I}$ are the means of the uncertainty and investment variable, respectively. When the means of the variables are not given in the study we contacted the authors of the studies and asked them for the relevant information. Further, failing to control for differences in effect size precision by weighting the data with standard errors of the effect sizes may lead to substantial bias and increase in the meta-estimator variance (Koetse et al., 2009b). Obtaining standard errors of the elasticity estimates is therefore crucial. For double-log elasticities the standard errors are provided in the studies. For point elasticities they have to be computed externally. In the transformation of $\hat{\lambda}$ in equation (7.9) the investment variable is stochastic, so in order to compute the standard error of the point elasticity, which is done by applying the Delta method, we would require the entire model covariance matrix. Since this matrix is not available for any of the studies, a simplifying procedure is used, based on the assumption that investment is not stochastic. In this case, the standard error of the point elasticity in equation/ (7.9) is approximated by:

$$SE(\eta^{pe}) = SE(\hat{\lambda}) \cdot \frac{\overline{U}}{\overline{I}} \tag{7.10}$$

where $SE(\hat{\lambda})$ is the estimated standard error of $\hat{\lambda}$.

5.2 Measurement of Investment and Uncertainty

The level of aggregation of the investment variable is fairly homogeneous across studies. Although some studies use a very specific type of investment, such as investment in producer durables or investment in fixed machinery, the bulk of the studies use an aggregate measure of investment. The existing aggregation differences were few and too diverse to control for in our meta-analysis. Regarding the measurement of investment there are three distinct ways in which investment is specified in studies: as unscaled investment (I), as investment scaled by capital (I/K), and as investment scaled by some measure of income, such as sales or GDP (I/S).[8] It is a priori unclear whether uncertainty affects these measures of investment differently, so our meta-regression analysis will have to shed more light on this issue.

The source of uncertainty in studies is even less homogeneous. We distinguish seven sources of uncertainty: output price uncertainty, input price uncertainty, sales uncertainty, stock price uncertainty, profit rate uncertainty, inflation rate uncertainty and other sources of uncertainty.[9] There is no clear consensus in the literature on how to create a good measure of uncertainty. The main reason is that underlying the method of measuring uncertainty are often implicit assumptions regarding the expectation formation process of firms and decision-makers. As a consequence, several measures were used in the underlying studies. Most studies use historical data on the variable under investigation to derive their measure of uncertainty. They either take the unconditional standard deviation of a series, or they use a more complicated prediction model in order to arrive at their uncertainty proxy. In the latter case, usually a (general) autoregressive conditional heteroskedasticity model (ARCH or GARCH) is applied to take out the 'predictable' autoregressive part of a series (Episcopos, 1995; Huizinga, 1993). Not surprisingly, these models are often used in the finance literature as a method to model volatility.[10] Ultimately, the choice for a specific model depends on the assumptions of the expectation formation process by investment decision-makers. Koetse et al. (2006) provide a more extensive, critical evaluation of different measures of uncertainty.

An important criticism with respect to the use of historical data to measure uncertainty is that uncertainty is essentially a forward-looking phenomenon. Since historical data are by definition backward looking, they are not optimal for measuring uncertainty. A first option to create a more forward-looking uncertainty measure is to use market measures of risk, such as the risk premium embedded in the term structure of interest rates (Ferderer, 1993). Another, more popular approach is to ask entrepreneurs or economists for their subjective evaluations of uncertainty. Six of the articles in our database use such a subjective uncertainty measure, thereby avoiding the inherent theoretical problems with historical data. A first example is the approach taken by Ferderer (1993). He uses a so-called forecaster discord as the uncertainty measure, which is based on forecasts made by economists participating in a monthly survey conducted by a forecasting service for economic indicators. Each month, forecasts are made with respect to the value of several variables. Forecaster discord is measured as the standard deviation of the forecast made by each individual in the group.[11] Since the target date of the forecasts changes only once a year, the length of the forecast horizon changes each month. Therefore, on average, standard deviations of forecasts over longer periods should be higher.[12] Alternatively, Guiso and Parigi (1999), Pattillo (1998) and Lensink et al. (2000) use a survey in which entrepreneurs are asked to provide a probability distribution of the development of expected sales over some period. This measure comes closest to the ideal-typical measure of individually perceived uncertainty. Ultimately, we distinguish between two ways of measuring uncertainty: a backward-looking or 'objective' uncertainty measure, and a forward-looking or 'subjective' uncertainty measure.

5.3 Theoretical Issues

Two specific theoretical models can be distinguished in the empirical investment literature. The first model, discussed extensively in Jorgenson

(1971), is the accelerator model of investment. In this model investment spending is driven by income or sales. This type of model has either sales or GDP as an explanatory factor among the explanatory variables, depending on the level of data aggregation. The second distinct model is the Q-model of investment. In this model an investment opportunity is related to Tobin's Q (Cuthbertson and Gasparro, 1995; Tobin, 1969). This type of model, which may in practice overlap with accelerator models, usually has a measure of Tobin's average Q among the explanatory variables. Besides these clearly distinctive investment models there are various variables that are important in explaining investment behaviour, in particular relative prices of capital and labour. Omission of these four explanatory variables may lead to a bias in the estimate of the investment–uncertainty relationship. In our meta-analysis we therefore include distinctive dummy variables depending on whether the model incorporates Tobin's Q, an accelerator variable, wages and/or the price of capital.

In addition one should note the potentially differential impact of industry-specific and firm-specific uncertainty on the outcome of a study, due to differences in the degree of investment irreversibility at these levels. In the meta-analysis, this effect will largely be picked up by the distinction between studies that use data with different levels of aggregation. Unfortunately, this distinction may reveal other effects of data aggregation as well, thereby obscuring the effect of the differences mentioned above. Still, in our empirical analysis we distinguish three levels of data aggregation: the country, industry and the firm levels.

5.4 Operational Issues

There are a number of empirical issues in the investment–uncertainty relationship that we can investigate in our meta-analysis. In some cases it will not be straightforward to determine why the estimates differ along these lines, largely because the variables mentioned below reflect a multitude of issues that may all influence the effect size. For instance, regional differences may reflect such different aspects as sectoral composition, degree of competition and institutional settings. The most important operational and observable differences between studies are: geographical origin (Europe or the USA; developing countries were excluded), data type (cross-section, time series or panel data), data frequency (annual, quarterly or monthly), and estimation method (ordinary least squares, general methods of moments, instrumental variables, and other estimation methods).

Next to the differences in model specification discussed above, many other specification differences exist. We distinguish between specifications that include a lagged dependent variable (to control for autocorrelation), a time trend, governmental expenditures and debt. Finally, in a number of studies the influence of more than one uncertainty source is being used in the same estimation, for instance wage uncertainty combined with sales uncertainty. We include a dummy variable when this is the case.

5.5 Unobserved Sources of Variation

It is clear that some sources of heterogeneity cannot be accounted for in our meta-analysis. First, among the moderators of the relationship identified in the literature are the degree of irreversibility of investment, the degree of risk aversion, and assumptions on production factor substitution (see section 3). Unfortunately, most of the empirical studies do not provide explicit information with respect to these issues. Investment is generally measured as an aggregate, which makes a direct distinction between different levels of irreversibility impossible. Furthermore, the degree of risk aversion is unobservable, at least in a direct way, in studies that use aggregate data. In studies using disaggregate firm-level data it is at least possible to distinguish between different levels of risk aversion, but only if firms are explicitly asked to reveal such information. In none of the studies in our sample this is the case. Finally, different levels of substitution between input factors, for example, between capital and labour, may influence the investment–uncertainty relationship. Also with respect to this distinction the sampled studies do not provide any explicit information. The only possible way to account for these sources of heterogeneity is by including fixed effects for specific sectors, or even countries, which may differ with respect to the three issues discussed above. The problem, however, is that a sector-specific fixed effect measures the impact of all three sources of heterogeneity. Furthermore, the effect may obviously be caused by various other sources of heterogeneity, arguably underlying market structure and differences in firm size. Although the meta-analysis would pick up part of the heterogeneity in the sample of effect sizes, we would not be able to attribute it to a specific source. Moreover, a practical difficulty is that very few studies actually distinguish between sectors at a low level of aggregation. Most studies that use sector level data do so for the entire manufacturing sector, implying that very little sectoral variation is present in the underlying set of studies. Especially because of the latter, sector-specific fixed effects are not included in the meta-analysis. As a result, part of the heterogeneity in the sample of elasticities is likely left unexplained. However, the fact that empirical studies generally use aggregate manufacturing data has a potential upturn as well. Although within each study the differences with respect to the three issues set out above may be substantial, the comparability across studies is warranted because almost all studies include the same intra-study variation.

Another potentially important moderator variable is the underlying market structure. As discussed in section 4, this issue is investigated in empirical studies by analysing whether the relationship between uncertainty and investment is different for different degrees of competition. However, some studies also use firm size as a proxy for underlying market structure. Unfortunately, these studies are scarce, implying that dummy variables for market structure and firm size may pick up study specific effects that are not at all related to these two issues. Therefore, we do not include these dummies in our empirical analysis. Because effect sizes from the studies that explicitly distinguish between different degrees of competition in the underlying product market and between firms of different sizes are still included in our meta-sample, some potential heterogeneity is unaccounted for. However, since these elasticities are scarce, the impact of this unobserved heterogeneity is likely small.

6 EMPIRICAL ANALYSIS

6.1 Descriptive Statistics

In this chapter we are interested in the direction and magnitude of the investment–uncertainty relationship, so we chose to use the elasticity as our effect size.[13] Our meta-sample contains 51 double-log elasticities with associated standard errors from three studies. Furthermore, for 288 of the linear and log-linear estimates from 13 studies we were able to calculate point elasticities and standard errors at the mean value of the investment and/or uncertainty variables in the studies (see section 5). This results in a sample of 339 double-log and point elasticities with associated standard errors. We put some further restrictions on the elasticity sample, because some of the point elasticities are extremely large. Specifically, from the study by Bo (2002) we obtained extremely large point estimates, ranging from −111 to +13. In order to prevent these extreme elasticity values from causing spurious outcomes in the meta-regression analysis, we excluded the entire study (16 elasticity estimates) from our sample. In the remaining sample of 323 elasticities, obtained from 16 different studies, there are still relatively large elasticity values. It is impossible to judge whether these values are due to random (sampling) error or whether they represent a real effect. Therefore, we analyse two different elasticity samples: a full sample of 323 elasticities, and a restricted sample that excludes elasticities outside of the 95 per cent confidence interval around the mean (10 point

Table 7.1 Descriptive statistics for several samples of elasticities[a]

	DL-elasticities	Full sample		Restricted sample	
		P-elasticities	All elasticities	*P*-elasticities	All elasticities
No. of observations	51	272	323	262	313
Minimum	−1.51	−10.98	−10.98	−2.23	−2.23
Maximum	0.04	5.16	5.16	1.77	1.77
Median	−0.21	−0.05	−0.08	−0.04	−0.07
Mean	−0.37	−0.24	−0.26	−0.11	−0.15
Standard deviation	0.41	1.16	1.07	0.48	0.48
Percentage negative	84%	63%	66%	62%	65%

Note: [a] *DL* refers to double-log elasticities, and *P* to point elasticities.

elasticities in total). Descriptive statistics for both samples are presented in Table 7.1.

As we can see from the table, point elasticities display wider variation than double-log elasticities. The minimum and maximum values of point elasticities, both in the full and in the restricted sample, are considerably higher than their counterparts in the double-log sample. This pattern is confirmed by the differences in the standard deviations of the samples. Furthermore, the mean and median of the point elasticity sample are considerably smaller than the mean and median of the double-log sample. Finally, the minimum and maximum elasticity values in the full sample are substantially larger than those in the restricted sample.

6.2 Fixed and Random Effect Estimates

The simple statistics provided above do not take into account the standard errors of the elasticity estimates. For this purpose we revert to fixed and random effect models. These models are univariate and do not yet account for sources of systematic effect size variation. We furthermore calculate the *Q*-test statistic, which measures whether the effect sizes in the sample represent a single population or multiple different underlying populations. The *Q*-statistic is given by (Sutton et al., 2000, p. 38):

$$Q = \sum_{i=1}^{I} \frac{1}{\sigma_i^2}(\eta_i - \hat{\eta})^2 \tag{7.11}$$

where η_i are the estimated effect sizes, σ_i^2 the squared standard errors of the effect sizes, $\hat{\eta}$ is the meta-estimate from the fixed effect model, and i (= 1, 2,

Table 7.2 Fixed and random effect estimates, and the Q-*statistic on effect-size heterogeneity*[a]

	Multiple sampling (Full sample)		Single sampling (Full sample)	
	DL-elasticities	*P*-elasticities	*DL*-elasticities	*P*-elasticities
Fixed effect estimate	−0.137*	−0.0003	−0.091	−0.002
	(0.009)	(0.0003)	(0.061)	(0.001)
Q-statistic (FE)	215.5*	882.6*	2.12	27.0*
	(50)	(271)	(2)	(12)
Random effect estimate	−0.177*	−0.035*	−0.091	−0.008
	(0.025)	(0.008)	(0.050)	(0.007)
Root of between-study variance	0.125*	0.075*	0.000	0.006
	(0.024)	(0.008)	(0.094)	(0.005)

Note: [a] For the fixed and random effect estimates standard errors are given in parentheses. For the *Q*-statistic degrees of freedom are presented in parentheses. The results refer to the multiple sampling case, and the case of single sampling of the median of the effect sizes per study. *DL* refers to double-log elasticities, and *P* to point elasticities. Statistical significance at the 5 per cent level is indicated with an asterisk.

. . ., I) indexes the observations. The Q-statistic follows a χ^2 distribution with I–1 degrees of freedom under the null hypothesis of homogeneity. Multiple sampling refers to the situation where studies present more than one effect size, which is potentially a problem because of lacking independence. A possible way to account for this problem is to use a single sampling setup. The difficulty with this approach is that one has to select a single effect size from each study providing multiple effect sizes. Several options are available. We can draw the effect size randomly, which has the potential problem that mere sampling issues drive the results. An alternative is to take the average effect size, which prevents the results from being driven by sampling issues, but this has the disadvantage that the standard error is not readily available. Therefore, when more than one effect size is drawn from a single study, we take the median of the effect sizes. This approach prevents sampling issues driving the results and has the clear advantage that effect size standard errors are readily available. When the amount of effect sizes obtained from a study is an even number, we take the effect size with the lowest absolute value in order to ensure that we do not overestimate the meta-effects.

Fixed and random effect estimates and the Q-statistic for the multiple sampling and the single sampling framework are provided in Table 7.2. Note that for both approaches we only present results for the full elasticity sample ($N = 323$); the results for the restricted sample are very similar. All estimates in the table are negative but they differ in terms of magnitude and statistical significance. The random effect estimates in the multiple sampling approach are statistically significant at the 5 per cent level, and the overall double-log elasticity estimate is substantially larger than the point elasticity estimate, in absolute value. In the single sampling approach the double-log estimates are again larger in absolute value, but all estimates are now statistically insignificant. Based on these univariate estimates, the insights on the impact of uncertainty on investment spending remain ambiguous. However, the Q-statistics associated with the fixed effect model clearly indicate that there is substantial effect size heterogeneity, especially in the sample of point elasticities. Moreover, the fact that the Q-statistics from the random effect model are statistically significant at the 5 per cent level suggests that this heterogeneity is not caused solely by purely random variation of the true underlying effect across studies, implying that the samples contain systematic variation in elasticity estimates. Stated differently, there is more than one underlying population effect size. Since the estimates on the investment–uncertainty relationship provided in Table 7.2 assume a single underlying population, they are not very informative. In order to analyse the systematic variation in the elasticity sample, and to uncover the empirically relevant sources underlying this variation, we estimate a multivariate meta-regression model in the next subsection.

6.3 Meta-Regression Analysis

With respect to the specification of the meta-regression model it is important to note that in the sample of double-log elasticities the amount of variation in study characteristics is very limited, in contrast to the amount of variation contained in the sample of point elasticities. Furthermore, preliminary estimation results for the pooled sample (that is, a sample of both double-log and point elasticities) show that the estimated coefficients are very sensitive, both in sign and in magnitude, to the inclusion of a dummy variable for point elasticities in the meta-regression specification. The problem is that it is impossible to assess whether the systematic variation is caused by point elasticities or by other factors. In order to circumvent this problem we separately estimate regression models for the samples of double-log and point elasticities. In both cases we use a mixed effect model to allow for fixed and random variation of the true underlying effects.

In addition to the limited variation in study characteristics within the sample of double-log elasticities, many of the dummy variables are perfectly correlated, thereby putting further restrictions on the meta-model specification. Moreover, because of the perfect correlation between the explanatory variables, it is impossible to assess what specific study characteristic causes the variation in the sample. A simple bivariate analysis of variance shows that the variation in the sample is due to several study characteristics. However, when a multivariate meta-model is estimated, none of these characteristics appear to be relevant, judging from the statistical significance of the regression coefficients. It is therefore difficult, if not impossible, to say which factor causes the systematic variation in the sample of double-log elasticities.[14]

The variation in study characteristics in the sample of point elasticities is far more substantial. The procedure for the determination of the model specification is as follows. We only include dummy variables that contain sufficient variation, in order to prevent dummy variables from exhibiting study-specific effects rather than relevant study characteristics. We also estimate two separate models. In the first model, referred to as the full model, we include dummy variables for:

- investment measured as *I*/*S* (because the *I*/*K* dummy contains too little variation the *I* and *I*/*K* dummies are pooled and included in the constant),
- input price uncertainty, sales uncertainty, stock price uncertainty, inflation rate uncertainty (the main uncertainty measures included in the constant are output price and profit rate),
- theoretically relevant study characteristics (dummy on wages and capital prices are excluded because they contain too little variation),
- data related characteristics (specifically those that are statistically significant in an analysis of variance), and
- econometric and operational characteristics (those that are statistically significant in an analysis of variance).

The second model serves as a robustness check, since it is not unlikely that the results in the first model are influenced by multicollinearity between the explanatory variables. The specification of the second model is determined by an iterative approach. From the first model we exclude the dummy variable with the highest associated *p*-value, and subsequently re-estimate the model. We repeat this procedure until all estimates are statistically significant at the 10 per cent level. This model is referred to as the restricted model. We also estimate a full and a restricted model using a restricted sample of point elasticities (see section 6.1) in order to analyse whether the potential outliers in the full sample of elasticities influence the results. The results from the meta-regression analyses are presented in Table 7.3.

The results show that restricting the model specification has only a

Table 7.3 Mixed effect model estimations on the full and the restricted sample of point elasticities[a]

Sample/model	Full sample		Restricted sample	
	Full model	Restricted model	Full model	Restricted model
Constant	−0.108	−0.087**	−0.118**	−0.054*
	(0.070)	(0.011)	(0.070)	(0.028)
I/S	0.074**	0.066**	0.073**	0.085**
	(0.017)	(0.013)	(0.017)	(0.015)
Input price uncertainty	−0.073**	−0.052**	−0.075**	−0.068**
	(0.026)	(0.022)	(0.026)	(0.024)
Sales uncertainty	−0.027	–	−0.028	–
	(0.024)		(0.023)	
Stock price uncertainty	−0.110**	−0.095**	−0.093*	−0.098**
	(0.051)	(0.044)	(0.048)	(0.048)
Inflation rate uncertainty	−0.007	–	−0.001	–
	(0.031)		(0.031)	
Tobin's *Q* excluded	0.069	–	0.077	–
	(0.067)		(0.065)	
Accelerator variable excluded	0.032*	–	0.031	0.033*
	(0.019)		(0.021)	(0.017)
Firm-level data	−0.002	–	0.004	–
	(0.032)		(0.032)	
Cross-section data	−0.007	–	−0.007	–
	(0.014)		(0.013)	
USA data	0.061*	0.056**	0.069*	0.061**
	(0.036)	(0.013)	(0.036)	(0.013)
No IV estimation	−0.040	–	−0.043	−0.053*
	(0.034)		(0.035)	(0.031)
Lagged dependent variable	−0.019	–	−0.021	–
	(0.032)		(0.032)	
Root of between-study variance	0.049**	0.055**	0.048**	0.051**
	(0.008)	(0.008)	(0.008)	(0.008)
N	272	272	262	262

Note: [a] Standard errors are in parentheses, and statistical significance at the 5 and 10 per cent level is indicated by ** and *, respectively.

moderate impact on the magnitude of the estimates and, in general, leads to a decrease in the estimated standard errors. Besides differences in the magnitude and the statistical significance of some of the coefficients, the results for the restricted sample of elasticities are similar to the results for the full elasticity sample. The results suggest that there are several study characteristics that have a systematic effect on the estimated elasticity. Some of these characteristics refer to clear shortcomings of the studies. For instance, not controlling for endogeneity in estimating the model and ignoring the impact of sales acceleration on investment spending may bias the estimated elasticity. For other study characteristics it is more difficult, or simply unnecessary, to assess whether they refer to positive or negative traits of studies. For instance, measurement of investment and the source of uncertainty included in the model may have a substantial impact on the estimated elasticity. Specifically, our results suggest that both stock price and input price uncertainty have a more negative impact on investment spending than other sources of uncertainty. Also, while in European countries the effect is negative, as shown by the constant, the impact of uncertainty on investment spending appears to be statistically indistinguishable from zero in the USA.[15]

7 CONCLUSION

The effect of uncertainty on investment spending has been a heavily debated issue since the early 1970s. However, the many theoretical insights developed over the years provide an ambiguous picture on the direction of the effect. In this chapter we analysed the heterogeneity in the outcomes of empirical studies by means of a meta-analysis. Since we are interested in both the direction and the magnitude of the relationship between uncertainty and investment we use an elasticity as the metric for the effect size. For most studies we computed elasticities with information from the underlying studies, ultimately resulting in a sample of 323 elasticities from 16 different studies.

The results from univariate fixed and random effect models differ with respect to the magnitude and the statistical significance of the overall estimate, but the estimates are negative without exception. However, since tests indicate that elasticities in the sample are likely drawn from different populations, the insights from these univariate models are probably misleading. We therefore estimate a meta-regression model in which we control for difference between studies to uncover the empirically relevant sources of variation. More specifically, we estimate a mixed effect model in order to control for inherent heteroskedasticity and to allow for random variation of the true underlying elasticity across studies. Our results suggest that, in general, the relationship between uncertainty and investment spending is negative. Although there are several empirically relevant sources of systematic variation in the sample of elasticities, none of them appear to influence the outcome of a study in such a way that a positive and statistically significant elasticity results. However, our findings suggest that the relationship may be zero under certain circumstances.

With respect to the energy efficiency paradox discussed in the introduction, our findings provide little direct insight into the impact of uncertainty on specific energy-saving technology investments. The underlying reason is that this issue has received relatively little attention in the empirical literature so far. However, it is clear that many important sources of uncertainty surrounding the uptake of energy-saving technologies exist, such as uncertainty about the development of energy prices, uncertainty about the performance and prices of future energy-saving technologies, and uncertainty about future government policies in view of the uptake of these technologies. Although only limited evidence is available so far (de Groot et al., 2001; Koetse et al., 2006), our results seem to corroborate the inference that increased uncertainty has a negative impact on energy-saving technology adoption.

ACKNOWLEDGEMENTS

We would like to thank Tom Stanley and Herman Vollebergh for useful suggestions and comments on an earlier draft of this chapter.

NOTES

1. See, among others, Bo (2001) for an extensive discussion and empirical investigation on this issue.
2. This theory is therefore also referred to as the real options theory of investment.
3. The assumption that the only cost of delaying an investment is forgone revenue during the period of delay is obviously not always true. Regarding the adoption of new technologies, first mover advantages, learning-by-doing and learning-by-using may be important fruits of early adoption.
4. Although in the real options theory the long run impact of uncertainty on the level of investment is uncertain, postponing investment in an uncertain world obviously implies lower investment levels in the short run.
5. This, in turn, is in contrast with the results attained by Caballero (1991).
6. See Aizenman and Marion (1999) for an empirical analysis of this issue.
7. An issue related to market structure is the possibility that differences in degree of market competition are associated with differences in the magnitude of uncertainty or volatility (Ghosal, 1995). This argument does not necessarily address the relationship between investment and uncertainty itself, but rather it addresses how, how much and why the amount of uncertainty may differ between markets with different degrees of competition.
8. We exclude studies that estimate the impact of uncertainty on the capital–labour ratio (*K*/*L*). The reason is that uncertainty may affect both capital and labour, implying that the net effect of uncertainty on the level of investment is ambiguous.

9. For empirical analyses on the differential impact of different sources of uncertainty, see Koetse et al. (2006) and Huizinga (1993).
10. A third option is to estimate an autoregressive moving average (ARMA) of a series (see, for instance, Goldberg, 1993).
11. The implicit assumption with this measure of uncertainty is that greater interpersonal differentiation of expectations is an indication of higher uncertainty.
12. Driver and Moreton (1991) create their uncertainty measure in a similar fashion.
13. In Koetse et al. (2009a) we perform a meta-analysis on sign and statistical significance, that is, excluding the magnitude of the relationship.
14. Results from the bivariate analysis of variance are available upon request from the authors.
15. The studies that are included in the constant are studies that use data from the UK and Japan, but mostly these are studies that use a cross section of countries belonging to the Organisation for Economic Co-operation and Development.

REFERENCES

(Studies labelled with an * are included in the meta-analysis.)

Abel, A.B. (1983), 'Optimal investment under uncertainty', *American Economic Review*, **73**, 228–33.

Abel, A.B. and J.C. Eberly (1999), 'The effects of irreversibility and uncertainty on capital accumulation', *Journal of Monetary Economics*, **44**, 339–77.

Aizenman, J. and N.P. Marion (1999), 'Volatility and investment: interpreting evidence from developing countries', *Economica*, **66**, 157–79.

Bar-Ilan, A. and W.C. Strange (1999), 'The timing and intensity of investment', *Journal of Macroeconomics*, **21**, 57–77.

Bateman, I.J. and A.P. Jones (2003), 'Contrasting conventional with multi-level modeling approaches to meta-analysis: expectation consistency in U.K. woodland recreation values', *Land Economics*, **79**, 235–58.

*Bell, G.K. and J.M. Campa (1997), 'Irreversible investments and volatile markets: a study of the chemical processing industry', *Review of Economics and Statistics*, **79**, 79–87.

Bo, H. (2001), 'Corporate Investment under Uncertainty in the Netherlands', dissertation, University of Groningen, The Netherlands.

Bo, H. (2002), 'Idiosyncratic uncertainty and firm investment', *Australian Economic Papers*, **41**, 1–14.

*Bo, H. and R. Lensink (2005), 'Is the investment–uncertainty relationship nonlinear? An empirical analysis for the Netherlands', *Economica*, **72**, 307–31.

*Brunetti, A. and B. Weder (1998), 'Investment and institutional uncertainty: a comparative study of different uncertainty measures', *Weltwirtschaftliches Archiv*, **134**, 513–33.

Caballero, R.J. (1991), 'On the sign of the investment–uncertainty relationship', *American Economic Review*, **81**, 279–88.

*Calcagnini, G. and E. Saltari (2000), 'Real and financial uncertainty and investment decisions', *Journal of Macroeconomics*, **22**, 491–514.

Card, D. and A.B. Krueger (1995), 'Time-series minimum-wage studies: a meta-analysis', *American Economic Review*, **85**, 238–43.

Carruth, A., A. Dickerson and A. Henley (2000), 'What do we know about investment under uncertainty?', *Journal of Economic Surveys*, **14**, 119–53.

Chenery, H.B. (1952), 'Overcapacity and the acceleration principle', *Econometrica*, **20**, 1–28.

Clark, J.M. (1917), 'Business acceleration and the law of demand: a technical factor in economic cycles', *Journal of Political Economy*, **25**, 217–35.

Cuthbertson, K. and D. Gasparro (1995), 'Fixed investment decisions in UK manufacturing: the importance of Tobin's *Q*, output and debt', *European Economic Review*, **39**, 919–41.

*Darby, J., A.H. Hallett, J. Ireland and L. Piscitelli (1999), 'The impact of exchange rate uncertainty on the level of investment', *Economic Journal*, **109**, c55–67.

de, Groot, H.L.F., E.T. Verhoef and P. Nijkamp (2001), 'Energy saving by firms: decision-making, barriers and policies', *Energy Economics*, **23**, 717–40.

Dixit, A.K. and R.S. Pindyck (1994), *Investment under Uncertainty*, Princeton, NJ: Princeton University Press.

*Driver, C. and D. Moreton (1991), 'The influence of uncertainty on UK manufacturing investment', *Economic Journal*, **101**, 1452–59.

Episcopos, A. (1995), 'Evidence on the relationship between uncertainty and irreversible investment', *Quarterly Review of Economics and Finance*, **35**, 41–51.

Fawkes, S.D. and J.K. Jacques (1987), 'Problems of adoption and adaptation of energy-conserving innovations in UK beverage and dairy industries', *Research Policy*, **16**, 1–15.

Ferderer, J.P. (1993), 'Does uncertainty affect investment spending?', *Journal of Post Keynesian Economics*, **16**, 19–35.

Ghosal, V. (1995), 'Input choices under price uncertainty', *Economic Inquiry*, **33**, 142–58.

*Ghosal, V. and P. Loungani (1996), 'Product market competition and the impact of price uncertainty on investment: some evidence from US manufacturing industries', *Journal of Industrial Economics*, **44**, 217–28.

* Ghosal, V. and P. Loungani (2000), 'The differential impact of uncertainty on investment in small and large businesses', *Review of Economics and Statistics*, **82**, 338–49.

*Goel, R.K. and R. Ram (1999), 'Variations in the effect of uncertainty on different types of investment: an empirical investigation', *Australian Economic Papers*, **38**, 481–92.

*Goel, R.K. and R. Ram (2001), 'Irreversibility of R&D investment and the adverse effect of uncertainty: evidence from the OECD countries', *Economics Letters*, **71**, 287–91.

Goldberg, L.S. (1993), 'Exchange rates and investment in United States industry', *Review of Economics and Statistics*, **75**, 575–88.

Goldfarb, R.S. (1995), 'The economist-as-audience needs a methodology of plausible inference', *Journal of Economic Methodology*, **2**, 201–22.

Görg, H. and E. Strobl (2001), 'Multinational companies and productivity spillovers: a meta-analysis', *Economic Journal*, **111**, 723–40.

Guiso, L. and G. Parigi (1999), 'Investment and demand uncertainty', *Quarterly Journal of Economics*, **114**, 185–227.

Hartman, R. (1972), 'The effects of price and cost uncertainty on investment', *Journal of Economic Theory*, **5**, 258–66.

*Huizinga, J. (1993), 'Inflation uncertainty, relative price uncertainty, and investment in U.S. manufacturing', *Journal of Money, Credit, and Banking*, **25**, 521–57.

Hunter, J.E. and F.L. Schmidt (1990), *Methods of Meta-Analysis: Correcting Error and Bias in Research Findings*, London: Sage Publications.

Jorgenson, D.W. (1963), 'Capital theory and investment behaviour', *American Economic Review*, **53**, 247–59.

Jorgenson, D.W. (1971), 'Econometric studies of investment behavior: a survey', *Journal of Economic Literature*, **9**, 1111–47.

Jorgenson, D.W. and J.A. Stephenson (1967), 'Investment behaviour in U.S. manufacturing, 1947–1960', *Econometrica*, **35**, 169–220.

Koetse, M.J., H.L.F. de Groot and R.J.G.M. Florax (2008), 'Capital-energy substitution and shifts in factor demand: a meta-analysis', *Energy Economics*, **30**, 2236–51.

Koetse, M.J., H.L.F. de Groot and R.J.G.M. Florax (2009a), 'A meta-analysis of the investment–uncertainty relationship', *Southern Economic Journal*, **76**, 283–306.

Koetse, M.J., R.J.G.M. Florax and H.L.F. de Groot (2009b), 'Consequences of effect size heterogeneity for meta-analysis: a Monte Carlo study', *Statistical Methods and Applications*, published online, DOI 10.1007/s10260-009-0125-0.

Koetse, M.J., A.J. van der Vlist and H.L.F. de Groot (2006), 'The impact of perceived expectations and uncertainty on firm investment', *Small Business Economics*, **26**, 365–76.

Koyck, L.M. (1954), *Distributed Lags and Investment Analysis*, Amsterdam: North Holland.

Kulatilaka, N. and E.C. Perotti (1998), 'Strategic growth options', *Management Science*, **44**, 1021–31.

*Leahy, J.V. and T.M. Whited (1996), 'The effect of uncertainty on investment: some stylized facts', *Journal of Money, Credit, and Banking*, **28**, 64–83.

*Lensink, R. (2002), 'Is the uncertainty–investment link non-linear? Empirical evidence for developed economies', *Weltwirtschaftliches Archiv*, **138**, 131–47.

Lensink, R., P. van Steen and E. Sterken (2000), 'Is size important for the investment–uncertainty relationship? An empirical analysis for Dutch firms', SOM Research Report 00E03, Groningen: University of Groningen.

*Lensink, R., P. van Steen and E. Sterken (2005), 'Uncertainty and growth of the firm', *Small Business Economics*, **24**, 381–91.

Meadows, D., D.L. Meadows, J. Randers and W. Behrens (1972), *The Limits to Growth*, New York: Universe Books.

Meyer, J. and E. Kuh (1957), *The Investment Decision*, Cambridge, MA: Harvard University Press.

Nakamura, T. (1999), 'Risk aversion and the uncertainty–investment relationship: a note', *Journal of Economic Behavior and Organization*, **38**, 357–63.

Nakamura, T. (2002), 'Finite durability of capital and the investment–uncertainty relationship', *Journal of Economic Behavior and Organization*, **48**, 51–6.

*Ogawa, K. and K. Suzuki (2000), 'Uncertainty and investment: some evidence from the panel data of Japanese manufacturing firms', *Japanese Economic Review*, **51**, 170–92.

Pattillo, C. (1998), 'Investment, uncertainty, and irreversibility in Ghana', *IMF Staff Papers*, **45**, 522–53.

*Peeters, M. (2001), 'Does demand and price uncertainty affect Belgian and Spanish corporate investment?', *Louvain Economic Review*, **67**, 119–39.

Pindyck, R.S. (1982), 'Adjustment costs, uncertainty, and the behavior of the firm', *American Economic Review*, **72**, 415–27.

Pindyck, R.S. (1988), 'Irreversible investment, capacity choice, and the value of the firm', *American Economic Review*, **78**, 969–85.

Pindyck, R.S. (1993), 'A note on competitive investment under uncertainty', *American Economic Review*, **83**, 273–77.

Rosenthal, R. and M.R. DiMatteo (2001), 'Meta-analysis: recent developments in quantitative methods for literature reviews', *Annual Review of Psychology*, **52**, 59–82.

Saltari, E. and D. Ticchi (2005), 'Risk aversion and the investment–uncertainty relationship: a comment', *Journal of Economic Behavior and Organization*, **56**, 121–25.

Schipper, L. and S. Meyers (1992), *Energy Efficiency and Human Activity*, Cambridge: Cambridge University Press.

Smith, V.K. and J.C. Huang (1995), 'Can markets value air quality? A meta-analysis of hedonic property value models', *Journal of Political Economy*, **103**, 209–27.

Stanley, T.D. (2001), 'Wheat from chaff: meta-analysis as quantitative literature review', *Journal of Economic Perspectives*, **15**, 131–50.

Sutton, A.J., K.R. Abrams, T.A. Sheldon and F. Song (2000), *Methods for Meta-Analysis in Medical Research*, New York: John Wiley and Sons.

Tinbergen, J. (1938), 'Statistical evidence on the acceleration principle', *Economica*, **5**, 164–76.

Tobin, J. (1969), 'A general equilibrium approach to monetary theory', *Journal of Money, Credit, and Banking*, **1**, 15–29.

Weichselbaumer, D. and R. Winter-Ebmer (2005), 'A meta-analysis of the international gender wage gap', *Journal of Economic Surveys*, **19**, 479–511.

8. The effects of uncertainty on investments: analysing the environmental impact of energy market liberalization

Daan P. van Soest and Henri L.F. de Groot

1 INTRODUCTION

One of the main factors affecting an entrepreneur's decision to invest in new, energy-saving technologies is uncertainty about future government policies, changes in energy prices, and the arrival of new technologies. Indeed, accounting for uncertainty about the future has been found to contribute substantially to explaining variation in the intensity and timing of investments (see Chapter 7). In this chapter, we empirically test the validity of the predictions of one specific model of investment behaviour. The model was developed by Dixit and Pindyck (1994) and emphasizes the combined importance of uncertainty and irreversibility of investment decisions. Given the dynamic characteristics of investments, uncertainty about future prices can have an important (additional) impact on the decision whether or not to invest. Higher uncertainty may induce firms to postpone the decision to invest, awaiting the arrival of new information. Thus, firms mitigate the likelihood of investing in a project that *ex post* turns out to be unprofitable due to fallen (relative) prices of inputs. Entrepreneurs' propensity to postpone is higher the larger the (sunk) cost involved when investing, the higher the uncertainty they face, and the higher the degree of persistence in the uncertain variable.[1]

In this chapter, we investigate the quantitative importance of such predictions by assessing the impact of the liberalization of natural gas markets in the USA on investments in energy-saving technologies. More specifically, we look at a special case of an energy-saving technology in the paper production industry for which detailed data on energy-saving potential are available, as well as data on investment costs and the associated increase in operating and maintenance costs.

The main reason why we focus our attention on US data is because liberalization of the US gas market started some 15 years before the liberalization wave hit Western Europe. Price ceilings were removed in 1979 while the phasing out of price controls started in 1989. The liberalization was completed in January 1993 (EIA, 1999a, 1999b). Analysing these data gives insight into how energy market liberalization has affected energy price developments in the USA, both in the short run as well as in the longer run. Experiences in the USA can be used to predict developments in Europe. We explore the implications of the price developments associated with liberalization to investigate the impact on the adoption of energy-saving technologies, and hence on the amount of energy used. This implies that the analysis is not only of academic interest, but also yields valuable insights for policy-makers. To that effect, we simulate how the US experience would affect the Dutch paper industry. We do so by assessing how the investment behaviour of the Dutch paper industry would change if the impact on energy prices, in terms of levels as well as volatility, of energy market liberalization in the Netherlands would be similar to the USA. We take the paper industry as a case because it is energy intensive, and hence energy price changes as well as changes in volatility are likely to play an important role in this sector's investment behaviour. The availability of data has been relevant for our choice as well.

This chapter's approach is simple in that it focuses on just one type of uncertainty, specifically energy price uncertainty associated with energy market liberalization. Also, we ignore complicating factors like market power (especially monopsony power), and technological change. Ignoring these matters is warranted as long as the state of technology, the amount of market power, and other sources of uncertainty are unaffected by the liberalization process itself. The extent to which that is true is an open question. For example, liberalization may well result in increased regulatory uncertainty. However, at this point no information is available about the impact of liberalization on these aspects, and therefore we leave these issues for future research.

The set-up of this chapter is as follows. In section 2, we develop a simple model to illustrate the effects of energy price levels and volatility on investment behaviour. In section 3, we analyse the development of the natural gas price in the USA during the period 1984–98. We thereby intend to assess the effect of liberalization on the level and volatility of gas prices. In section 4 the empirical results are applied to the investment decision concerning an energy-saving technology in the paper production process to give a crude approximation of the effects of the liberalization of the US gas markets on investment behaviour. We conclude in section 5.

2 AN ILLUSTRATIVE MODEL

This section develops a straightforward theoretical model to illustrate the basic issues arising in an environment where firms make irreversible investment decisions under uncertainty. Suppose that, currently, firms are applying an energy-intensive technology to produce a certain amount of output, but there is also an energy-saving technology available that can produce the same amount of output, albeit at the expense of higher operating and maintenance costs. Define ΔE as the change in energy used to produce a unit of output if the current technology is replaced by an energy-saving technology, and ΔOC as the associated increase in operating and maintenance costs.[2] Furthermore, assume that the costs associated with the adoption of the (new) energy-saving technology are C_A.

Uncertainty is introduced into the model in a simple way. Investment costs and the increase in operating and maintenance costs (OC) are assumed to be known with certainty, but future energy prices (P_E) are subject to uncertainty. Consistent with reality, we assume a time trend for prices, but disturbances can shift the energy price away from its trend path. More specifically, the energy price is assumed to follow a Brownian motion:

$$dP_E = \alpha P_E dt + \sigma P_E dz \tag{8.1}$$

In this equation, α is the trend parameter, and $dz = \varepsilon\sqrt{dt}$, where ε is a normally distributed independent variable with a zero mean and a standard deviation of one. This implies that the expected change of the energy price level over a short period dt is equal to $\alpha P_E\, dt$ with variance $\sigma^2 P_E^2 dt$ (Dixit and Pindyck, 1994, pp. 70–1). Therefore, the expected energy price at time t equals:

$$\mathrm{E}(P_E(t)) = P_{E0}e^{\alpha t} \tag{8.2}$$

In this simple framework, we can now determine the costs and benefits of switching from the energy-intensive technology to the energy-saving technology. The change in technology results in savings on energy expenses, but requires additional expenditures in terms of operating and maintenance costs. Taking into account the adjustment or installation costs (C_A),[3] the expected value of switching to the energy-saving technology equals:

$$\Omega(P_E) = \int_0^\infty \mathrm{E}(P_E)\Delta E e^{-rt} dt - \frac{\Delta OC}{r} - C_A = \frac{P_E \Delta E}{r - \alpha} - \left(\frac{\Delta OC}{r} + C_A\right) \tag{8.3}$$

where r is the exogenously given discount rate.[4]

The question now is at what energy price will the firm decide to switch to the energy-saving technology? In each period, the firm compares the benefits of undertaking the investment, in terms of cost reductions achieved, with the benefits of postponing the decision one period. The latter include access to more information about energy prices in the next period. Given the uncertainty that the firm faces, postponing the decision reduces the probability of investing in a project that turns out to be unprofitable *ex post*. In mathematical terms, the firm maximizes:

$$F(P_E) = \max\left\{\Omega(P_E), \frac{1}{1 + rdt}\mathrm{E}(F(P_E) + dF(P_E))\right\} \tag{8.4}$$

The value $\Omega(P_E)$ is labelled the 'termination value'. When the firm decides to undertake the investment, its expected return is known. The expected return of waiting, represented by the second term within the large brackets, is usually referred to as the 'continuation value'. The firm's optimal decision maximizes the net present value of the investment option (F). As soon as the termination value exceeds the continuation value, the investment is undertaken. The energy price at which this point is reached will from here on be referred to as the critical energy price (P_E^*).

Solving the model, the following differential equation is obtained:[5]

$$rFdt = \mathrm{E}\left[F_P dP_E + \frac{1}{2}F_{PP}(dP_E)^2\right] = \left[F_P \alpha P_E + \frac{1}{2}F_{PP}\sigma^2 P_E^2\right]dt \tag{8.5}$$

Try $F = AP_E^{\beta}$ as a solution to this differential equation. Solving the differential equation, two roots can be found:

$$\beta_{1,2} = \left(\frac{1}{2} - \frac{\alpha}{\sigma^2}\right) \pm \sqrt{\left(\frac{1}{2} - \frac{\alpha}{\sigma^2}\right)^2 + \frac{2r}{\sigma^2}} \tag{8.6}$$

This term essentially captures the impact of price uncertainty on the critical energy price level at which the switch towards the energy-saving technology will be carried out. The general solution is of the form $F(P_E) = A_1 P_E^{\beta_1} + A_2 P_E^{\beta_2}$ where β_1 and β_2 represent the positive and negative roots, respectively. The higher the energy price, the higher the value of the energy-saving investment option will be. This implies that the term with the negative root can be ignored: A_2 equals zero.

Now the critical value of the energy price can be determined by using two additional conditions (Pindyck, 1991; Dixit and Pindyck, 1994). First, in the optimum it must hold that at the critical energy price level, the value of the investment project is equal to the termination value $F(P_E^*) = \Omega(P_E^*)$. Given the fact the investment is undertaken, waiting apparently no longer has a positive net value; see equation (8.4). Second, optimality requires that the option value function $F(P_E)$ and the termination value function $\Omega(P_E)$ meet tangently at the critical price level $F_P(P_E^*) = \Omega_P(P_E^*)$. Using these two additional conditions, it can be found that the critical energy price equals:

$$P_E^* = \left(\frac{\beta_1(\alpha, \sigma, r)}{\beta_1(\alpha, \sigma, r) - 1}\right)\left(\frac{r - \alpha}{\Delta E}\right)\left[\frac{\Delta OC}{r} + C_A\right] \tag{8.7}$$

where

$$\beta_1(\alpha, \sigma, r) = \left(\frac{1}{2} - \frac{\alpha}{\sigma^2}\right) - \sqrt{\left(\frac{1}{2} - \frac{\alpha}{\sigma^2}\right)^2 + \frac{2r}{\sigma^2}}$$

The last two terms in expression (8.7) reveal the price at which the investment in the energy-saving technology would be undertaken in the absence of uncertainty, according to a standard net present value rule.[6] Firms put a mark-up of $\beta_1/(\beta_1 - 1)$ on this price. This mark-up depends negatively on β_1 and therefore positively on σ; see equation (8.6). Hence, a higher variability results in an increase of the critical price level above which the investment will be undertaken. Furthermore, the mark-up is lower the higher the discount rate r, and the lower the trend rate of growth α. With respect to the discount rate, the intuition is that heavier discounting gives lower weights to future developments and therefore leads to less prudent behaviour. A higher trend rate of growth increases the value of the option to invest (the continuation value). This means that the opportunity costs of investing increases and hence the mark-up will be higher.[7] Thus, appropriate consideration of uncertainty results in a critical price that exceeds the critical NPV price, P^{NPV}, by a factor $\beta_1/(\beta_1 - 1)$.

Using equations (8.2) and (8.7), the expected period in which the investment in energy-saving technology is undertaken can now be calculated. From equation (8.2), we know that $E(P_E(T^*)) = P_{E0}e^{\alpha T^*}$, so the expected investment lag is given by:

$$T^* = \frac{1}{\alpha}\ln\left(\frac{P_E^*}{P_{E0}}\right) \tag{8.8}$$

Taking the first derivatives of T^* with respect to P_{E0} and σ, it is clear that a lower initial energy price implies that the investment is expected to be postponed into the future while a decrease in uncertainty implies that the investment is expected to take place sooner. If increased

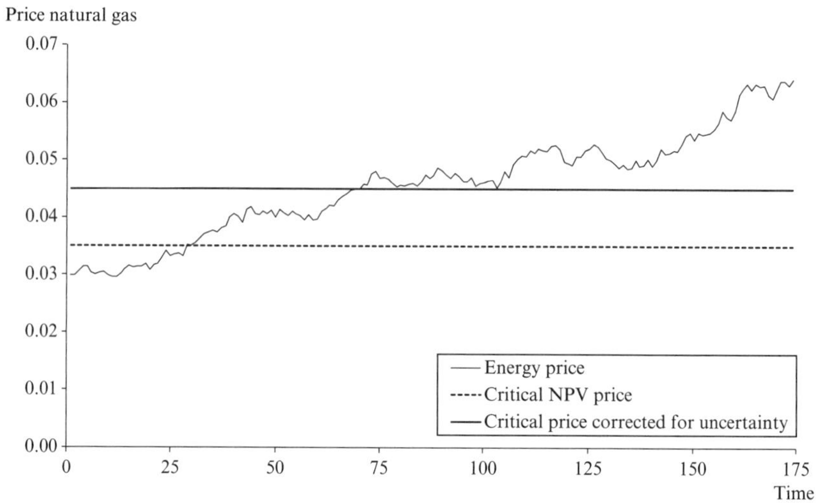

Figure 8.1 *The timing of investment in hypothetical energy-saving technology*

competition results in lower energy prices and higher variance, both effects will induce the firm to postpone the investment in energy-saving technology. If, however, lower prices are associated with less volatility (as argued by, for example, Ferderer, 1996), energy price liberalization will have an ambiguous effect on the propensity to invest in energy-saving technologies.

To further illustrate the model, we refer to Figure 8.1. In this figure, we have drawn a hypothetical development path of the energy price. In addition, we have drawn the critical energy price that results from a standard net present value decision rule and the critical energy price that holds when controlling for the effects of uncertainty on investment behaviour. The figure reveals the moments at which adoption of the new technology takes place according to the alternative investment decision rules. The difference in the timing of adoption is caused by the uncertainty regarding energy prices: under uncertainty and irreversibility, firms will postpone adoption to reduce the likelihood of making the wrong decision.

In the remainder of this chapter, we empirically assess the effects of liberalization of the US gas market on the level and uncertainty of gas prices. Using the simple framework developed in this section, we subsequently assess the likely impact of this liberalization on investments in energy-saving technologies.

3 EMPIRICAL APPLICATION

Let us take a closer look at data on natural gas prices in the US over the period 1984–98. The data were taken from the Energy Information Administration (EIA, 1999a). In this period, the national gas market was gradually liberalized, with liberalization being completed in 1993. Figure 8.2 depicts the US natural gas price deflated with the producer price index[8] and the long-term trend[9] in the gas price over the entire sample period.

Inspection of the data gives rise to two observations. First, the deflated natural gas price fell continuously during the liberalization process, but by and large stabilized after 1993 when market liberalization was completed. Second, price volatility has increased considerably over the liberalization period, and continued to do so after 1993. This seems to indicate that the market becomes increasingly responsive to demand and supply imbalances. Of course, the fluctuations shown in Figure 8.2 arise partly because of seasonal variation. In order to correct for that, we plotted the 12-month percentage change in the natural gas price and its trend, which was again obtained by applying a Hodrick-Prescott filter, in Figure 8.3.

This figure yields the same observations as Figure 8.2. Up until 1993 the natural gas price fell, showing a slight recovery afterwards, while volatility increased over the entire 1984–98 period.[10]

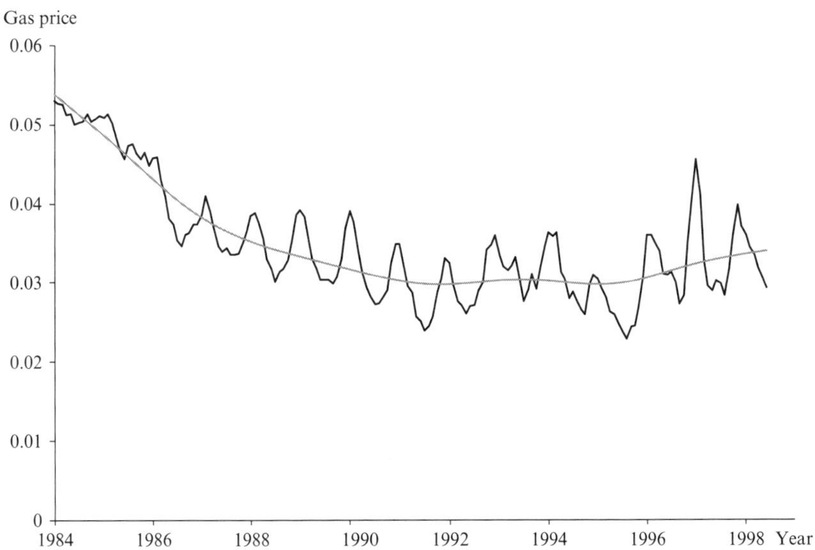

Figure 8.2 The deflated US natural gas price index and its trend

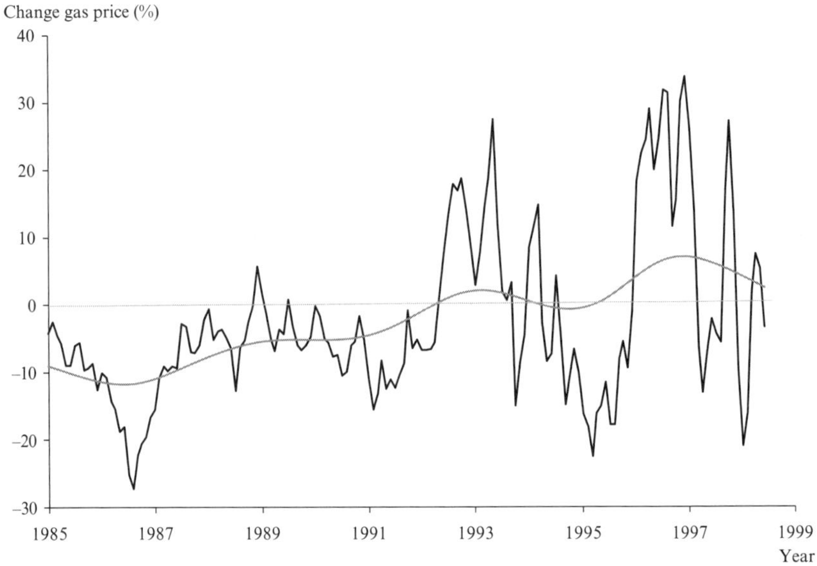

Figure 8.3 *The 12-month percentage change in the US natural gas price and its trend*

To determine the trend and volatility of US gas prices, we have checked whether the series can be described by a geometric Brownian motion (see Conrad, 1997). The details of the estimation procedure can be found in the Appendix to this chapter. The analysis shows that US natural gas prices indeed follow a geometric Brownian motion. Splitting the entire time span in two periods, pre- and post-1993, we find a significantly negative trend in the gas price in the pre-1993 period and an insignificantly positive trend in the post-1993 period. Uncertainty turns out to have increased from 16 to 20 per cent, on an annual basis. Obviously, apart from liberalization, other factors may have caused these developments such as technological developments and instability in the Organization of the Petroleum Exporting Countries (OPEC). Nevertheless, it seems fair to conclude that the liberalization of the US gas market has indeed resulted in lower gas prices and that it has at the same time resulted in an increase in the volatility of these prices, strengthening the uncertainty surrounding investments in energy-saving technologies.

The previously described liberalization of energy markets is often argued to have detrimental consequences for the environment (see, for example, Elliott and Pye, 1998). By lowering energy prices, liberalization is expected to result in increased energy consumption and fewer invest-

Table 8.1 Pre- and post-liberalization mark-ups on the critical energy price[a]

	Pre-liberalization		Post-liberalization	
	$r = 0.01$	$r = 0.02$	$r = 0.01$	$r = 0.02$
β_1	3.48	4.78	2.91	3.93
Mark-up	1.40	1.26	1.52	1.34

Note: [a] Pre-liberalization information is based on $\alpha = 0.0004$ and $\sigma = 0.045$; post-liberalization information is based on $\alpha = 0.0004$ and $\sigma = 0.056$

ments aimed at improving energy efficiency.[11] What is often neglected in these arguments is that consumption and investment decisions are not only driven by the level of energy prices, but also by their volatility. Liberalization can be expected to have important consequences for the volatility of energy prices, although the impact is ambiguous. On the one hand, price levels and volatility are often positively correlated in energy markets: high energy price levels have been observed to have a higher variability than low prices (see, for example, Ferderer, 1996). On the other hand, in comparison to the situation of administered prices, liberalization creates a more dynamic economic environment where energy prices react quickly to changes in demand and supply. Thus, lower prices may also be associated with higher volatility. Whether one or the other effect dominates is an empirical question.

As a first attempt to assess the effects of liberalization on investment behaviour, we use the estimates for the volatility of energy prices to determine the mark-up that firms put on top of the critical price that would hold in the absence of uncertainty. Results are presented in Table 8.1, where monthly values of the discount rate and trend and uncertainty parameters are used.

In Table 8.1, we abstain from including the effect of a change in the structural growth rate of energy prices. The reason for this is that we see no a priori reason for why the structural development of energy price would be affected by liberalization.[12] We do account for a change in the level of the energy price as a result of liberalization, as will become apparent in the next section. However, such a change leaves unaffected the mark-up over the critical price that holds in the absence of uncertainty. From this table it is clear that the mark-up has increased by approximately 7.5 per cent as a result of liberalization, the increase depending on the discount rate (see also section 2).

4 ASSESSING THE IMPACT OF LIBERALIZATION ON ENERGY-SAVING INVESTMENTS

Having established that during the liberalization process the general energy price level has fallen while uncertainty has increased, we can now assess the impact of these developments on investment behaviour. For this analysis we take the simple model laid out in section 2 as a starting point. According to theory, both the lower energy prices and the increased uncertainty about future energy prices will lengthen the period in which no investment in cleaner technologies takes place.

To illustrate the quantitative effect of liberalization, we consider an energy-intensive industry where substantial reductions in energy use can be achieved, namely the paper and pulp industry. In a research report issued by the Dutch paper industry (Versluijs et al., 1993), several potential energy-saving technologies are listed. It has been estimated that 22 per cent of the energy currently used can be reduced cost effectively; that is, with a payback period of less than seven years. By means of an example, we focus on one specific energy-saving technology associated with the drying process. This part of the production process is highly energy intensive. Substantial reductions can be achieved by introducing the so-called long nip press, which increases the fibre concentration and thus lowers the water content of the raw material so that substantial savings can be achieved in the drying phase in terms of the amount of steam needed. The resulting reduction in energy consumption is estimated to be approximately 16 per cent.

The analysis in this section is based on the costs and benefits of the new technology per tonne of paper produced. It takes about 5 Giga Joule (GJ) to produce one tonne of paper.[13] Using the estimate of a 16 per cent reduction in energy use, annual energy savings of 0.75 GJ per tonne of paper can be achieved. Using the international average calorific content of natural gas (1000 m^3 yields 31.7 GJ), the amount of gas annually saved equals 23.5 m^3 per tonne of paper produced. The costs associated with this new technology comprise investment costs, and operating and maintenance costs. The investment costs associated with the adoption of a long nip press are US$18.75 per tonne of paper, and the additional operating and maintenance costs are $US0.50/tonne (all figures are derived from de Beer et al., 1994, pp. 53–6).

Using these data, we determine critical gas prices at which investment is predicted to take place according to a standard net present value rule and according to the modified net present value rule accounting for uncertainty. The results are given in Table 8.2.

To get a sense for the impact on the investment timing, we performed

Table 8.2 Critical energy price in the paper case[a]

	$\sigma = 0.045$	$\sigma = 0.056$
β_1	3.82	3.21
Mark-up	1.35	1.45
P_E^{NPV}	0.030	0.030
P_E^*	0.041	0.044

Note: [a] Prices are derived on the basis of $\alpha = 0.0004$, $r = 0.0125$, $C_A = 18.75$, $\Delta OC = 0.5$, and $\Delta E = 23.5$.

a Monte Carlo analysis. We took three different combinations of parameter values for the initial energy price level and its volatility. These were subsequently used to generate 1000 time paths for the gas prices for each set of parameter values. Confronting these time paths with the critical energy price corresponding to each parameter constellation, we determined the time at which the switch from the energy intensive to the energy extensive technology is made. The initial natural gas price index for the pre-liberalization period (the 1984.1 value) was about 0.04 while the monthly price volatility in that period was about 0.045. At the end of the liberalization period, the natural gas price index had fallen to a level of about 0.03, while the volatility in the post-liberalization period was about 0.056. Using these figures in the Monte Carlo simulations yields the results as represented in Figure 8.4, where Case I ($P_{E0} = 0.04$, $\sigma = 0.045$) and Case II ($P_{E0} = 0.03$, $\sigma = 0.056$) represent the pre- and post-liberalization situations, respectively.

In Figure 8.4, the cumulative likelihood of adoption is depicted on the vertical axis for a certain period depicted on the horizontal axis. It is clear from this figure that the implementation lag increased substantially since the completion of the natural gas market liberalization; compare Cases I and II. To separate out the price level effect and the uncertainty effect, an intermediate third case is also shown in the figure. Case IIa is based on the pre-liberalization energy price level ($P_{E0} = 0.04$) and on the post-liberalization volatility level ($\sigma = 0.056$). As a result, the difference between Cases I and IIa reflects the pure impact of higher uncertainty, while the difference between Cases II and IIa reflects the pure impact of a lower energy price level. Loosely stated, the experiment indicates that higher uncertainty decreases the probability that adoption of the new technology takes place within 10 years (120 months) from 90 per cent to 79 per cent (compare Case I and IIa), whereas the lower energy price further reduces this probability to 41 per cent (compare Case IIa and II).

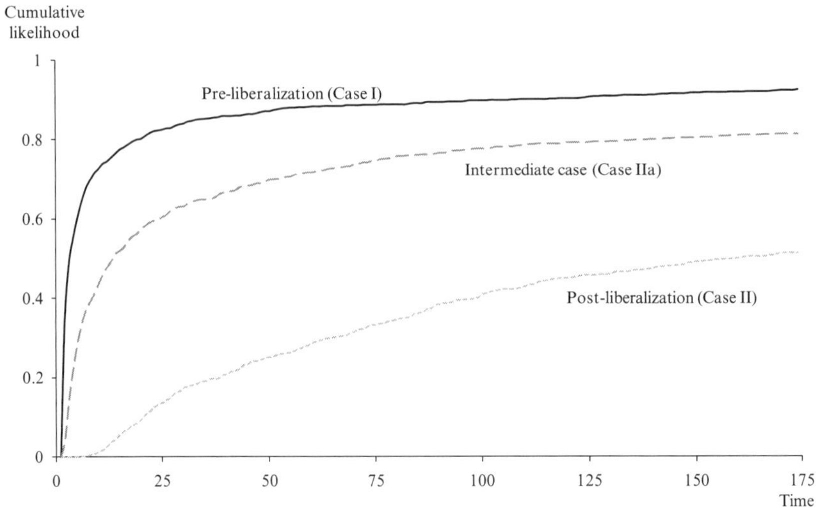

Figure 8.4 The cumulative likelihood adoption of the energy-saving technology over time

5 CONCLUSIONS

Energy market liberalization is often suspected to have adverse environmental consequences, as the resulting decrease in energy prices would reduce the attractiveness of investing in energy-saving technologies. However, lower price levels are not the only consequence of liberalization: the variability of energy prices may go up because markets become more responsive to differences in energy demand and supply, or the variability goes down as there is often a positive correlation between energy price levels and volatility. Higher volatility, implying higher uncertainty, reduces the attractiveness of energy-saving investments because of the irreversibility of the investment decision: if prices are highly volatile, implying that they may increase or decrease substantially between periods, the likelihood of making the wrong decision (that is, *ex post*) increases, and firms are therefore likely to become more prudent in their investment behaviour. The higher the variability, the higher the critical energy price at which an energy-saving technology will be adopted *ceteris paribus*.

Using US data, we established that the liberalization of the natural gas market will have a substantial detrimental impact on the willingness to invest in more energy efficient technologies in the Dutch paper industry, provided that the effects of the Dutch liberalization process are not too dissimilar from those in the USA. Not only did the price level fall in the US gas market, but also its variability increased substantially resulting in a large increase in the implementation lag. Applying these insights to the investment decision with respect to one specific technology, the adoption of the long nip press in the paper production process, we find that there is indeed a substantial impact of both the lower energy price as well as the increased volatility in energy prices as a consequence of the liberalization of gas markets.

This means that from a social point of view, there is a trade-off between energy market liberalization and environmental protection for two reasons. The first reason is that, obviously, lower energy prices depress the return on investment in energy-saving technologies. However, our analysis shows that there is a second reason for the trade-off to occur: the volatility of the energy price has also increased over the liberalization period. The increased variance results in a higher mark-up over the standard expected net present value criterion. As a result, liberalization results in a higher critical energy price to trigger investments in energy-saving technologies. In assessing the desirability of energy market liberalization, not only the environmental consequences of a lower energy price should be taken into account but also its higher volatility.

Whether the environmental consequences of energy market liberalization in Europe will turn out to be as detrimental as predicted in this chapter is yet to be determined. It is too early to tell and the impact will depend on many factors, of which the development of energy prices in response to liberalization is but one. However, policy-makers concerned about the probability of Europe meeting its Kyoto targets, do wise to consider additional measures to curb energy use. More specifically, policy-makers should consider the impact of policy measures on uncertainty of economic actors since the quantitative impacts of increases in uncertainty can be substantial.

ACKNOWLEDGEMENTS

We are grateful to the Netherlands Organisation for Scientific Research (NWO) for financial support, and also to two anonymous reviewers for their useful comments on an earlier version of this chapter.

NOTES

1. Note that we focus on the impact of uncertainty (in combination with irreversibility) on the investment decisions of energy users, and not of energy suppliers.

2. In other words, we model firms as producing according to a Leontief production technology. This assumption is reasonable since substitution possibilities for inputs needed to operate technologies are limited once the technology has been installed. Generalizations could allow for (imperfect) substitutability among the various inputs but are not likely to affect the main results. Also note that adoption of an energy-saving technology does not necessarily result in an increase in operating and maintenance costs. The investor will always be confronted with a real trade-off as long as investment costs are positive.
3. Important for the argument to follow is that at least a significant part of the investment in the new technology is irreversible. For example, it may be impossible to sell the technology at its purchase price; second-hand markets may even be absent. Other important parts of the investment costs can generally also not be recouped, such as reorganizational efforts and expenses aimed at incorporating the new technologies in the production process and acquisition of information. Even if returning to the old technology is feasible at low or zero costs, the costs of installing the new technology cannot be recouped and are therefore irreversible.
4. The results have been derived using dynamic programming, which is based on the assumption that the price risk cannot be spanned by constructing an appropriate market portfolio. If we would have dropped this (implicit) assumption, contingent claims analysis could be used which would have enabled us to derive a risk-adjusted discount rate. Using the capital asset market pricing approach, this discount rate would be equal to $r + \phi\rho_{PM}\sigma$, where ϕ is the market price of risk, and ρ_{PM} the correlation coefficient between market risk and the riskness of the energy price (see Dixit and Pindyck, 1994, p. 185).
5. Throughout this chapter, G_i and G_{ii} denote the first and second partial derivatives of the function G with respect to the variable i.
6. This price is derived from the simple net present value rule of the investment according to which the investment should be undertaken as soon as:

$$P_E^{NPV} > \left(\frac{r - \alpha}{\Delta E}\right)\left[\frac{\Delta OC}{r} + C_A\right]$$

7. This does not imply that the investment will actually be postponed. A higher trend also increases the termination value. As this increase generally dominates the increase in the continuation value (see equation 8.4), the critical price in equation (8.7) tends to fall.
8. Producer prices were taken from the International Financial Statistics (IFS) database of the International Monetary Fund. Annual data were converted to monthly data using linear intrapolation. Using the price index of capital as a deflator does not alter the qualitative results of the analysis.
9. The trend has been derived by applying a Hodrick-Prescott filter to the deflated price of natural gas series. Another way of establishing the same result is to run regressions of the seasonally adjusted data on a time trend.
10. Of course, not all observed changes can be attributed exclusively to the liberalization process; for example, shifts in global demand and supply of energy also affect US gas prices. Independent of the source of fluctuation, these data allow us to measure the impact of uncertainty on investment decisions (the mark-up as discussed in section 2).
11. Concerns about the (persistently) low energy prices and their associated difficulties of fostering energy saving were recently forcefully expressed in the Dutch policy plans for meeting the Kyoto targets (VROM, 1999).
12. Theoretically, one could argue that liberalization leads to a steeper time profile of energy prices as it results in lower initial prices but faster exhaustion of resources. Alternatively, one could argue that liberalization speeds up technological progress in the sector and thereby lowers costs of gas recovery and thereby flattens the time profile. For the purpose of this chapter, in which we are mainly interested in the effects of liberalization-induced uncertainty, we abstain from these issues.

13. Energy use may be somewhat higher or lower, depending on the desired quality of paper.

REFERENCES

Conrad, J.M. (1997), 'On the option value of old-growth forest', *Ecological Economics*, **22**, 97–102.

de Beer, J.G., M.T. van Wees, E. Worrell and K. Blok (1994), 'ICARUS-3: the potential of energy efficiency improvement in the Netherlands up to 2000 and 2015', Utrecht: Department of Science, Technology and Society, Utrecht University.

Dixit, A.K. and R.S. Pindyck (1994), *Investment under Uncertainty*, Princeton: Princeton University Press.

Elliott, R.N. and M. Pye (1998), 'Investing in industrial innovation: a response to climate change', *Energy Policy*, **26**, 413–23.

Energy Information Administration (EIA) (1999a), 'Why do natural gas prices fluctuate so much?', Washington, DC: Energy Information Administration, US Department of Energy.

Energy Information Administration (EIA) (1999b), 'Natural gas 1998 issues and trends', Washington, DC: Energy Information Administration, US Department of Energy.

Ferderer, J.P. (1996), 'Oil price volatility and the macroeconomy', *Journal of Macroeconomics*, **18**, 1–26.

Pindyck, R.S. (1991), 'Irreversibility, uncertainty, and investment', *Journal of Economic Literature*, **29**, 1110–48.

Stewart, J. (1991), *Econometrics*, New York: Philip Allan.

Versluijs, R.P., J.E. van Oostwaard, B.M.P. Mulder, J.G. Singels, L.A. Wattimena, J.M. van Wolferen and F. de Haas (1993), 'Energieonderzoek: Mogelijkheden tot Vermindering van het Specifiek Energieverbruik in de Nederlandse Papier- en Kartonindustrie', TNO Report PK 92.1134, Delft: TNO.

VROM (1999), *Uitvoeringsnota Klimaatbeleid*, The Hague: Ministry of Housing, Spatial Planning and the Environment (VROM).

APPENDIX: ESTIMATING THE GROWTH RATE AND VARIANCE OF THE ENERGY PRICE

This appendix describes the procedure used to determine whether a geometric Brownian motion can describe US natural gas prices. After establishing this result, we continue by determining the parameter values best describing the price development.

Determining whether indeed the energy price follows a geometric Brownian motion boils down to establishing $dP_E = \alpha P_E dt + \sigma P_E dz$. Although the trend and uncertainty parameters will be slightly different, testing whether the natural logarithm of the energy price follows an ordinary Brownian motion is econometrically more convenient (see Conrad, 1997). In general, the first derivative of the natural logarithm of a specific variable, in this case the energy price, P_E, equals:

$$d\ln(P_E) = \frac{1}{P_E} dP_E - \frac{1}{2P_E^2}(dP_E^2) \tag{8A.1}$$

Substituting $dP_E = \alpha P_E dt + \sigma P_E dz$ in (8A.1) and applying Ito's Lemma (Dixit and Pindyck, 1994, p.81), we find:

$$d\ln(P_E) = \alpha dt + \sigma dz - \tfrac{1}{2}\sigma^2 dt = (\alpha - \tfrac{1}{2}\sigma^2)\, dt + \sigma dz = adt + sdz \tag{8A.2}$$

Hence, after testing whether the log of the energy price follows an ordinary Brownian motion, we can determine the trend and uncertainty parameters by using the relationship between a, s, α and σ as presented in equation (8A.2).

Testing whether the natural logarithm of the energy price follows an ordinary Brownian motion boils down to jointly testing the significance of a trend parameter and the level of the natural log of the energy price in the previous period. This requires running two regressions, an unrestricted and a restricted version (Stewart, 1991, pp. 199–203). Denoting the natural logarithm of the energy price as p_E, the unrestricted regression equation is given by:

$$p_{Et} - p_{Et-1} = \gamma d + \beta t + (\rho - 1)p_{Et-1} + \lambda(p_{Et-1} - p_{Et-2}) + \varepsilon_t \tag{8A.3}$$

where d is the vector of dummies to capture differences in monthly growth rates, and γ the associated vector of coefficients. The null hypothesis that p_E follows a Brownian motion is H_0: $\beta = 0, \rho = 1$. This implies that the following restricted version should also be estimated:

$$p_{Et} - p_{Et-1} = \gamma d + \lambda(p_{Et-1} - p_{Et-2}) + \xi_t \tag{8A.4}$$

By calculating the F-statistic using the sums of squared residuals of these two equations, the joint null hypothesis can be tested.

The results for the two sub-periods are presented in Table 8A.1. The F-test is given by:

$$F = \left(\frac{\widetilde{T} - k}{q}\right)\left(\frac{SSR_R - SSR_U}{SSR_U}\right)$$

where $\widetilde{T}$ is the number of observations corrected for the monthly dummies, k and q are the number of variables in the unrestricted and restricted regressions (ignoring the dummy variables), and SSR_R and SSR_U are the sum of squared residuals of the restricted and unrestricted regressions, respectively. The F-values for the distinguished periods are 2.6805 for the 1984–92 period and 5.1974 for the 1993–98 period. The Dicky-Fuller critical F-value at the 5 per cent level (for $\widetilde{T} > 50$) is 6.49, and hence the null hypothesis of a Brownian motion cannot be rejected (Stewart, 1991, p. 203).

Now the appropriate α and σ of the energy price series can be found by calculating the mean (a) and standard deviation (s) of the $d\ln(P_E)$ series and applying the correction indicated in equation (8A.2). Indeed, for the 1984.2–1992.12 period, a is found to be -0.00585 and s^2 equals 0.00200, which implies that α equals -0.00485 and σ equals 0.0447. For the period 1993.1–1998.6 a is found to be 0.00717 and s^2 equals 0.003187; that means that α equals 0.008764 and σ equals 0.05646. Note that the uncertainty parameters are much higher (in absolute terms) than the trend parameters. Converting these figures to yearly trends and uncertainty parameters, the annual trends are approximately zero while the annual uncertainty parameters are 0.1549 and 0.1956 for the first and second sub-period, respectively.

Table 8A.1 Regression results of the restricted and unrestricted models for both time periods[a]

Variable	Unrestricted 1984.3–1992.12	Restricted 1984.3–1992.12	Unrestricted 1993.1–1998.6	Restricted 1993.1–1998.6
January	−0.278	0.006	−0.864	0.015
	(−2.16)	(0.49)	(−3.12)	(0.56)
February	−0.306	−0.026	−0.918	−0.050
	(−2.42)	(−2.10)	(−3.35)	(−1.89)
March	−0.329	−0.049	−0.913	−0.043
	(−2.61)	(−4.24)	(−3.33)	(−1.63)
April	−0.312	−0.031	−0.931	−0.048
	(−2.46)	(−2.43)	(−3.35)	(−1.80)
May	−0.328	−0.042	−0.906	−0.009
	(−2.54)	(−3.39)	(−3.20)	(−0.32)
June	−0.296	−0.006	−0.937	−0.029
	(−2.26)	(−0.47)	(−3.28)	(−1.13)
July	−0.302	−0.008	−0.926	−0.012
	(−2.28)	(−0.68)	(−3.22)	(−0.40)
August	−0.281	0.015	−0.939	−0.017
	(−2.11)	(1.28)	(−3.24)	(−0.60)
September	−0.282	0.013	−0.901	0.027
	(−2.12)	(1.16)	(−3.08)	(0.94)
October	−0.264	0.030	−0.916	0.012
	(−1.99)	(2.56)	(−3.13)	(0.44)
November	−0.255	0.036	−0.805	0.119
	(−1.94)	(3.01)	(−2.77)	(4.17)
December	−0.256	0.031	−0.871	0.036
	(−1.98)	(2.53)	(−3.04)	(1.09)
$p_{Et-1} - p_{Et-2}$	0.357	0.292	0.391	0.278
	(3.46)	(2.95)	(3.07)	(2.11)
t (trend × 10^3)	−0.503		0.441	
	(−1.75)		(1.10)	
p_{Et-1}	−0.095		−0.241	
	(−2.19)		(−3.22)	
Adjusted R^2	0.53	0.52	0.45	0.36
Sum squared residuals	0.104	0.110	0.178	0.214
No. of observations	106	106	66	66

Note: [a] Student's t-values are in parentheses.

PART III

Energy Policy Design

9. Energy investment behaviour: firm heterogeneity and subsidy design

Daan P. van Soest and Herman R.J. Vollebergh

1 INTRODUCTION

Over the past three decades, much attention has been paid to the potential of public policies to reduce the emission of polluting substances that give rise to environmental problems, such as global warming and acid rain. A wide range of policy instruments is available, ranging from pollution taxes, tradable emission permits, abatement subsidies, quotas and performance standards, to covenants. The main criterion by which the economic profession has evaluated the instruments is that of static efficiency; that is, the extent to which these instruments can achieve a specified environmental objective at minimum cost. A necessary condition for static efficiency is that marginal abatement costs are equalized across firms, and in this respect instruments like emission taxes or tradable emission permits are generally preferred to quotas, performance standards and investment subsidies (Baumol and Oates, 1988).

However, static efficiency is just one of many criteria on which to base instrument choice, and it is not necessarily the most important one (Jaffe and Stavins, 1995; Jaffe et al., 2003). With ongoing economic growth and the associated pressure on the natural regenerative capacity of the environment, ever more efficient technologies should be developed and adopted to ensure that environmental protection and economic growth are not incompatible in the long run. These technologies should be able to produce the same amount of output with less pollution, or to abate pollution at lower marginal cost. Indeed, Kneese and Schulze (1975, p. 38) already argued that 'over the long haul, perhaps the most important single criterion on which to judge environmental policies is the extent to which they spur new technology toward the efficient conservation of environmental quality'.

In this chapter, we review the recent theoretical and empirical literature on the effectiveness of investment subsidies to induce adoption of new, energy-saving (or energy-efficient) technologies. We focus on subsidies as opposed to the other instruments listed above because they have received relatively little attention in the environmental economics literature. Indeed, from two recent surveys of the adoption literature (Nelissen and Requate, 2007; Requate, 2005) and market-based instruments (Stavins, 2003), it is clear that the economics profession has focused predominantly on pollution taxes, tradable pollution permits and quotas. Despite the fact that the subsidy instrument is not popular among economists, it is actually used quite often in practice. Subsidies appear in many forms; not just explicit investment subsidies are widely used, but also tax deductibility schemes are very common. The latter include investment credits, accelerated depreciation, partial expensing, and exemptions (Jenkins and Lamech, 1992; OECD, 2006; Price et al., 2005). This popularity can at least partly be explained by the fact that subsidies temper the increase in average cost associated with environmental policies, and hence they have a less detrimental effect on the domestic industry's international competitiveness, as compared to, for example, environmental taxes or quotas (Dietz and Vollebergh, 1999). In addition, distributional fairness considerations may give rise to the use of subsidies rather than taxes, for example because it is especially the relatively poor who purchase the products produced by the more pollution-intensive industries (Abrego and Perroni, 2002).

This chapter is organized as follows. In the next section we review the arguments against subsidy schemes and argue that the economics profession tends to overestimate the disadvantages of subsidies, and ignores some of the disadvantages of the alternative instruments. We conclude that the subsidy instrument deserves a second look, and hence the question arises as to how subsidy schemes should be designed to achieve maximum effectiveness per dollar spent. To do so, we describe the decision making process of managers in section 3, and infer how investment decisions are likely to be affected by subsidy design. In the fourth section we review empirical and experimental evidence about optimal subsidy design, and we present our conclusions in section 5.

2 ADVANTAGES AND DISADVANTAGES OF ENVIRONMENTAL SUBSIDIES IN THEORY

As stated in the introduction, recent literature reviews confirm that subsidies as an instrument of environmental policy have not received much attention from economists (Nelissen and Requate, 2007; Requate, 2005). This holds for both abatement and investment subsidies. Abatement or Pigovian subsidies typically reward reductions in emissions as compared to a certain baseline level, whereas investment or adoption subsidies reduce the cost of adopting new technologies. In this section we first discuss the main reasons why subsidies are not looked upon very favourably in economics. Next, we show that at least some of these reasons are less compelling than they appear at first sight, and that the environmental economics literature may be too optimistic about the effectiveness of alternative instruments such as environmental taxes. The conclusion of this section is that especially investment subsidies deserve a second look.

2.1 Arguments Against the Use of Environmental Subsidies

Economists usually have a negative view on environmental subsidies for at least three reasons. First, subsidies would have adverse effects on entry and exit decisions of firms. Second, subsidies require distortional taxes at the margin. Finally, with asymmetric information, improper design of subsidies typically contributes to ineffective incentives, and therefore to inefficiency. Let us discuss these three arguments in turn.

We start with the well-known objection that providing investment or abatement subsidies results in an undesirable expansion of polluting industries. Profits in the regulated industries are higher when subsidies are used instead of taxes or quotas, and hence the number of firms in the industry is also likely to be higher (Baumol and Oates, 1988). The argument is as follows. From a static efficiency perspective (that is, for a given technology), abatement subsidies and environmental taxes have the same impact on the behaviour of incumbent firms if the per-unit emission tax rate is equal to the per-unit pollution reduction subsidy. The intuition behind this result is straightforward. Abatement subsidies reward the reduction of emissions as compared to the baseline, and hence *not* reducing emissions is costly at the margin. In this sense, abatement subsidies and environmental taxes are identical in terms of the incentives they provide to existing firms to reduce abatement. For equal per-unit tax and subsidy rates, the profit-maximizing amount of pollution is the same under both instruments, and hence both instruments are equally effective in reducing emissions in the short run.

In the longer run, however, when firms can enter or leave the industry, the consequences of the two instruments are not equal, at least if the barriers to enter or exit are not too high. Firms enter (leave) the industry in case of abatement subsidies (taxes) as the instrument increases (decreases) profitability in the industry. Thus the aggregate pollution level unambiguously falls in the case of a tax, but it may possibly increase in the case of subsidies. With subsidies industry-wide emissions go up if the percentage decrease in per firm pollution production is smaller than the percentage increase in the total number of firms in the industry.

The second objection comes from optimal tax theory. The argument is that raising and distributing subsidies is usually not costless to society because of the way in which they are financed. The government needs to raise funds by imposing (higher) taxes on for example labour (that is, income taxes), capital, or value-added, and thus affects marginal decisions of agents elsewhere in the economy (Parry, 1998). Indeed, value-added taxes affect individuals' consumption and savings decisions, and income taxes affect their decisions with respect to employment and leisure. So, environmental subsidies correct a market failure (excessive pollution) but tend to do so at the expense of introducing additional distortions elsewhere in the economy.[1] Taxes, on the other hand, not only correct the environmental market failure, but they also raise funds that, in turn, can be spent elsewhere in the economy. Indeed, environmental taxes can be used to lower distortionary taxation in output or labour markets, and hence potentially produce a so-called double dividend.[2]

The third major objection is that agents may claim subsidies while the subsidies do not affect their behaviour. Less enigmatically stated, subsidies are not just claimed by agents for whom the investment opportunity is not profitable without the subsidy, but also by agents who would have adopted the technology anyway, even with zero subsidies (Wirl 2000; Wirl and Orasch, 1998). Indeed, for some firms the costs of purchasing an energy-saving technology may be smaller than the benefits it provides in terms of, for example, a reduced energy bill, and hence for these firms the subsidy would just be a windfall profit. For other firms the subsidy is not high enough to offset their net investment costs. For those firms the subsidy is not sufficient to induce them to adopt, and hence their behaviour is also unaffected by the scheme. Only those firms will be induced to change their decision and adopt the new technology for which the costs exceed their private benefits by an amount that is less than the subsidy. However, depending on the cost structure of firms in the industry, this third type of firm may actually be a small fraction of all firms in the industry, whereas the first type may be dominant. That means that a large amount of money is spent without inducing much additional investment. In combination with the observation that raising funds for subsidies is costly to society, the so-called windfall profits associated with subsidies may be substantial.

2.2 The Disadvantages of Subsidies Reconsidered

The efficiency arguments against the use of subsidies seem compelling. Indeed, it is typically thought that the only reason why subsidies are so widespread is because of distributional aspects or considerations concerning international competitiveness. Although the case against subsidies seems strong, some recent literature suggests that it is less strong than typically thought. Therefore, we first present some recent insights that challenge some of the arguments against subsidies mentioned in the previous section. Next we present some recent claims favouring subsidies because they would have additional welfare improving effects.

2.2.1 Challenging the traditional arguments

Some of the disadvantages discussed above can be mitigated, at least in theory, or they may be less problematic than suggested. First, several of the negative side effects of subsidies can be easily reduced, in particular in a general equilibrium setting. For instance, the undesirable side effect of subsidies inducing entry into the industry can be solved by using an additional policy instrument, such as quotas or output taxes (Eskeland and Devarajan, 1996). The subsidy stimulates adoption of cleaner production technology, and therefore makes dirty inputs relatively more expensive, causing substitution away from the dirty input. Quotas directly limit the aggregate level of activity, and output taxes do so indirectly by compensating the (relative) price reduction of the polluting good due to the abatement or investment subsidy. As has been shown by Fullerton and Wolverton (1999), such a two-part instrument exactly matches the incentive effects of a direct tax on pollution or waste. Note that these taxes also raise revenues that finance the subsidies and therefore reduce the negative effect of the marginal cost of public funds.[3]

Also the problem of windfall profits associated with subsidies can be addressed, at least in theory. Arguedas and van Soest (2009) show how governments may be able to separate those firms that require a subsidy from those that do not, even if the government does not have detailed information about each specific firm's production process or cost structure. Governments are able to do so by constructing a menu of combinations of investment subsidies and pollution taxes. In that case, firms that are most efficient in abatement will tend to choose those combinations that have high subsidies, but also relatively high pollution taxes. The reason is that for a one dollar increase in subsidies, the more efficient firms are willing to accept a higher marginal tax rate than less efficient firms, and still be better off. This is because taxes are paid over unabated pollution, and hence more efficient firms have a smaller tax base. As a result, for a given set of tax/subsidy combinations, the more efficient firms (weakly) prefer the tax/subsidy arrangements with the largest subsidies (and highest tax rates). If these firms are also those that actually need the largest subsidies to be induced to invest, the government is able to reduce the amount of money spent on subsidies as compared to the case of just introducing a single subsidy level. Therefore, the windfall profit argument against subsidies loses at least some of its bite if investment subsidies are used in combination with pollution taxes.

2.2.2 Welfare improving subsidies

Arguments have been put forward that actually favour the use of investment subsidies as compared to the alternative policy instruments, including abatement subsidies. One such argument is that the alternative instruments are not as effective as they may appear. Suppose the regulator imposes a pollution tax to induce adoption of more efficient abatement technologies. Upon adoption, the industry's marginal abatement cost function shifts down. Keeping the tax rate fixed, industry-wide abatement increases substantially as compared to the initial abatement level, and marginal environmental damage falls below marginal abatement cost – at least if the marginal damage function is increasing in the industry's pollution level, as is the case with most local pollutants. Therefore, the regulator has an incentive to decrease the tax rate *ex post*, to again equate marginal damage and marginal abatement cost. If firms know this in advance, the incentive to adopt is weaker than the initial pollution tax rate would suggest (Kennedy and Laplante, 1995).

Obviously, this so-called time inconsistency problem is relevant for all instruments that can be changed after firms have adopted the new technology, for instance, taxes, tradable emission permits, quotas, and abatement subsidies. But the advantage of investment subsidies is that they are sunk as soon as the firm adopts the new technology. The government may renege on the announced tax rates, but it is typically not possible to reclaim investment subsidies paid out in the past. Hence, investment subsidies are the only instrument that does not suffer from time inconsistency. This argument is spelled out in detail by Abrego and Perroni (2002). Indeed, to achieve the desired penetration rate of a new technology, the regulator should estimate the industry's marginal abatement cost function, find the level of pollution at which this function intersects the marginal environmental damage function, and infer the associated level of taxes or quotas. Announcing this level of stringency is credible, and the regulator then only needs to establish how much investment subsidies should be offered to implement the desired share of technology adoption in the industry under consideration.

Finally, there is increasingly more attention for the advantageous role subsidies may play when non-environmental externalities, such as knowledge and adoption externalities, are present as well. If only such positive externalities are present, the regulator actually wishes to expand such an industry, and hence subsidies are the appropriate instrument. However,

Newell et al. (2005, p. 169) argue that this conclusion carries over to a world with two interacting market failures, that is, negative environmental and positive technology spillovers. They argue that the rate of investment in such a technology is below the socially optimal level in cases where environmental externalities have not been fully internalized. Moreover, they maintain that it is unlikely that environmental policy alone creates sufficient incentives for first-best levels of adoption. Like in the case of the two-part instrument, however, it is essential to use investment subsidies not as a substitute, but as a complement to environmental policy.

3 THE IMPACT OF SUBSIDIES ON THE ADOPTION DECISION BY FIRMS

In the previous section we have argued that environmental subsidies no longer seem to be the ugly duckling in the set of environmental policy instruments. However, the available empirical evidence on the effectiveness and efficiency of subsidies in general and subsidy design in particular seems to support the economists' belief that subsidies are inefficient. Some studies report that the numbers of agents whose behaviour is not affected by the subsidy, tends to be high (for example, Malm, 1996), although the evidence is mainly confined to the demand side management (DSM) programme by electric utilities in the USA during the 1990s (see also Wirl 2000; Wirl and Orasch, 1998). Only Hassett and Metcalf (1995) provide some counter-evidence showing that energy-conservation credits given to households have been effective in stimulating the penetration of modern energy-saving technologies. Using much better (panel) data at the household level, they were able to reproduce previous results showing no, or even adverse effects. However, upon properly controlling for unobservable individual characteristics, the estimated effects suggest that subsidies have been effective after all. Nevertheless, Hassett and Metcalf (1995) also admit that their result adds little to the knowledge of the severity of the windfall profits problem associated with this subsidy scheme.

So we conclude that the issue whether subsidies can be an effective let alone efficient instrument in practice is still largely unresolved. Moreover, we conclude that this issue is likely to depend on how subsidies are designed in practice. Pigovian subsidies typically provide a positive price per unit of emissions saved, but subsidies usually have different designs in practice. Accordingly, we ask ourselves the question whether and how subsidy schemes should be designed to achieve maximum effectiveness per dollar spent. Before turning to two recent contributions to the empirical analysis of subsidy design and firm behaviour in the next section, we briefly review the standard view on the economic evaluation of investments and the role investment subsidies are supposed to play. Therefore, this section first explains how standard economics assumes that rational agents, like firms, organize the (economic) information (including the subsidy) on potential investments to choose between them. Next, we discuss some recent empirical evidence on how such decisions are 'framed' in practice.

The information used for evaluating investment decisions by firms is typically related to the cost and benefit streams associated with specific investment projects. According to economic theory, the choice of a particular investment project, like the adoption of an energy-saving technology, follows from an explicit comparison of the (expected) costs and benefits involved. After potential investments in energy-saving technologies are ranked according to their net present value (*NPV*), neoclassical economic theory suggests that any rational firm would invest in the technology with the highest *NPV*.

To illustrate, consider the case where the decision-maker in the firm contemplates implementing a specific investment project for which there are three alternative technologies available, labelled X, Y and Z. The minimum amount of (economic) information required about each technology i (= X, Y, Z) would include: its costs of adoption (usually at the beginning of the project, I_i), the (expected) life time of the technology (T_i), the user and maintenance cost during the (economic) lifetime of the project (M_{it}), the technology's contributions to output or the stream of benefits of the project (B_{it}), and the appropriate discount rate for this project r_i. Note as well that the agent may obtain some kind of subsidy (S_{it}) that, depending on its type and time of payment, adds to the (net) benefit of investment. Then the decision-maker would compute the *NPV* of each technology i according to:

$$NPV_i = \sum_{t=0}^{T_i} \frac{-I_{it} - M_{it} + B_{it} + S_{it}}{(1 + r_i)^t}, \; i = X, Y, Z \tag{9.1}$$

Her decision is then likely to be guided by the decision rule to select the alternative that generates the highest *NPV*.

According to this *NPV* framework, the (predicted) effect of a subsidy is rather straightforward. Defining the net adoption costs as the actual adoption costs minus the investment subsidy provided, technologies with lower net adoption costs are likely to obtain a higher ordering in the ranking of available alternative technologies. Therefore the subsidized technology will be adopted more readily than in absence of a subsidy. If firms are heterogeneous, for instance with respect to the technology currently in use or with respect to capital (or borrowing) constraints, the adoption

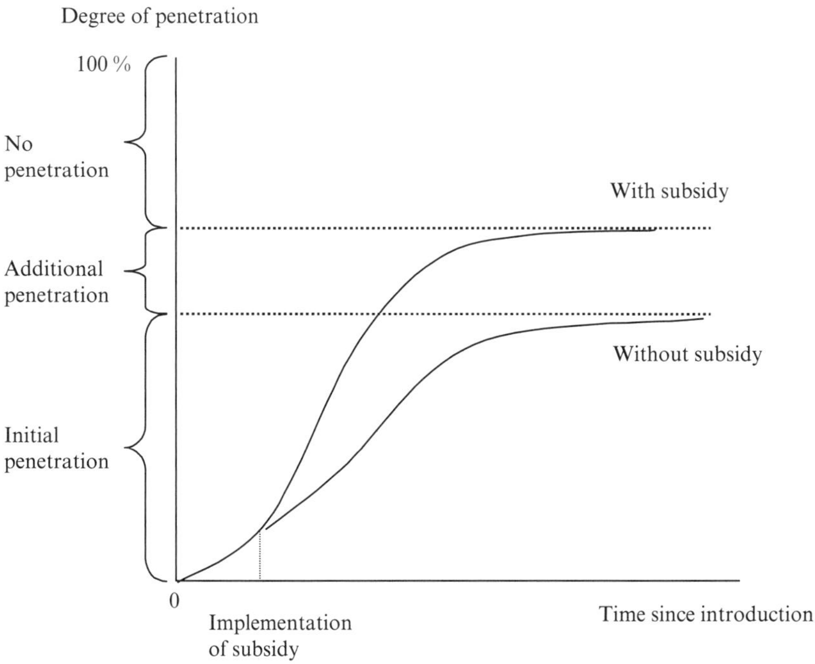

Figure 9.1 The effect of a subsidy on the penetration of a particular technology

of a particular technology is likely to follow a gradual pattern over time or a so-called penetration curve (see Figure 9.1). A subsidy will shift this penetration curve to the left, whereas the overall penetration will be lifted somewhat due to lower net cost of this particular technology. Note that this line of reasoning holds even if firms belong to different risk classes, that is, if they differ with respect to the discount rate they apply to their investment decision.

This logic applies for different subsidy schemes as well, even if schemes differ as to how they provide incentives for adoption. Although any scheme lowers the net adoption costs, the extent to which, and hence their effectiveness, depends on the type of investment subsidy scheme. For instance, incentives from an investment subsidy, a depreciation allowance or an investment tax credit are very different. First, an investment subsidy reduces the cost of investment up to a maximum of 100 per cent. This subsidy, however, might be provided as a lump sum or as a percentage of the investment. Moreover, if the full investment subsidy is obtained at $t = 0$, the subsidy is provided entirely 'up front'. In this case the *NPV* of the investment is reduced with the full amount of subsidy provided, and the net benefit stream is reduced with exactly the amount of the subsidy.

The effect of depreciation allowances or investment tax credits is quite different. These are facilities that reduce the user cost of capital by allowing for accelerated fiscal depreciation or a tax deduction (up to 100 per cent) at the time of the investment. Both the depreciation allowance and the investment tax credit work through a reduction of income or corporate tax payments. To understand how these facilities affect firm's investment decisions, we should first note that the income or corporate payment usually allows for deductions based on a so-called 'normal' or 'true' rate of depreciation, that is, the actual decrease in the machine's market value. Because such (second-hand) markets are often not well developed, governments use simpler procedures, like linear depreciation. For instance, the tax authority would allow one-fifth deduction for machines with a lifetime of five years, and one-tenth for machines with a ten-year lifetime. A depreciation allowance is a subsidy that, compared to this linear or other standard scheme, typically allows for accelerated depreciation up to 100 per cent in the year of investment. This yields a tax saving equal to the marginal tax times the amount of depreciation allowed earlier. Usually, firms may also choose the moment for application themselves, which allows them to use this facility at times when they make a real profit and have to pay tax accordingly. In this case the additional term in the *NPV* formula that captures the subsidy effect is equal to the tax savings to which it gives rise, and it is likely to differ across agents depending on the tax rate applied to them and their financial circumstances. An investment tax credit is also linked to these (income or corporate) tax payments, but typically allows for a reduction of the investment cost with some percentage and reduces tax payments accordingly.[4]

So far we have illustrated the effect of different subsidy schemes for investors using the *NPV* method assuming implicitly that all decision-makers only base their decisions on a clinical evaluation of the (expected) investment cost and benefits of adopting. However, this information only explains part of the behaviour of decision-makers in practice. It should be noted that still little is known about how precisely subsidies influence investment behaviour of (small) firms when they decide about investing in, or adopting a new technology. The recent rise of behavioural economics suggests that non-standard assumptions on consumer choice might be more important than previously thought (Bernheim and Rangel, 2005). The same is likely to hold for the behaviour of firms. Indeed, the actual evaluation of the flows of costs and benefits may well differ between decision-makers, depending on their norms, habits and personal characteristics (Tversky and Kahneman, 1981, p. 453). Recent evidence suggests that investors evaluate similar investment options differently depending on their own circumstances. In particular, the findings by DeCanio and Watkins (1998) are challenging. They report that for a similar project, a programme in green lightning, 'exogenous' characteristics of firms matter, even if one controls for differences in risk between firms (and sectors).

One source of variation that may result in different 'frames' is how the financial evaluation process itself is organized within firms. In practice, firms often do not follow the cost–benefit approach based on *NPV* considerations. Graham and Harvey (2001) provide evidence that firms may indeed differ as to how they make their investment decisions. They show that large firms rely heavily on present value techniques and the capital asset pricing model, while small firms are more likely to use the payback criterion. This criterion requires the calculation of the number of years after which all costs of the technology have been recouped, and then imposes that only projects with payback times smaller than a certain threshold are eligible for implementation. This implies that the timing associated with costs and benefits is important: projects that yield high benefits in the distant future are less likely to be adopted under a payback criterion than under the *NPV* rule. Aalbers et al. (2009b) provide additional evidence that such decisions indeed differ across firms, even within the same industry. They report that firms with, on average, a much smaller turnover than those in the Graham and Harvey (2001) sample, either apply the payback criterion or an even simpler evaluation scheme. Firms that do not use any formal capital budgeting method for an energy-saving investment project still rely on a rule of thumb that more or less mimics payback computations based on a simple comparison of investment cost with physical energy savings.

It is obvious that if investors differ with respect to the type of capital budgeting method they apply, it is very likely that they will evaluate the same investment project differently. However, other sources of 'framing' may exist as well, and they are likely to be more important if the evaluation is not checked using capital budgeting methods. This typically makes investors sensitive to other signals than traditional capital budgeting methods in deciding about their investments. For example, the decision-maker might be sensitive to the signal that a technology-specific subsidy may provide about the importance of a particular investment. In addition, subsidies may also draw the attention of decision-makers to the existence of (new) technologies, thus mitigating search costs. This demonstration effect of subsidy programmes, or alternatively, the attention value effect, may work like the effect of information dissemination for technology adoption.[5] Koomey (2000) estimates the demonstration effect of a tax credit on the adoption of specific technologies (that is, an effect that is independent of the size of the subsidy and only requires the provision of a subsidy per se) and finds that the signalling effect alone accounts for about 24–30 per cent of the adoption triggered.

Differences in technology characteristics may provide another source of 'framing'. Even if firms would frame their capital budgeting evaluation similarly, they may adopt different technologies because not all of the characteristics of these technologies could be subsumed in financial terms. Also production technologies differ and not all options fit equally well at different places. One well-known source of variation between technologies is whether they are add-on or integrated. Indeed, firms may evaluate an add-on technology that can typically be installed without much impact on its primary production process differently from integrated technologies that have a much larger effect on its primary production.

4 OPTIMAL SUBSIDY DESIGN IN PRACTICE

The previous section claims that subsidy design may affect evaluation of investment decisions by agents and therefore matters for the effectiveness and, by implication, the efficiency of using this instrument. It draws attention to the role of heterogeneity between firms and decision-makers, and suggests that firms may respond differently to the same incentive even if they are in the same factual circumstances. This section illustrates and discusses the implications of different subsidy designs for the variation in effectiveness and efficiency of subsidies given to firms. We discuss two recent empirical studies, one that evaluates a subsidy programme currently in place in the Netherlands (Aalbers et al., 2009b), and another aimed at identifying the impact of subsidy design in a more controlled environment (that is, using economic experiments; Aalbers et al., 2005, 2009a).

4.1 The Importance of Subsidy Design: Evidence from a Dutch Subsidy Programme

Aalbers et al. (2009b) evaluate the rents that accrue to firms adopting new energy-saving technologies by employing unique micro-data from a Dutch tax rebate scheme, launched in 1997. This scheme consists of two programmes aimed at stimulating investments in energy-saving technologies, the EIA (Energy Investment Credit)[6] for for-profit firms and the EINP (Energy Investments Subsidy in Non Profit sectors) for not-for-profit firms.[7] Firms investing in an energy-saving technology may apply for a tax deduction (EIA) or a subsidy (EINP) provided that the technology appears on the list of subsidized technologies, the so-called Energy

List. Once a year, this list is updated at the request of firms that supply energy-saving technologies. Criteria for inclusion are concerned with the fact that the use of the technology should result in a substantial reduction in the consumption of energy, and that the technology is not common. If technologies no longer meet these criteria, they are removed from the list. The Energy List differs for the EIA and EINP programmes, because not-for-profit firms typically invest in different technologies than for-profit firms. Indeed, only a small number of technologies, like insulation, energy-saving lighting systems and frequency converters, appear on both lists. The total number of techniques on the Energy List has varied from year to year. For example, the number of techniques on the EIA Energy List has varied from 80 to 110 between 1997 and 2001.[8]

It turns out that rents are related to characteristics that represent (potential) heterogeneity across firms, like differences in financial evaluation or background (profit versus non-profit), as well as characteristics of the design of the subsidy, in particular what type of technologies are subsidized. Using a bivariate probit model Aalbers et al. (2009b) analyse 862 subsidized investment decisions in 20 technologies across 57 sectors. Interestingly, they find that a considerable percentage of agents claim that the subsidy has had no effect on their behaviour, but *also* that the way in which firms make their investment decisions significantly affects the likelihood of this claim. More specifically, the extent to which energy-saving characteristics of technologies affect their adoption crucially depends on *how* firms evaluate these characteristics Firms indeed use different rules of thumb when evaluating investments decisions (see also Sarnat and Levy, 1969). In addition, Aalbers et al. (2009b) find that the way in which firms frame their investment decision is correlated with observable characteristics like the size of the firm and the sector in which they are active. Larger firms are more likely to use the payback period as a capital budgeting technique, but the use of the payback period also differs extensively between sectors. Agricultural and transport firms are much less likely to use a capital budgeting technique than are horticulture firms. Finally, technology characteristics are also important in the distribution of rents. Firms investing in technologies with high payback periods are less likely to be a free-rider compared with firms investing in technologies with low payback periods. Also, firms investing in expensive technologies are less likely to be a free-rider compared with firms investing in cheap technologies.

These findings illustrate that eligibility of subsidy schemes can be used as an instrument to influence their effectiveness and therefore their efficiency. The better the subsidy is targeted to 'marginal' firms changing their behaviour because of the subsidy, the smaller is the welfare loss of using subsidies to change investment decisions. Moreover, one would also expect differences in windfall profits across subsidy schemes as they affect investment behaviour differently.

However, the study of Aalbers et al. (2009b) is restricted to only one type of subsidy scheme. Moreover, this study only uses information about firms that applied for the subsidy and actually adopted energy-saving technologies. Therefore, no information is available about the firms that decided not to apply. It may well be that an inherent bias exists between firms that applied and those that did not apply. In the next subsection, we discuss a paper that is potentially able to address these two issues.

4.2 The Importance of Subsidy Design: Experimental Evidence

A second study by Aalbers et al. (2005) analyses the effectiveness of different subsidy designs using a controlled comparison of such schemes within and across different economic and technical decision environments (see also Aalbers et al., 2009a). By using an economic experiment, the authors are able to construct the decision environment of agents in such a way that they all essentially face the same investment decision under slightly different circumstances, such as differences in subsidy design or other characteristics that may affect the ultimate adoption decision. Moreover, the experimental setting is such that the decision-makers also face binding time constraints to mimic the decision situations that managers in small and medium enterprises find themselves in. In these smaller firms, there is insufficient managerial time available to both search for the best available technology, and, at the same time, follow the output market sufficiently close to maximize profits (DeCanio, 1998). Therefore, searching for a new technology is costly and potentially reduces profits.

Accordingly, subjects in the experiment had to make two decisions in every round of the experiment: either to search for a new technology (and if one is offered, whether they wish to adopt it), or to focus on the output market and to decide how much output they would like to produce given information about the state of the output market. Technologies can be ranked according to the amount of energy saved, as compared to a default technology. The price of energy is assumed constant, so that savings on the firm's energy bill increase linearly with the amount of energy savings of the technologies. Investment costs, however, go up with the amount of energy saved. Parameters were chosen such that the worst and the best technology in terms of energy savings have the lowest *NPV*, and technologies that yield an intermediate amount of savings have the highest *NPV*. Accordingly, the function of the *NPV* of all technologies is pyramid-shaped when technologies are ranked according to the savings they provide; see Figure 9.2, where *e* denotes the amount of per-period

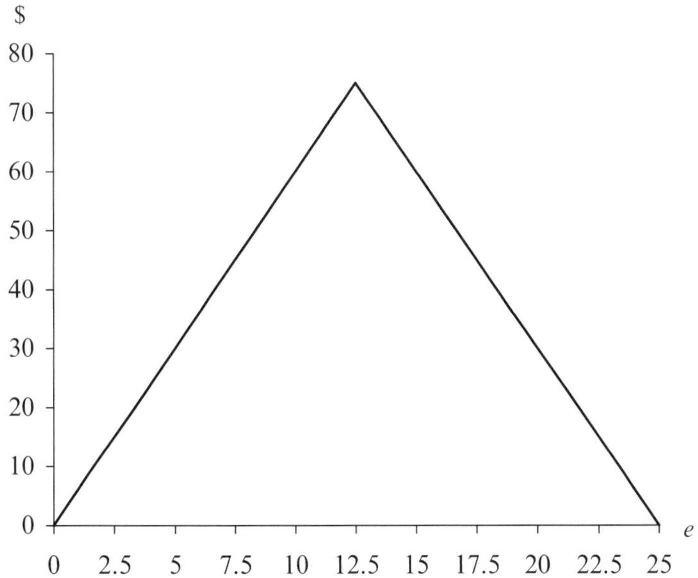

Figure 9.2 Net present value for different energy-saving technologies (e)

savings a technology yields. Note that the 'best' technologies from an environmental point of view are those that yield the highest per-period savings, which are also the most expensive ones. Indeed, one would expect firms to adopt technologies that yield intermediately large savings if they aim to maximize profits. When choosing to search for a new technology, subjects received a random technology offer, and subsequently were given the choice to purchase or not.

In each decision environment, subsidies were either absent or the most energy-saving technologies were subsidized. Variations in subsidy design studied include low versus high investment subsidies (6 versus 13 per cent), small versus large relatively very efficient technologies, and add-on versus integrated technologies. Comparing adoption decisions across treatments with and without subsidy as well as across different subsidy designs yields insight into the effectiveness of subsidies in general as well as in terms of their design. The analysis of Aalbers et al. (2005, 2009a) focused on differences across treatments in terms of the level of new technologies adopted, adoption speed (time needed to find a technology that the agent decides to adopt) as well as the actual physical energy savings of the chosen technology. Both students and firm managers participated in the experiments.

The experimental results confirm that subsidies indeed are highly effective. Introducing subsidies on the investment costs of the best technologies, in terms of physical energy savings, induces subjects to purchase technologies that yield higher savings than in the absence of such subsidies. This is the case for all treatments and irrespective whether subjects were students or managers. Concerning the subsidy level, the study reports an interesting difference between students and managers. Whereas the positive impact of a subsidy on technology choice is independent of the percentage of investment subsidies offered to students, managers are very sensitive to a higher subsidy level. So it seems that a higher subsidy level is indeed more effective in stimulating adoption in practice. Note, however, that its desirability does depend on the expected size of the windfall profits (see section 2).

Another important issue is the extent to which subsidies disclose information about the availability of energy-saving technologies. If the regulator offers a list of subsidized technologies, she basically signals agents to search for a limited number of technologies. This is an advantage for the agents because many low-efficiency technologies are thus excluded from the search, and hence agents save time and money. However, this list also excludes the technologies with the highest *NPV* (without subsidies) because they have intermediate levels of energy efficiency. So, apart from the financial incentive effect, providing a subsidy also has a signalling effect, that is, it affects the search method. Statistical analysis of two treatments where agents may either explicitly limit their search to technologies that are subsidized, or only learn whether the technology comes with a subsidy or not, reveals that more energy-saving technologies are adopted in the situation where subjects can specifically search for subsidized technologies.[9] This implies that the signalling effect of subsidies is in fact very important, and that the amount of subsidies provided is of secondary importance.

Having established that signalling is important, the next question is whether subjects can be induced to invest in even more energy-saving technologies by only subsidizing the very best technologies (the top 12 per cent in terms of energy savings), or by being less exclusive (by subsidizing the top 36 per cent). The results reveal that the more exclusive list is in fact most effective in inducing adoption of more energy-saving technologies. This is surprising as by being more exclusive, more of the intermediately efficient technologies fall outside the range of subsidized technologies, whereas these technologies do yield the highest private benefits in terms of *NPV*.

Related to the issue of optimal subsidy design is whether up-front investment subsidies are preferred to so-called abatement subsidies. This has been studied by comparing adoption decisions under a scheme where the subsidy is paid out upon adoption, and a scheme in which the subsidy is paid out in every period in which energy savings are achieved, as compared to the default technology. Interestingly, the two subsidy schemes were found to be equally effective in inducing adoption.[10] This is remarkable in the sense that if agents are at least to some extent risk averse, they would prefer the up-front payment to the abatement subsidy scheme. And if subjects use a payback criterion rather than an *NPV* analysis, we would also expect them to prefer the up-front investment subsidy.

A final issue is to what extent investment subsidies are more effective for technologies that affect the firm's competitiveness at the output market (by reducing marginal cost), or for technologies that do not affect competitiveness. So, the relevant comparison is between integrated technologies and add-on technologies. Aalbers et al. (2005) address this issue by again designing two situations, one where the new technologies merely yield a fixed stream of energy savings (that is, the add-on technology, such as double glassing or a more energy-efficient boiler), and one where the new technologies actually decrease marginal production costs, thus resulting in higher profits on the output market. The parameters were framed such that the expected discounted flow of cost savings were exactly the same, but subsidies were clearly more effective in case of integrated technologies. This implies, perhaps surprisingly, that subsidies are more effective in affecting the adoption behaviour of firms that contemplate replacing integrated technologies than the behaviour of firms that consider adopting a new add-on technology. The difference in effectiveness is, however, relatively small.

5 CONCLUSION

Subsidies as an instrument of environmental policy have not received a substantial amount of attention in the economics literature. One of the main reasons is the conviction of many economists that subsidies are dominated in terms of effectiveness as well as in terms of efficiency by measures such as environmental taxes and tradable permits. In this chapter, we have shown that this view only holds to some extent if the objective is to induce firms to adopt new, environmentally friendly technologies. Announced levels of stringency, in terms of tax rates, or total amounts of permits distributed, may not always be credible: as soon as a new technology is adopted on a fairly large scale, industry-wide emissions of polluting substances fall substantially, and politicians may be induced to set taxes below (or distribute permits up and above) the announced level. Investment subsidies do not suffer from such time-inconsistency problems, and hence may be preferable to other instruments. In addition, subsidies may be the appropriate instrument in the case where positive externalities with respect to adoption exist. For instance, efficiency of technologies might increase if agents obtain more experience with respect to their use and the externalities involved in this learning process justify subsidies. In addition, we argue that some of the traditional arguments against subsidies may be less damaging as they appear on first sight. Recent advances in economic theory suggest that the ideas that subsidies may lead to an increase in industry-wide emissions because they increase profitability of the regulated industry, and subsidies will be claimed by firms that would have adopted the subsidized technology anyway and thus obtain windfall gains, may have lost some of their bite.

Furthermore, the discussion of more recent empirical studies demonstrates the importance of a proper subsidy design, that is, the choice of the subsidy base and the rate. Indeed, it has been illustrated for a Dutch energy investment subsidy scheme that windfall gains may vary with the type of firm involved as well as with the financial characteristics of the technologies. This raises the question whether subsidies influence behaviour and to what extent this effect depends on the type of subsidy. Key determinants here are the level of subsidy, whether eligibility is directly associated with the investment itself or with the actual amounts of emissions abated, and whether a subsidized technology directly contributes to the firm's competitiveness. In an experimental setting it has been shown that subsidies matter, both because of their financial incentive and their help in limiting the firm managers' search for new technologies by eliminating the inefficient technologies. No differences in adoption have been reported as far as financially equivalent adoption and abatement subsidies are concerned. Finally, subsidies are marginally more effective in inducing adoption of technologies that are directly associated with the firm's production process as opposed to just being an add-on technology. Perhaps the fact that subsidies rank high on the list of most frequently used policy instruments in practice is less cumbersome than economists usually think.

NOTES

1. Lump-sum taxes are non-distortionary, obviously, but there are hardly any examples of such taxes in the real world.
2. See de Mooij (1999) for an overview of the double dividend literature.
3. In a subsequent paper, Fullerton and Wolverton (2005) also consider how a revenue requirement for the government affects the level of the tax-subsidy schemes. Interestingly, the negative welfare effect of financing subsidies is also somewhat reduced if one allows for interactions with other distortions. Parry (1998) has shown that in the

presence of a (pre-existing) labour tax, the cost of raising revenue to finance a (revenue-neutral) abatement subsidy is partly offset by an indirect tax-interaction effect. In the case of subsidies, this is the welfare gain from an increase in labour supply owing to the increase in the real household wage because the price of subsidized consumption falls.
4. The net effect of both the depreciation allowance and investment tax credit is only similar across firms if they do not face any differences in borrowing constraint, and their capital asset profile is similar. However, this is usually not the case in practice, and this causes considerable differences with respect to the user cost of capital across firms.
5. Note that such demonstration effects are absent in case of taxes or quotas, because they are targeted at pollution and do not provide information about the (relative) desirability of particular technologies to mitigate pollution.
6. In Dutch 'Energie InvesteringsAftrek' (EIA).
7. Dutch 'Energie Investeringsaftrek Non-Profit sector en bijzondere sectoren' (EINP).
8. The current Energy List (in Dutch) can be downloaded from www.senter.nl.
9. For this purpose, the subjects' adoption environment has been limited to settings where the subsidy percentages were equal and where the same technologies were subsidized.
10. Note that the parameters in the two treatments are such that the expected discounted flow of abatement subsidies was equal to the amount of subsidies received up front.

REFERENCES

Aalbers, R., E. van der Heijden, J. Potters, D.P. van Soest and H.R.J. Vollebergh (2009a), 'Technology adoption subsidies: an experiment with managers', *Energy Economics*, **31**, 431–42.

Aalbers, R.F.T., H.L.F. de Groot and H.R.J. Vollebergh (2009b), 'Rents from energy technology subsidies', mimeo, Netherlands Bureau for Economic Policy Analysis (CPB), The Netherlands.

Aalbers, R.F.T., E.C.M. van der Heijden, A.G.C. van Lomwel, J.H.M. Nelissen, J.J.M. Potters, D.P. van Soest and H.R.J. Vollebergh (2005), 'Naar een Optimaal Design van Investeringssubsidies in Milieuvriendelijke Technieken', OCFEB Studies in Economic Policy no. 15, Rotterdam: OCFEB.

Abrego, L. and C. Perroni (2002), 'Investment subsidies and time-consistent environmental policy', *Oxford Economic Papers*, **54**, 617–35.

Arguedas, C. and D.P. van Soest (2009), 'On reducing the windfall profits in environmental subsidy programs', *Journal of Environmental Economics and Management*, **58**, 192–205.

Baumol, W.J. and W.E. Oates (1988), *The Theory of Environmental Policy*, Cambridge: Cambridge University Press.

Bernheim, B.D. and A. Rangel (2005), 'Behavioral public economics: welfare and policy analysis with non-standard decision makers', NBER Working Paper 11518, available at: SSRN, http://ssrn.com/abstract=776006.

De Mooij, R. (1999), 'The double dividend of an environmental tax reform', in J.C.J.M. van den Bergh (ed.), *Handbook of Environmental and Resource Economics*, Cheltenham, UK and Northampton, MA, USA: Edward Elgar Publishing.

DeCanio, S.J. (1998), 'The efficiency paradox: bureaucratic and organizational barriers to profitable energy-saving investments', *Energy Policy*, **26**, 441–54.

DeCanio, S.J. and W.E. Watkins (1998), 'Information processing and organizational structure', *Journal of Economic Behavior and Organization*, **36**, 275–94.

Dietz, F.J. and H.R.J. Vollebergh (1999), 'Institutional barriers for economic instruments', in J.C.J.M. van den Bergh (ed.), *Handbook of Environmental Economics*, Cheltenham, UK and Northampton, MA, USA: Edward Elgar Publishing.

Eskeland, G. and S. Devarajan (1996), 'Taxing bads by taxing goods: towards efficient pollution control with presumptive charges', in L. Bovenberg and S. Cnossen (eds), *Public Economics and the Environment in an Imperfect World*, Boston, MA: Kluwer.

Fullerton, D. and A. Wolverton (1999), 'The case for a two-part instrument: presumptive tax and environmental subsidy', in A. Panagariya, P. Portney, and R. Schwab (eds), *Environmental and Public Economics: Essays in Honor of Wallace E. Oates*, Cheltenham, UK and Northampton, MA, USA: Edward Elgar Publishing.

Fullerton, D. and A. Wolverton (2005), 'The two-part instrument in a second-best world', *Journal of Public Economics*, **89**, 1961–75.

Graham, J.R. and C.R. Harvey (2001), 'The theory and practice of corporate finance: evidence from the field', *Journal of Financial Economics*, **60**, 187–243.

Hassett, K.A. and G.E. Metcalf (1995), 'Energy tax credits and residential conservation investment: evidence from panel data', *Journal of Public Economics*, **57**, 201–17.

Jaffe, A.B. and R.N. Stavins (1995), 'Dynamic incentives of environmental regulations: the effects of alternative policy instruments on technology diffusion', *Journal of Environmental Economics and Management*, **29**, 43–63.

Jaffe, A.B, R.N. Stavins and R.G. Newell (2003), 'Technological change and the environment', in K.G. Mäler and J.R. Vincent (eds), *Handbook of Environmental Economics*, Vol. 1, Amsterdam: Elsevier Science.

Jenkins, G.P. and R. Lamech (1992), 'Fiscal policies to control pollution: international experience', *Bulletin for International Fiscal Documentation*, **46**, 483–502.

Kennedy, P.W. and B. Laplante (1995), 'Equilibrium incentives for adopting cleaner technology under emissions pricing', World Bank Policy Research Paper 1491, Washington, DC: World Bank.

Kneese, A.V. and C.L. Schulze (1975), *Pollution, Prices, and Public Policy*, Washington, DC: Brookings Institute.

Koomey, J. (2000), 'Avoiding the "Big Mistake" in forecasting technology adoption', paper presented at the 8th International Energy Forum ENERGEX 2000, Las Vegas, 23–28 July.

Malm, E. (1996), 'An actions-based estimate of the free-rider fraction in electricity utility DSM programs', *Energy Journal*, **17**, 41–8.

Nelissen, D. and T. Requate (2007), 'Pollution-reducing and resource-saving technological progress', *International Journal of Agricultural Resources, Governance and Ecology*, **6**, 5–44.

Newell, R.G., A.B. Jaffe and R.N. Stavins (2005), 'A tale of two market failures', *Ecological Economics*, **54**, 164–74.

Organisation for Economic Co-operation and Development (OECD) (2006), Database on instruments used for environmental policy and natural resources management, http://www2.oecd.org/ecoinst/queries/index.htm.

Parry, I. (1998), 'A second-best analysis of environmental subsidies', *International Tax and Public Finance*, **5**, 157–74.

Price, L., C. Galitsky, J. Sinton, E Worrell and W. Graus (2005), 'Tax and fiscal policies for the promotion of industrial energy efficiency: a survey of inter-

national experience', paper LBNL-58128, Berkeley, CA: Lawrence Berkeley National Laboratory.
Requate, T. (2005), 'Dynamic incentives by environmental policy instruments – a survey', *Ecological Economics*, **54**, 175–95.
Sarnat, M. and H. Levy (1969), 'The relationship between rules of thumb to the internal rate of return: a restatement and generalization', *The Journal of Finance*, **24**, 479–90.
Stavins, R. (2003), 'Experience with market based environmental policy instruments', in K.G. Mäler and J.R. Vincent (eds), *Handbook of Environmental Economics*, vol. 1, Amsterdam: Elsevier Science.
Tversky, A. and D. Kahneman (1981), 'The framing of decisions and the psychology of choice', *Science*, **211**, 453–8.
Wirl, F. (2000), 'Lessons from utility conservation programs', *Energy Journal*, **21**, 87–108.
Wirl, F. and W. Orasch (1998), 'Analysis of United States' utility conservation programs', *Review of Industrial Organization*, **13**, 467–86.

10. Effectiveness of energy policies in the service sector

Kornelis Blok, Suzanne Joosen and Mirjam Harmelink

1 INTRODUCTION

Many governments have implemented policies to limit energy use in response to the threat of climate change. International agreements with quantified emission limitation and reduction targets for greenhouse gas emissions force governments to be critical about the actual effects of such policies. In addition, limited government budgets force governments to be cautious in terms of the introduction of policies associated with a substantial financial burden. Therefore, when evaluating climate policies, the two following questions have become important. First, are the policies effective, that is, to what extent do the policies lead to emission reduction? Second, are the policies cost-effective; in other words, do the policies lead to emission reduction at a level of cost that is acceptable to the government?

Policy evaluation in the area of energy and environmental issues is widely applied, but by and large it focuses more on the analysis of policy processes than on the determination of the quantitative impact of policies. In this analysis, the evaluation approach we select is the so-called theory-based evaluation (see, for instance, Rossi et al., 2004), an approach that has been widely applied for policy analysis in general, but as of now not very frequently in the context of the analysis of energy and climate policies (Blumstein et al., 2000). The main advantage of theory-based evaluation is that it goes through the full cause–effect chain. This makes it possible not only to determine whether policies are effective or not, but also, in the case where they are not effective, what the causes of failure are. Policy evaluation can take two different forms by focusing on an *ex post* (looking back) or an *ex ante* (looking ahead) assessment. Theory-based evaluation is most suitable for *ex post* analysis. The results of *ex post* analysis can be used to adapt and improve existing policies, but also for the design of new policies.

In this chapter, the above-mentioned questions will be discussed for energy policies directed towards the service sector in the Netherlands. The service sector includes both the tertiary and the quaternary sector, and hence commercial as well as public services are included. Policies directed at direct emissions (through fuel use) as well as at indirect emissions (through electricity use) are covered. The period of evaluation is 1995–2002. Joosen et al. (2004) provides a more extended coverage of this work, albeit in Dutch.

The organization of this chapter is as follows. In section 2 we describe the energy policies that have been targeted at and implemented in the service sector in the Netherlands. Methodological aspects of the method of theory-based evaluation are set out in section 3 for the various policy instruments that have been used. Section 4 analyses the impact of the energy policies, and results are presented for the various policy instruments in section 5. The overall impact is then provided in section 6. Section 7 contains conclusions, including a discussion of the results and associated policy recommendations.

2 ENERGY POLICIES FOR THE SERVICE SECTOR IN THE NETHERLANDS

The development of the consumption of natural gas and electricity in the Netherlands' service sector is depicted in Figure 10.1. More recent data are not available with the same amount of detail. Owing to the increase of energy use, carbon dioxide emissions caused by the Netherlands' service sector increased as well: from 24 million tonnes in 1995 to 26 million tonnes in 2001. These emissions include both direct emissions (for example, through the combustion of natural gas) and indirect emissions (for example, through the purchase of electricity and the associated emissions by power plants). The service sector emissions represent 11–12 per cent of the total carbon dioxide emissions in the Netherlands.

As part of the Dutch energy and climate policy, the following targets were set for the service sector:

- an energy efficiency improvement, which amounts to a reduction of average specific energy use, of 23 per cent between 1989 and 2000 (Ministry of Economic Affairs, 1993);
- a reduction of carbon dioxide emissions compared to a non-intervention level of 1.0 Mtonne in the year 2010, for building-related energy only with the base-year 1999 (Ministry of Housing, Spatial Planning and Environment, 1999).

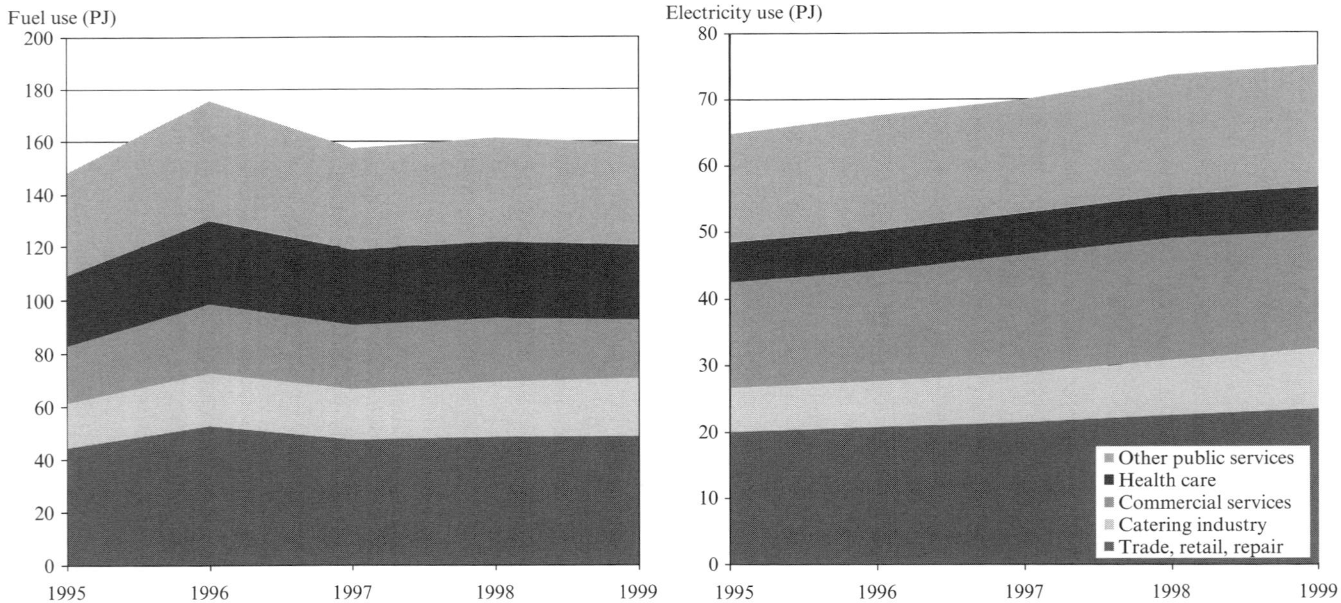

Note: The high fuel consumption in 1996 is due to temperature effects. Data after 1999 are not available at this level of detail.

Figure 10.1 Development of fuel and electricity consumption in the Netherlands' service sector from 1995–99

These targets cannot directly be translated to an energy use or carbon dioxide emissions reduction target for our research period. We will come back to this point later.

Also, note that both targets do not necessarily mean that emissions caused by the service sector should be reduced in absolute terms. In order to reach these targets, a variety of policy instruments has been put in place for the service sector in the Netherlands. An overview is provided in Table 10.1.

2.1 Policy Instruments with a Normative Character

There are three types of normative policy instruments, referred to as the Energy Performance Standard, long-term agreements and regulations based on the Environmental Management Act.

The Energy Performance Standard for new buildings was introduced in 1996. Up till then regulation existed for individual building components. These regulations still exist but they are now complemented with a standard for the integrated building. The Energy Performance Standard gives the building companies the opportunity to reach the overall target in a flexible way. The standard is set for most categories of buildings in the service sector. The requirements were made stricter in 1998 and 2000.

Long-term agreements on energy efficiency negotiated between the national government and industrial sectors were already introduced in the early 1990s (Farla and Blok, 2002; Rietbergen et al., 2002). Later on, these were extended to some sub-sectors within the service sector, specifically banks, insurance companies, supermarkets, higher education institutes and some individual companies, like the Dutch Railways and Schiphol Airport. These agreements followed more or less the same approach as those in manufacturing industry. They include quantified energy efficiency improvement targets and an annual progress monitoring system.

The new Environmental Management Act of the Netherlands (Wet Milieubeheer), introduced in 1993, allowed setting requirements for energy efficiency as part of the environmental permit system. The government has introduced so-called general administrative orders specifying such requirements for a number of sectors. Typical requirements are to take all energy conservation measures with a payback time of less than three or five years, depending on the type of measure.

2.2 Economic Policy Instruments

There are also three types of economic policy instruments. One type of economic regulation is related to the Environmental Action Plan, and the

Table 10.1 Overview of instruments of energy policy for the service sector in the Netherlands[a,b]

Instrument	Segment			Period								
	New buildings	Existing buildings	Appliances	1995	1996	1997	1998	1999	2000	2001	2002	2003–10
EPN												
MAP												
EIA/VAMIL												
EINP												
REB												
LTA												
WM												

Notes:

[a] The abbreviations for the policy instruments are derived from the Dutch name of the regulation. Their meaning is as follows:
EPN = Energy Performance Standard,
MAP = Environmental Action Plan of the Netherlands' energy distribution companies,
EIA/Vamil = Fiscal regulations for energy investments,
EINP = Energy Investment subsidy for Non-Profit organizations,
REB = Regulatory Energy Tax,
LTA = Long-term agreements, and
WM = Energy requirements through General Administrative Orders (Environmental Management Act).

[b] The table indicates during which period the policies were in force and whether they are valid for new buildings, existing buildings or (electric) appliances. If a box is shaded, this means that the instrument only applies to a part of that category. As indicated, MAP and EINP are no longer in force, Vamil is no longer applicable for energy investments, and REB is no longer regulatory and now called EB.

other two types are fiscal policy measures, taking the form of either fiscal discount regulations or a regulatory energy tax.

The Environmental Action Plan of the Netherlands' utilities was introduced in 1990. The energy distribution companies committed themselves to invest in energy efficiency and renewable energy. The activities were made financially possible by a small levy on the electricity and natural gas tariffs. The utilities provided subsidies for investments in energy conservation. In addition, a facility for cheap loans was introduced (the latter is not treated in this study).

A fiscal discount for energy conservation investments ('energie-investeringsaftrek', abbreviated EIA in Dutch) was introduced in 1997. The regulation is valid for companies that pay income tax or corporate tax. The benefit is equivalent to 15–20 per cent of the investment sum, depending on the tax regime that companies are in. In addition, companies can make use of the so-called Vamil regime (the abbreviation comes from the Dutch '*vrije afschrijving milieu-investering*') that allows for early depreciation of all investments with an environmental benefit. By applying early depreciation, taxation can be shifted to a later year. As the EIA is only valid for companies, in parallel a subsidy was introduced for non-profit organizations, which is referred to as the EINP. The subsidy rate varied between 14.5 and 18.5 per cent.

The regulatory energy tax was introduced in 1996. Large consumers of electricity and natural gas were exempted. The regulatory energy tax is especially relevant for the smaller consumers in the service sector. Tax rates were gradually increased. In the year 2002, typical tax rates most relevant for the service sector were 2 cents per kWh for electricity and 5.8 cents per cubic metre of natural gas,[1] which is equivalent to about €30 per tonne of carbon dioxide.

3 THEORY-BASED EVALUATION

The method of theory-based evaluation is not new and has been used numerous times to evaluate policies. The method is extensively described and illustrated by Rossi et al. (2004). The examples mentioned in Rossi et al. (2004) are, however, all outside of the energy sector. In the energy sector the method was used to design, evaluate and adapt 'market transformation' programmes in the field of energy efficiency in California (Blumstein et al., 2000). This example notwithstanding, the application of the theory-based method in evaluating energy efficiency policies has been limited and the method has so far not been applied in a systematic way in the policy implementation process.

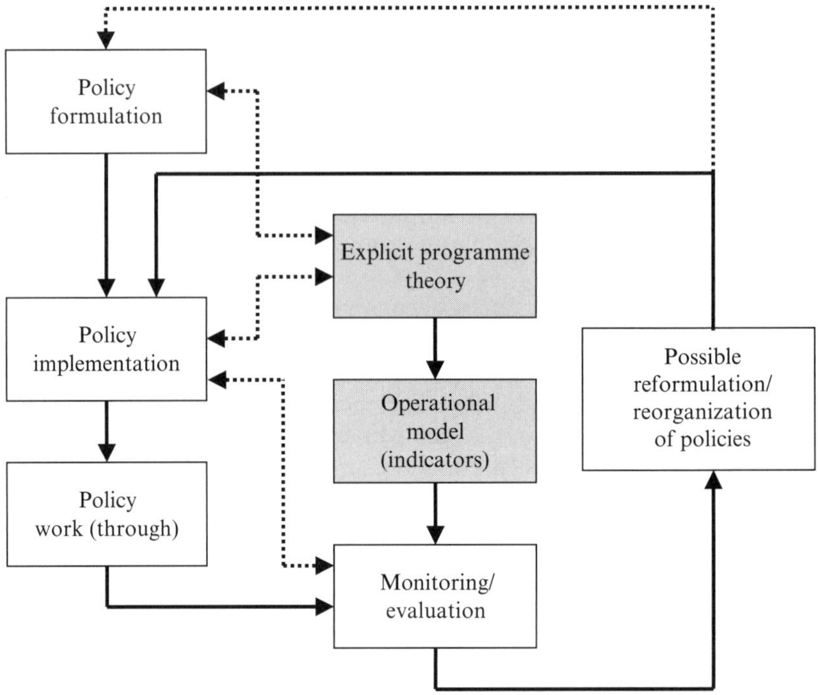

Figure 10.2 Outline of the policy cycle and the role of the programme theory in the policy cycle

3.1 General Principles

In practice, theory-based policy evaluation boils down to establishing a plausible theory on how a policy instrument (or a package of instruments) is expected to lead to energy efficiency improvements. Application of the theory-based approach in *ex post* policy evaluation means that the whole policy implementation process is unravelled to evaluate the effectiveness and cost-effectiveness of the different steps of the implementation process. The process of unravelling provides insight into where something went wrong in the process of policy design and implementation, and where the keys are for improving the effectiveness and cost effectiveness. See Figure 10.2 for an illustration of the role theory-based evaluation can play in the policy cycle.

3.2 Practical Framework to Evaluate Individual Policy Instruments

For the assessment of policy instruments the theory-based policy evaluation includes six different steps.

In the *first* step, the policy instrument is characterized. This is a description of the policy instrument including, among other things, targets, the period during which the policy instrument was enforced, target groups, policy implementing agents, available budget, available information on initial expected effectiveness and cost effectiveness of the instrument.

In the *second* step, a policy or programme theory is drawn up. A policy or programme theory includes all the assumptions with respect to the way policy instruments should reach their targeted effect. Sometimes the policy theory is clearly described in official documents and is well known by the policy-makers. In these cases we speak of an 'explicit policy or programme theory'. In most cases, however, the policy theory for a specific instrument is not clearly described and in these cases the programme theory is drawn up based on experiences with similar instruments and we speak of an 'implicit policy or programme theory'. Drawing up a policy theory in practice includes documenting all implicit and explicit assumptions in the policy implementation process, and mapping the cause–impact relationships, including the relationship with other policy instruments.

In the *third* step, the programme theory is translated to concrete (quantifiable) indicators. This means that for each assumed cause-impact relation an indicator is drawn up to 'measure' whether the cause–impact relation actually took place and to 'measure' whether the change (or part of the change) that took place is due to the implementation of the policy instrument. In other words, it is assessed whether the policy instrument was the causal force for change. This step also includes the development of the necessary formulas to calculate the effectiveness and cost-effectiveness.

In the *fourth* step, the cause–impact relations and the indicators are visually reflected in a flowchart.

In the *fifth* step, the policy theory is verified and if necessary adjusted. In step 2, the policy theory is drawn up with the help of available (official) documents or on experiences with similar instruments. In the fifth step the policy theory is verified through interviews with policy-makers and implementing agents and other actors involved in the implementation and monitoring of the policy instrument.

In the *sixth* and final step, available information is gathered and analysed to draw up the indicators, conclusions are drawn on the effectiveness and cost effectiveness of the policy instrument using the formulas and indicators, analyses are made to investigate the success and failure factors attributed to the analysed instruments, and recommendations are formulated to improve the effectiveness and cost effectiveness of the policy instrument.

4 RESULTS FOR LONG-TERM AGREEMENTS ON ENERGY EFFICIENCY

In order to explain the theory-based approach of evaluation we present detailed results for the policy instrument referred to as long-term agreements. For the other policy instruments, key findings are presented in the following section.

4.1 Characterization of the Long-Term Agreements

Within the framework of energy efficiency policy in the Netherlands, the long-term agreements (LTA) on energy efficiency were introduced in 1992. This resulted in several agreements with branch organizations and individual organizations in the commercial and non-profit building sector. The overall aim of the government was to reach an energy efficiency improvement in the building sector of 25 to 30 per cent (on average) in 2000 as compared to 1989 (see Novem, various years). The LTAs were actually a package deal. Companies joining an LTA received support from the energy agency Novem (now this agency is merged into SenterNovem) and could apply for financial support for investments in energy savings utilizing various subsidy schemes and fiscal policy measures.

The starting-point of the government was to negotiate agreement with each sector consuming more than one petajoule (PJ) annually. This resulted in long-term agreements with branch organizations and large organizations listed in Table 10.2.

4.2 Programme Theory and Translation to Indicators

The policy theory as well as the assumptions regarding the way in which policy-makers and executors expect the long-term agreements to reach the targeted effects are included in Figure 10.3. This cause–effect chain is derived from the following basic procedure that was followed for the LTAs:

1. The government introduces the long-term voluntary agreement on energy efficiency with the aim to reach an energy efficiency improvement of 25 to 30 per cent in the commercial and non-profit building

Table 10.2 Overview of energy efficiency targets and energy savings for the different sectors and firms included in the long-term agreements

Sectors	LTA target	Reference year	End year	Monitored year	Realized energy saving in monitored year	Gross energy savings[a] 1995–2002	Energy savings that can be attributed to the LTA[a] 1995–2002	
	%				%	PJ_{prim}	PJ_{prim} low	PJ_{prim} high
Supermarkets	32	1995	2010	2001	−3.5	0.0	0.0	0.0
Banks	25	1995	2006	2002	10.0	0.5	0.1	0.3
Dutch Railways	11	1997	2010	2002	8.5	0.2	0.0	0.1
Insurance sector	23	1996	2006	2002	3.4	0.1	0.0	0.0
Schiphol Group	28	1989	2000	1999	21.0	0.3	0.1	0.2
Royal Dutch Airlines (KLM)	28	1989	2000	1999	35.0	0.6	0.1	0.3
Vocational training sector	30	1994			No monitoring information available			
School of higher vocational education	30	1994	2006	2001	−1.0	0.0	0.0	0.0
Foundation of universities	14	1996	2006	2002	−1.9	0.0	0.0	0.0
Hospital health care	30	1989	2001	1999	8.0	1.3	0.2	0.7
Sport and recreation sector	25	1998	2010	2002	0.0	0.0	0.0	0.0
Total						2.9	0.53	1.55

Note: [a] In the case of negative savings it is assumed that the LTA did not lead to energy savings.

Source: Novem (various years).

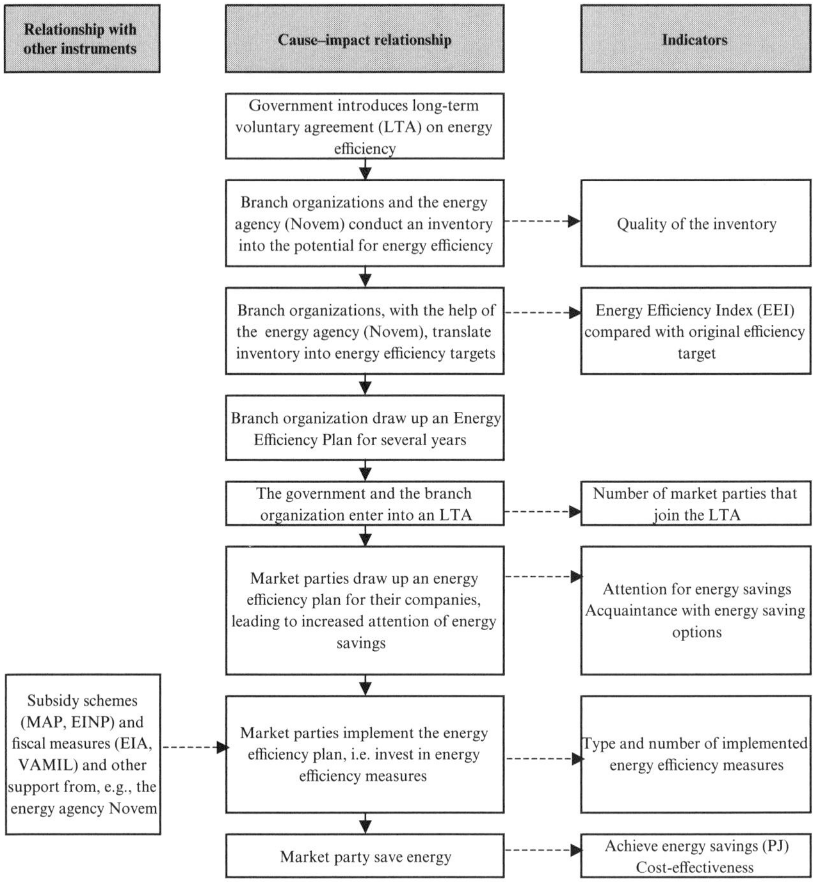

Note: The application shows how the long-term agreements (LTA) in the building sector should lead to energy savings and CO_2 reduction and indicators to 'measure' if assumed cause–impact relations actually occurred. Links with other policy instruments are indicated as well.

Figure 10.3 Policy theory application

sectors. Simultaneously, the government makes funds available to stimulate investments in energy-saving measures as part of the LTAs and to support the LTA process.

2. Branch organizations react by drawing up a report that includes an inventory of possible energy-saving measures and the energy-saving potential in their sector. The energy agency Novem supports branch organizations with this inventory.

3. The branch organizations, with support from Novem, translate the results of the inventory into quantitative targets for the sectors.
4. The branch organizations together with Novem draw up a long-range Energy Efficiency Plan (EEP), including concrete measures that can be implemented.
5. The government enters into a long-term agreement with branch organizations and individual firms.
6. Individual organizations and firms join the long-term agreements. They draw up an Energy Efficiency Plan for their individual organizations, leading to an increased attention to energy efficiency and insights on possibilities for saving energy.
7. Individual organizations and firms implement the Energy Efficiency Plan and save energy.
8. The individual organizations and firms monitor their energy use annually and report their energy efficiency index (EEI) to the energy agency Novem.
9. Novem systematically checks the submitted monitoring information and reports on the energy savings achieved at the sector level.

The LTAs are linked with several other policy instruments providing financial support for concrete investments. These were actually part of the package deal, and include subsidy programmes such as grants from the Environmental Action Plan (MAP) of energy companies, a subsidy scheme (EINP) and fiscal measures (EIA and VAMIL) established by the national government.

In Figure 10.3, the policy theory is summarized in the form of cause–impact relationships (middle column). Furthermore, the relationship with other policy instruments is included (left column) and in the right-hand side column the indicators are presented. The indicators are the factors that 'measure' whether the assumptions on the way the policy instruments would lead to changes actually occurred.

4.3 Evaluation of the Cause–Impact Relationships

Below, we give a detailed sequential account of the cause–impact relationships following the categorization in Figure 10.3.

4.3.1 Quality of the inventory

During the inventory possible energy efficiency measures were listed including good housekeeping, energy management, energy efficient heating, efficient lighting, insulation measures and efficient appliances. The starting-point of the inventory was that companies joining the long-term agreement should implement all measures with a payback time of less than five years. No information is available on the overall quality of the inventories, but during the interviews we did not get the impression that potentials were estimated either too high or too low.

4.3.2 Energy efficiency target compared with the initial overall targets of the government

Table 10.2 provides an overview of the targets that were agreed upon within the different long-term agreements at the sector or company level. These targets were based on the inventories executed per sector or company. Because the selected sectors and firms had a very diverse structure the sector-specific and firm-specific targets more or less deviate from the overall target of 25–30 per cent for the period 1989–2000. For some sectors, such as universities and Dutch Railways, the target is far below the overall indicative figures.

4.3.3 Number of parties that join the LTA

An analysis of the energy use of companies that joined an LTA shows that together these companies represent 15 per cent of the energy use in the building sector (Novem, various years; CBS, 2003). It is unknown which share of sectors using more than one PJ annually joined an LTA.

4.3.4 Attention to and familiarity with energy savings

The assumption was that through the introduction of LTAs in the built environment attention would be drawn to energy savings, and knowledge on energy-saving measures would increase. The Energy Efficiency Plan (EEP) at the firm level had to play a vital role in that respect. An evaluation of the LTA process in 2001 concluded that the LTAs have led to more attention for energy-saving measures at the company board level and among energy and environmental coordinators (Berenschot, 2001). The increased attention for energy-saving measures did not, however, lead to a higher priority for investment in energy-saving measures in sectors that entered into an LTA. Novem, for example, observed in the annual monitoring reports that, as firms and organizations are confronted with budget cuts, restructuring operations and sometimes high costs for new buildings and large-scale renovations, they gave energy efficiency measures less priority.

This was also visible in the very difficult process that led to an agreement on the monitoring methodology. Some sectors were very late in submitting monitoring information or even submitted no information at all.

4.3.5 Type and number of implemented energy efficiency measures

No public information is available on the type of energy efficiency measures that have been implemented by organizations that joined an LTA.

4.3.6 Achieved energy savings

Table 10.2 also provides an overview of the realized energy savings as they were monitored so far. It can be noted that the results are very diverse; one sector exceeded the target (Royal Dutch Airlines) whereas other sectors even showed negative savings. It is hard to attain a good overview of the achieved energy savings. As already mentioned some sectors did not deliver at all, or were too late delivering monitoring information. Furthermore, monitoring information is often not complete or evidently incorrect. For example, sectors did not always correct their Energy Efficiency Index for structural changes that took place. These included not correcting for longer opening hours of schools, changes in occupancy rates of hospitals or the increased use of information technology. This may be the cause for some sectors showing negative energy savings.

The reported energy savings cannot all be attributed to the implementation of the LTA. Part of the energy efficiency improvement would also have happened in the absence of LTAs, the so-called autonomous energy efficiency improvement. This autonomous efficiency improvement in the building sector was not systematically investigated. From research in the industry sector it is known that 20 to 50 per cent of the savings reported under LTAs can be attributed to the introduction of the LTA. In other words 50 to 80 per cent of the reported savings would also have been achieved in the absence of LTAs (Rietbergen et al., 2002). Assuming that these numbers can be applied to the building sector as well, primary energy savings in the period 1995–2002 achieved by the LTAs is 0.5 to 1.5 PJ, which corresponds to approximately 0.1 million ton of CO_2. It must, however, be noted that transferring results from the industry to the building sector must be interpreted cautiously because of the different structures of the sectors involved.

4.4 Cost-Effectiveness of the Long-Term Agreements

In order to determine the cost-effectiveness of the LTAs, government expenditures per unit of CO_2-reduction were determined. The government expenditures included capacity from Novem to support the LTA process, funding to support the execution of the LTAs (for example, to fund feasibility studies), and grants and fiscal support. Total government expenditure was estimated to be 53 to 76 million euro over the period 1995–2002, of which about 50 per cent was for subsidies and fiscal measures (Berenschot, 2001; Senter, 2003). This resulted in specific costs for the government of 60–190 euro per tonne of CO_2-emission mitigated. This shows that the cost-effectiveness of the LTA policy instrument in the building sector is not very good. We will provide a comparison with the cost effectiveness of other instruments in section 6 of this chapter.

4.5 Conclusions on Success and Fail Factors for the Long-Term Agreements

The LTAs in the building sector in the Netherlands have led to an increased attention for and knowledge on energy-saving measures. The LTAs, however, have not led to a higher priority in investments in energy-saving measures in sectors that entered an LTA. The low share of energy costs in overall costs, budget cuts for non-profit sectors, restructuring operations, high costs for new buildings and large-scale renovations resulted in a lower priority for energy efficiency measures. With some exceptions, this has led to disappointing results for the LTAs in the building sector in the Netherlands. The LTAs are far less effective and cost-effective than anticipated when they were introduced by the national government. The LTAs do not seem to add much to the generic instruments such as investment grants and fiscal measures available for the building sector as a whole.

5 KEY RESULTS FOR OTHER POLICY INSTRUMENTS

In this section we provide an overview of the effectiveness and cost-effectiveness for policy instruments except for the long-term agreements. The various policy instruments are listed in Table 10.1.

5.1 Energy Performance Standard

An important finding from the evaluation is that verification and enforcement is a weak point of the Energy Performance Standard. Permit requests were not sufficiently detailed to be assessed by the officials. Errors occurred regularly, and verification whether buildings were built according to the permit was virtually lacking. Nevertheless, the limited material available suggests that most likely the actual performance of the new buildings is close to that prescribed by the standard. This may be partly caused by the fact that initially the requirements of the performance standards were close to the then current building practices.

The actual effect of the energy performance standard is severely limited by the long time lag between permit request and building completion (on average nearly three years). There is only some effect of the introductory requirements in 1996, but the effect of the tightening of the standard in 1998 and 2000 is hardly or not at all visible yet in 2002.

5.2 Environmental Management Act

Municipalities can impose requirements with respect to energy use in the environmental permits of companies based on general administrative orders motivated through the Environmental Management Act. Note that this is only feasible for companies which are not participating in a long-term agreement on energy efficiency. Although the situation is improved compared to earlier research (Blok et al., 2004, pp. 65–70) the impact of these requirements is still very limited. About 80 per cent of the municipalities in 2002 include energy requirements in the permits, but in most cases these requirements are not binding and they are not very strict. Only 28 per cent of the permits are of good quality, and in only 10 per cent of the cases do the companies fully satisfy the requirements.

5.3 Environmental Action Plan

During the period 1995–2000 at least €75 million of subsidy was supplied from so-called MAP funds to the service sector. MAP refers to the Environmental Action Plan of the Netherland's energy distribution companies. The subsidy system did not tackle a wide variety of technologies in the service sector; over half the subsidies were for energy-efficient lighting systems.

5.4 Subsidy Schemes

To make use of a subsidy scheme it is first of all necessary that the scheme is known. This knowledge is moderate; for the EIA, it mainly depended on company size, ranging from 30–70 per cent for different categories of companies. Also, the knowledge on energy conservation possibilities is limited. For the service sector 70 per cent of the companies knew a range of options, but only 50 per cent had a good knowledge about the cost and benefits of the various options for their specific situation. The presence of a subsidy scheme had only a limited impact on the knowledge of energy conservation opportunities (that is, no 'attention effect'). Through fiscal and subsidy regimes a total of €450 million in energy investments in the commercial services and €725 million in the public services were financially

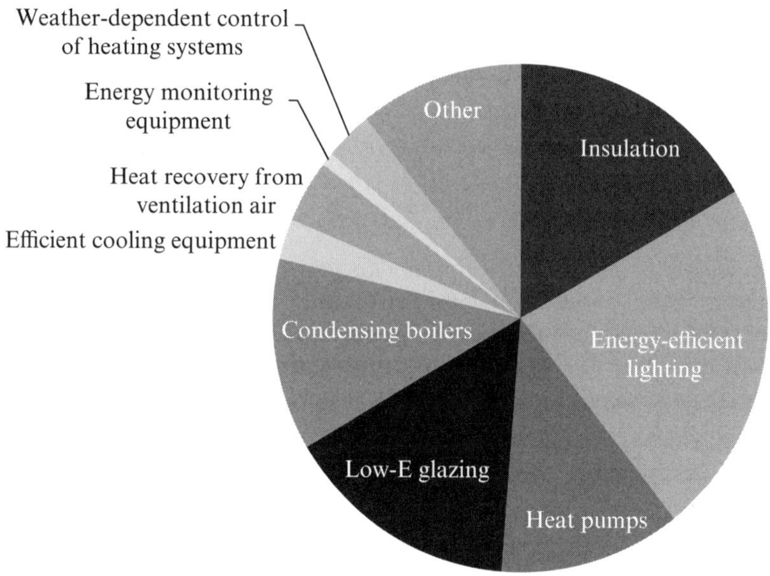

Note: This diagram only includes investments for which EINP or EIA subsidies were granted.

Figure 10.4 Distribution of investments in energy efficiency over the various technologies in the period 1997–2002 for the service sector

supported. The investments are fairly well spread across a range of different technologies (see Figure 10.4).

Critical for the effect of energy subsidies is the so-called free-rider effect. A free-rider is a firm or organization that takes the subsidy, but had planned to do the investment anyway. The share of free-riders depended on the type of technology and was estimated to range from 10–30 per cent for energy-efficient lighting systems, and between 30–60 per cent for condensing boilers and energy-efficient cooling equipment (de Beer et al., 2000; Blok et al., 2004, ch. 3).

5.5 Regulatory Energy Tax (REB)

Although a tax was introduced, the prices for electricity did not increase at all and the prices for natural gas hardly increased for the service sector (see Figure 10.5). The policy theory for energy taxes is complicated by the fact that there are various mechanisms that can lead to a variety of changes in

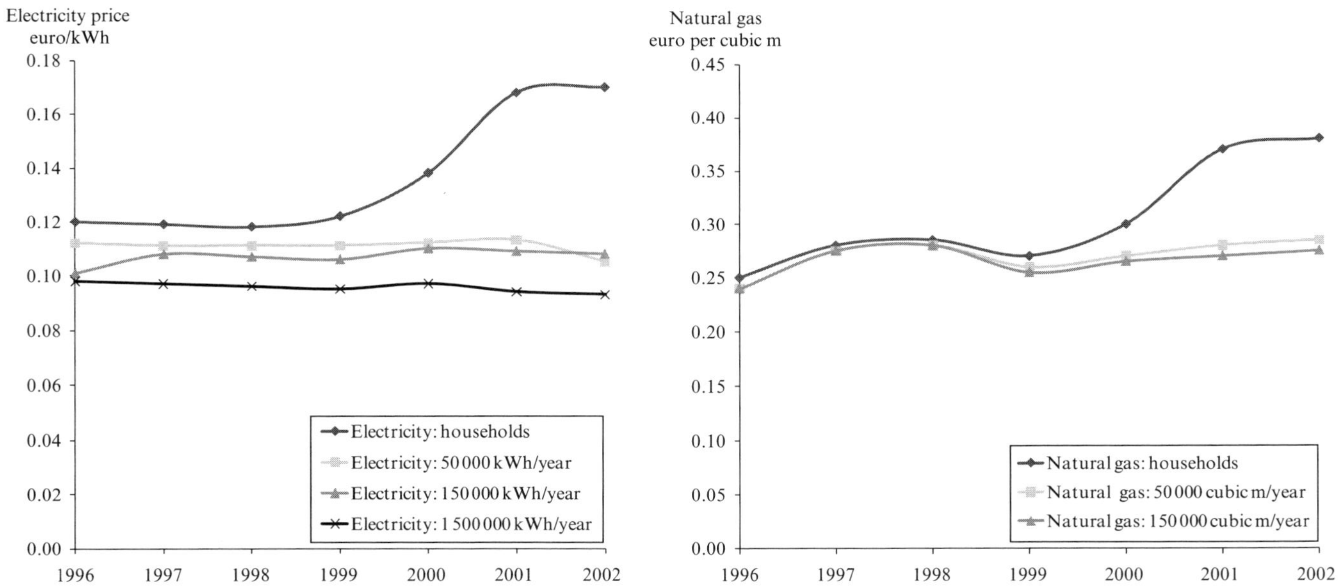

Note: For comparison, the electricity and natural gas prices for households are also shown.

Source: CBS (2003).

Figure 10.5 Prices in real terms (€$_{2002}$) for electricity and natural gas for service sector energy consumers with different volume sales per year behaviour (for example, buying more energy-efficient equipment, switching off lights more often, but also looking for a cheaper energy supplier). All these effects would be worth investigating, but this was beyond the scope of the current research project. Also, the changes in energy consumption caused by the energy tax are probably too small to be measured with standard econometric techniques. Therefore, the effect of the energy tax on energy consumption was estimated, using the assumption that the price elasticity for both electricity and natural gas will be between −0.02 and −0.10, based on Velthuijsen (1995), Koopmans et al. (1999) and EIM (2001).

6 OVERVIEW OF THE RESULTS FOR THE ENTIRE SERVICE SECTOR

The net effect of the various policy instruments is depicted in Figure 10.6. For each policy instrument uncertainty margins using a 95 per cent confidence interval are reported. The calculation of these uncertainty margins is done as follows. For all numbers that are required for the calculation of the net effect of a policy instrument (for example, the number of policy measures taken, the energy conservation per measure, the size of the free-rider effect) an estimate is made of the uncertainty range. Sometimes we were able to calculate the uncertainty range on the basis of the underlying data, sometimes we had to make our own assessment of the uncertainty range. Subsequently, the ultimate error in the net effect is calculated using standard error propagation rules.

The total carbon dioxide emission reduction in the service sector in the year 2002 amounts to 1.6 million tons, with an uncertainty range of 0.5–3.0 million tons (these aggregated figures are corrected for the overlap of the policy measures). The most important contribution comes from the various subsidy schemes (EIA/Vamil, EINP and MAP): these are estimated to make up nearly two-thirds of the total emission reduction achieved. Without the policies, emissions would have been 7 per cent higher, which implies that the rate of emission savings amounts to 1 per cent per year, with an uncertainty range of 0.3–1.9 per cent per year.

Of course, we ask ourselves how these emission reductions relate to the targets originally set. As mentioned above, the emission reduction targets were formulated in a different way and for different periods than the actual emission reductions we measured (see section 2 of this chapter). The first target set was an energy efficiency improvement of 23 per cent in 11 years.[2] Assuming an autonomous energy efficiency improvement of 1 per cent per year, this translates to an additional annual reduction of the energy

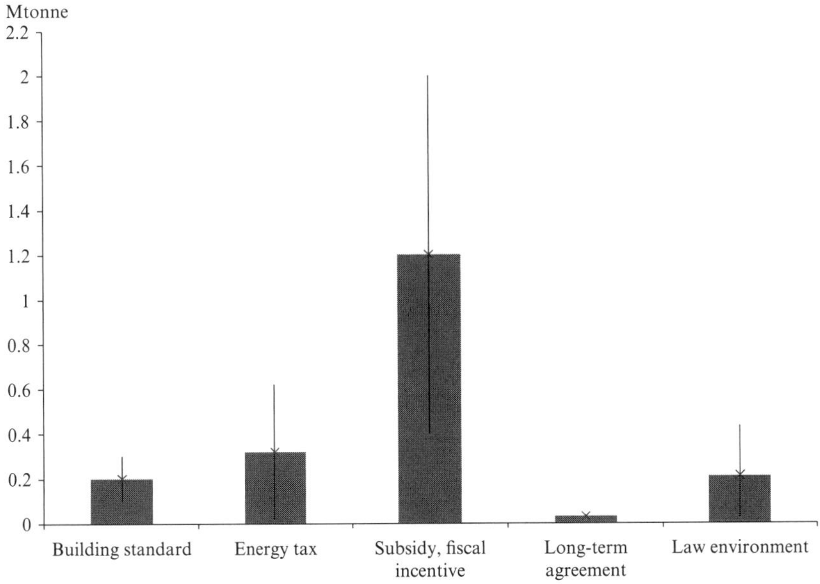

Note: The overlap of several measures is not taken into account for the individual measures. The error margins indicate a 95 per cent reliability interval in the end result.

Figure 10.6 Effect of the various policy instruments on carbon dioxide emissions in the service sector

consumption compared to the autonomous development of 1.4 per cent per year. The target of one Mtonne of emission reduction translates to an average emission reduction of 0.8 per cent per year. Our outcome for the emission reduction rate is around these targets. Given the uncertainty in our analysis, we can definitely not conclude that the target has not been reached.

Next to the effectiveness, we also investigated the cost effectiveness of the policies. To this end we made an overview of all the costs incurred by the government for the execution of the policies. The results are presented in Table 10.3. It turns out that building code regulations are relatively cheap for the government. As expected, investment subsidies are more expensive. A surprising result is that the long-term agreements are less cost effective than the subsidy schemes. For the industrial sector a different result was found (Blok et al., 2004, ch. 5). The disappointing cost effectiveness for long-term agreements in the service sector can be explained by two arguments. First, the long-term agreements in the

Table 10.3 Specific costs of the policies for carbon dioxide emission reduction in the Dutch service sector at the end of the study period in 2002[a]

Policy instrument	Specific costs of the policy instrument (€/tonne CO_2)	Cost categories included
Energy performance standard for new buildings	3–12	Financial support for transfer of knowledge and demonstration projects Maintenance cost for municipalities
EIA/Vamil (fiscal regulations for energy investments)	13–70	Personnel costs of energy agency Senter for support of the process Grants and fiscal support
Energy Investment subsidy for Non-Profit organizations (EINP)	8–50	Personnel costs of energy agency Senter for support of the process Grants and fiscal support
Long-term agreements (LTAs)	60–190	Personnel costs of energy agency Novem for support of the process Funding to implement LTAs (e.g., feasibility studies) Grants and fiscal support
Environmental Action Plan of the Netherlands' energy distribution companies	30–70	Energy companies' personnel costs for support of the process Grants and fiscal support

Notes: [a] Costs of policies were depreciated in 10 years against a social discount rate of 4 per cent. Prices in Dutch guilders with the price level of the year 2000 were converted to € at a rate of 2.2 guilders/€.

service sector, in terms of annual percent emission reduction, were less effective than those in the industrial sector. Second, sub-sectors in the service sector use a relatively small amount of energy as compared to sub-sectors of the industrial sector, and the programme costs are to a large extent fixed and only partly dependent on the size of the energy use. As a result, the programme costs per unit of energy use are higher for the service sector.

Note, that we did not provide figures on the cost effectiveness of the energy tax. In fact, the energy tax hardly causes costs for the government, but instead generates additional receipts. However, the introduction of the energy tax was accompanied by a reduction of other taxes (for example, the income tax and corporate taxes). Part of the energy tax income was also used to finance energy subsidy schemes. This makes it difficult to determine the net costs of the energy tax for the government.

One should also note that this line of reasoning is based on the perspective of the government only, and not those of the target group, nor those of society as a whole. An estimate of these two cost categories was also made and suggested that for both the benefits are larger than the costs. However, in this case as well the associated uncertainties are substantial (Joosen et al., 2004).

7 CONCLUSIONS

The analysis presented in this chapter used one consistent methodology, the method of theory-based evaluation, but for its use a substantive collection of data from primary sources was needed. These primary information sources, such as statistics, reports of government agencies and evaluations of individual policy instruments, were not set up along these methodological lines and often had completely different goals. In many cases, quantitative information could only be derived from such sources, but not without taking into account substantial uncertainty margins. These margins are reflected in the ultimate outcomes of our analysis as reflected in the previous section. They are often are larger than 50 per cent of the estimated effect of a policy.

It is tempting to compare our results with those found by Ramírez, Patel and Blok in Chapter 5 of this volume. These authors did not find evidence for a decrease of energy intensity in the Dutch service sector. This suggests that no energy efficiency improvements occur, although the effect of energy efficiency improvements might have been offset by a higher activity level (for example, more computers, more air conditioning). In our study, we find an additional improvement of energy efficiency, going beyond the unknown autonomous energy efficiency improvement. It should be noted that the period covered in Chapter 5 of this volume (starting between 1975 and 1985, and ending between 1997 and 1999) only partly overlaps with the period 1995–2002 that is covered in our study. Several of the policy instruments that we treated only became effective from 1996 or 1998 onwards.

One should also note that both in the work of Ramírez, Patel and Blok (Chapter 5 in this volume) as well as in our assessment, the uncertainties associated with the estimated policy impacts are substantial. We find that the main uncertainties are caused by the following data categories for which only information of insufficient quality is available: total building area of offices and other service sector buildings, energy use in sub-sectors of the service sector, actual energy use of new buildings, the actual performance and costs of energy-efficient equipment, and price elasticities of energy in the service sector.

The Netherlands government has applied a variety of policies to reduce carbon dioxide emissions in the service sector. We conclude that the combination of these policies has been effective in reducing carbon dioxide emissions. The biggest contribution to meeting the overall policy objectives comes from economic instruments (subsidies and taxes), but the use of normative instruments (standards and environmental permits) contributed to the results as well. The uncertainties associated with our policy impact analysis are, however, substantial. In order to reduce these uncertainties in future policy evaluations, it is necessary to have the necessary monitoring already in place while the policy instrument is in effect. This obviously requires upfront thinking about the appropriate evaluation procedure. In addition, target setting should be improved, particularly in view of making policy evaluations feasible and meaningful.

In 2003 the Netherlands government discontinued the Energy Investment subsidy for Non-Profit organizations (EINP). This implies a substantial disincentive to investing in energy-efficient technology, especially because within the service sector the subsidy scheme constituted a remarkably effective and also a relatively cost-effective policy instrument. Together with the fact that the energy tax will not have much effect, this implies that the future rate of energy efficiency improvement is likely to become substantially smaller. We expect the impact of alternative policy instruments (for example, the tightened energy performance standard for buildings) to be rather limited.

The interest in energy efficiency has revived in the Netherlands (Ministry of Economic Affairs, 2005), but it remains a crucial question as to how to increase future efficiency improvement rates, especially given the limitations on government budgets in economic downturns. One of the possibilities is to rely more on regulation. A building labelling system needs to be introduced as a result of the Energy Performance of Buildings Directive of the European Union. All buildings that are built, sold or hired out need to have an energy label that describes the energy quality of the building. This opens up possibilities to introduce new regulatory schemes associated with both the design performance and the actual performance of buildings. In addition, a scheme of tradable energy-saving certificates (the so-called 'white certificates') could be introduced on the basis of such a labelling system.

NOTES

1. These tax rates are valid for electricity consumption between 10 000 and 50 000 kWh and natural gas consumption between 5000 and 170 000 cubic metre of natural gas. The tax rates for households are two to three times higher.
2. We assume that energy efficiency improvement should be interpreted as a reduction of the average specific energy consumption.

REFERENCES

Berenschot (2001), *Evaluation Energy Program Services 1996–2000*, Utrecht: Berenschot.

Blok, K., H.L.F. de Groot, E.E.M. Luiten and M.G. Rietbergen (2004), *The Effectiveness of Policy Instruments for Energy-Efficiency Improvement in Firms*, Dordrecht: Kluwer Academic Publishers.

Blumstein, C., S. Goldstone and L. Lutzenhiser (2000), 'A theory-based approach to market transformation', *Energy Policy*, **28**, 137–44.

CBS (2003), Electricity, gas prices and price index for consumers, downloaded from StatLine on 25 August 2003, Statistics Netherlands (CBS), http://www.cbs.nl.

de Beer, J.G., M.M.M. Kerssemeeckers, R.F.T. Aalbers, H.R.J. Vollebergh, J. Ossokina, H.L.F. de Groot, P. Mulder and K. Blok (2000), *Effectiviteit Energiesubsidies: Onderzoek naar de Effectiviteit van Energiesubsidies en Fiscale Regelingen voor de Periode 1988–1999*, Utrecht: Ecofys.

EIM (2001), *National Energy Savings Monitor Building Sector 2001: Knowledge, Attitude, Behaviour and Penetration Degree of Energy Saving Measures in the Building Sector in the Netherlands*, Zoetermeer: EIM Business & Policy Research.

Farla, J.C.M. and K. Blok (2002), 'Industrial long-term agreements on energy efficiency in the Netherlands: a critical assessment of the monitoring methodologies and quantitative results', *Journal of Cleaner Production*, **10**, 165–82.

Joosen, S., M. Harmelink and K. Blok (2004), *Evaluatie van het Klimaatbeleid in de Gebouwde Omgeving 1995–2002*, Utrecht: Ecofys.

Koopmans, C.C., D.W. te Velde, W. Groot and J.H.A. Hendriks (1999), 'NEMO: Netherlands Energy demand MOdel: a top-down model based on bottom-up information', Research Memorandum 155, The Hague: Netherlands Bureau for Economic Policy Analysis (CPB).

Ministry of Economic Affairs (1993), *Vervolgnota Energiebesparing*, The Hague: Ministry of Economic Affairs.

Ministry of Economic Affairs (2005), *Nu voor later – Energierapport 2005*, The Hague: Ministry of Economic Affairs.

Ministry of Housing, Spatial Planning and the Environment (1999), *Climate Change Action Plan Part I: Domestic Measures*, The Hague: Ministry of Housing, Spatial Planning and the Environment.

Novem (various years), *Monitoring Reports Long-term Agreements in the Building Sectors*, Utrecht: Novem.

Rietbergen, M., J.C.M. Farla and K. Blok (2002), 'Do agreements enhance energy efficiency improvement? Analysing the actual outcome of long-term agreements on industrial energy efficiency improvement in the Netherlands', *Journal of Cleaner Production*, **10**, 153–63.

Rossi, P.M., H. Lipsey and H. Freeman (2004), *Evaluation: A Systematic Approach*, 7th edn, Thousand Oaks, CA: Sage Publications.

Senter (2003), *Information on the EIA, EINP and VAMIL*, Zwolle: Senter.

Velthuijsen, J.W. (1995), *Determinants of Investments in Energy Conservation*, Amsterdam: SEO Economic Research.

11. Energy model and policy advice: the effect of model choice

Willem J.H. van Groenendaal

1 INTRODUCTION

Models play an important role in policy evaluation and preparation. This chapter discusses if and what effects model formulation can have on the results of energy policy analysis. Reviewing the literature shows that a wide variety of models is used to evaluate energy policies. The type of model used for policy analysis normally depends of the type of question one wants to answer. However, energy policy problems, as most policy problems, are often complex and do not necessarily correspond to the variables used in a particular type of model. For example, the question on what pricing policy best promotes the use of particular energy-saving technologies is difficult to answer with an economic top-down model that utilizes price elasticities but does not identify these technologies. Bottom-up or engineering models that distinguish separate technologies, however, often lack the economic rigour to evaluate the economic implications of changes in prices appropriately. If each analysis leads to similar results there is no problem. However, if the results differ considerably we do have a problem.

For the formulation of its energy policy the Dutch national government may ask three institutes for advice: the Netherlands Bureau for Economic Policy Analysis (CPB), the Energy Research Centre of the Netherlands (ECN) and the National Institute of Public Health and Environment (RIVM). Each of these uses (among others) an energy model to formulate an advice. The CPB uses the Netherlands Energy demand MOdel (NEMO), a top-down model based on bottom-up information (Koopmans and te Velde, 2001; Koopmans et al., 1999), ECN uses SAVE, which stands for Simulation and Analyses of Virtual use in Energy scenarios (Daniëls and van Dril, 2007; van Dril et al., 1995). RIVM is developing the Model for Effectiveness of Policy Instruments for the Energy-Saving in Industry (MEI-Energy; van Wijk et al., 2001). Both SAVE and MEI-Energy, use detailed information on energy-saving options, but assess these options differently. MEI-Energy wants to be a tool that explicitly evaluates how the characteristics of policy measures affect the forces that drive energy use.

The three models are unique in the sense that all three are sector models using the technology options, and their direct cost and benefits, as reported in the ICARUS database (Michels, 2000) to evaluate investment options. Furthermore, all three are partial models, and economic growth and energy prices are determined outside the model. Each of the three models is equipped with particular diffusion mechanisms, which affect the policy analyses and thus the advice.

An interesting question is whether the government would receive the same advice from each of the three institutes when posing the same question. If so, the government can be sure that the advice has broad support among experts in the field of energy and economics, and has no reasons, other than political, to disregard the advice. However, if the policy analyses differ, the government has a problem, or maybe an opportunity, because she can use the advice that most suits her political agenda, or disregard all advices and do what suits her best for political reasons. A problem could be that, when available to the public, the opposition in parliament and/or non-governmental organizations (NGOs) pick another, conflicting, policy evaluation as compared to the policy option selected by the government.

In so far as policy advice is not dealt with by the government itself, but used by civil servants aware of the differences in policy preparation, there is a danger that they make use of these differences and ask advice of the institute that will suit their preferred policy best. If, on the other hand, the users are not aware of the differences in analysis, this may have serious consequences too. Policy based on the analysis of one institute could afterwards be reputed when evaluated by one of the others.

An experiment was performed to see if these dangers are real. Each of the three agencies was asked to independently evaluate three policy measures:

1. a CO_2-tax of €50 per ton,
2. a subsidy of 10 per cent on the investment cost of energy-saving technology, and
3. a command-and-control policy that requires companies to install energy-saving technologies with an internal rate of return (IRR) of 10 per cent or more.

These policies have been evaluated against a policy-free scenario in which energy prices grow by 2 per cent annually. The policies are evaluated for four sectors: food and beverages (SBI 15–16), textiles (SBI 17–19), other metal (SBI 28–35) and services (SBI 70–74).[1] All four sectors are energy-extensive sectors, since implementing policies for these sectors is less straightforward than for energy-intensive sectors, and the number of available energy-saving technologies is diverse. Contrary to energy-extensive sectors, energy-intensive sectors always pay attention to potential cost savings through energy saving. It is for energy-extensive sectors that the wide range of technologies available is not used to achieve energy savings. The main reason for this is that, since the savings in money terms are small, energy-extensive sectors need other than financial reasons to invest in energy saving (DeCanio, 1998; van Groenendaal, 1998b).

This chapter is organized as follows. Section 2 discusses different types of models used for policy analysis and their pros and cons. The type of analysis for which the different type models can be used will be discussed as well as potential drawbacks. This review is used in section 3 to identify what type of model each of the advisers of the Dutch government uses. The main differences and similarities of the Dutch models are discussed as well as their diffusion mechanisms. Section 4 uses the results of the experiment to compare the effects of the three policies for the four sectors analysed. Section 5 contains conclusions.

2 ENERGY DEMAND MODELLING

Energy demand modelling became popular after the first energy crisis in the 1970s when it became clear that cheap and sufficient energy was not obvious. Over the years different types of models, many of which are special-purpose models, have been developed (Bunn and Larsen, 1997; Kleinpeter, 1995). The type of model applied was closely related to the type of analyses one wanted to perform. In our case the goal is analysing the effects of policies to save energy.

A well-known distinction is that of models based on economic theory, for example, general or partial equilibrium models, as opposed to models with no explicit economic theory, such as vector autoregressive (VAR) models. Traditional structural or simultaneous equation (SE) models are somewhere in between in the sense that only parts of the model are normally based on economic theory and other parts are descriptive.

Another well-known distinction is that of top-down and bottom-up, where top-down models analyse aggregate behaviour without explicit formulation of energy technologies, whereas bottom-up energy models rely on detailed descriptions of production technologies (Hourcade, 1996). Because of the explicit characterization of the technologies used, bottom-up or engineering models better mimic the technical potential for energy savings as compared to models without explicit technologies.

The models do, however, require explicit long-term scenarios for existing technologies, as well as scenarios for the introduction of technologies that are already known, but have not yet reached the market. This implies that there is no room for new developments since all (future) technologies are explicit, and that the conditions for the introduction and growth of technology use are normally determined beforehand and outside the model. What varies is the speed with which technologies are diffused, which can (among others) be modelled using learning curves.

In the 1990s there has been a fierce debate about the inadequacy of the models used. This was a reaction to the ever larger econometric models built in the 1970s and 1980s. A good example is the HERMES modelling exercise (CEC, 1993). This type of very large structural equation type model was criticized as being inadequate because it did not take into account the Lucas critique, the most fundamental economic theoretical criticism on structural economic models, which states that any change in the rules of economic policy will have an impact on the behaviour of economic agents (Lucas, 1976). Therefore behavioural equations obtained by (then) standard econometric methods, such as least squares, will never adequately predict the result of policy changes.

Although the Lucas critique points at a serious theoretical flaw in many economic empirical models, the quantitative effect of this omission might not be that serious. For example, van Bergeijk and Berk (2001) show that the policy advice based on econometric analyses of key financial economic relations remained valid for many European countries despite a major institutional change, the introduction of the euro.

The Lucas critique has, however, led to changes in the way economic problems are modelled. One direction is general equilibrium (GE) modelling. General equilibrium models try to take into account all economic behaviour in an economic theoretically correct way. A second approach consists of vector autoregressive models, which are based on reduced-form equations and as such do not suffer from incorrectly specified economic behaviour, because they do not require any behavioural theory. Simultaneous equation models have some economic structure, but less stringent than (applied) general equilibrium models. They do, however, rely on statistical estimation methods to support equations, although data for sufficient statistical rigour are often lacking.

To analyse the long-term effect of energy policy on the Dutch energy sector, a model should allow for the evaluation of changes in policy, and the comparison of different scenarios over a time period of at least 20 to 40 years. Note that such a model is not a tool to get precise forecasts of energy use. Next, the three types of models distinguished above are discussed as well as their pros and cons.

2.1 General Equilibrium Models

General equilibrium or GE models are based on the rigorous theoretical economic concept of a multi-market equilibrium, where all markets clear due to price adjustments. The central idea underlying this work is to convert the Walrasian general equilibrium structure (formalized in the 1950s by Kenneth Arrow, Gerard Debreu, and others) from an abstract representation of an economy into a representative model of the actual economy. Empirically based general equilibrium models, also known as computable or applied general equilibrium (AGE) models, have been developed to evaluate concrete policy options. A well-known example is the ORANI project of MONASH University (see http://www.monash.edu.au/policy/oranig.htm). Shoven and Whalley (1992) describe all aspects of developing applied general equilibrium models, including developing an appropriate equilibrium structure, calibrating the model, compiling counterfactual equilibriums and interpreting results. They state that the Walrasian general equilibrium model provides an ideal framework for appraising the effects of policy changes on resource allocation, for assessing who gains and who loses, and that these policy impacts are not well covered by traditional simultaneous equation models.

Note that the econometric estimation of equations in AGE models has been replaced by calibration of the model. The typical structure of an AGE model is one of nested production functions of the Leontief, Cobb-Douglas and Constant (Relative) Elasticity of Substitution (CES) type (Adams et al., 1994; Naqvi, 1998). The coefficients used are not based on estimation, but (at best) on setting the parameters to meet the data in one or a few years, which is called calibration.

The functional forms used in many applied general equilibrium models have been seriously criticized by McKintrick (1998). Although correct from an economic theoretical point of view, the functional (and numerical) structure used in applied general equilibrium models strongly affects the results of policy simulations for small as well as large shocks. Based on a comparison between a constant elasticity of substitution based applied general equilibrium model and the normalized quadratic functional form, McKintrick concludes that better and more flexible functional forms for policy analysis in applied general equilibrium models need to be investigated. Also, the level of detail needs careful consideration (Fox and Fullerton, 1991).

A drawback of applied general equilibrium models is that they are not dynamic. They allow for a static comparative analysis only. For dynamic general equilibrium modelling a second type of general equilibrium model has been developed. A dynamic general equilibrium (DGE) model is based on a concise mathematical formulation of the long-term behaviour of all agents, and fully accounts for the effects of policy changes on model parameters. However, this type of model is not based on a complete description of the economy, but computes the equilibriums of fully specified but small artificial economies (Cooley and Dwyer, 1998). Many still see dynamic general equilibrium models as too stylized and too remote from fitting the data for actual policy analysis (Sims, 1996). Furthermore, dynamic general equilibrium models are also often calibrated instead of estimated, and they are therefore not based on a sound, complete and probabilistic assessment of model and data (Diebold, 1997).

2.2 Vector Autoregressive Models

Where general equilibrium models try to cover the economic theoretical criticisms on simultaneous equation models, vector autoregressive models originally ignored all economic theoretical aspects. Sims (1980) criticized simultaneous equation (and rational expectations) models because they rely on unrealistic identification restrictions. Traditional simultaneous equation models are often developed in parts. The restrictions introduced per part seem reasonable, but they become less reasonable when considering the complete system of equations. Furthermore, the assumptions on what is exogenous and what is endogenous are not based on economic theory.

Sims (1980) proposes to estimate models of the form:

$$C(L)\,Y_t = \mu_t \tag{11.1}$$

where $C(L)$ is a matrix polynomial with the identity matrix as the leading matrix, so the system is a reduced-form system, and the elements of μ_t are correlated, so the covariance matrix $\mathrm{E}(\mu_t\mu_t') = \Sigma_\mu$ is not a diagonal matrix. Equation (11.1) can be presented as a vector moving average model:

$$Y_t = C(L)^{-1}\mu_t \tag{11.2}$$

which can be used to calculate the impulse responses.

A clear distinction between the original vector autoregressive models and simultaneous equation models is the absence of exogenous variables. A vector autoregressive model is a closed system in which all variables are endogenous. This approach coincides with an important objection made by Sims (1980), specifically that the distinction between endogenous and exogenous variables in economic modelling is rather arbitrary. The system of equations given in equation (11.2) can be transformed via Choleski decomposition of the covariance matrix of the μ_t; that is, $T\mu_t = \varepsilon_t$ such that $E(\varepsilon_t\varepsilon_t') = T\Sigma_\mu T' = \Sigma_\varepsilon$, develops into a recursive system:

$$TY_t = TC(L)^{-1}\mu_t = \varepsilon_t + C^*(L)\varepsilon_{t-1} \tag{11.3}$$

where ε_t is white noise and the covariance matrix $E(\varepsilon_t\varepsilon_t') = \Sigma_\varepsilon$ is diagonal, T is the lower triangular matrix from the Choleski decomposition with ones on the diagonal.

An important difference between the simultaneous equation approach and the vector autoregressive approach is the interpretation of the error process. Where in the simultaneous equation approach these are interpreted as errors in equations due to the neglect of non-essential variables with minor influences, in vector autoregressive models the error terms are called structural innovations and are often considered to drive the stochastic dynamics of the equations (Gottschalk, 2001).

A problem many researchers have with the vector autoregressive approach is its a-theoretical character, resulting in impulse–response analysis that has no economic interpretation. As such this approach is hardly fit for policy analysis. Furthermore, the Choleski decomposition is not unique and leads to an arbitrary ordering of the variables and the results thus obtained can be sensitive to the ordering (Gottschalk, 2001).

This has lead to the development of the structural vector autoregressive (SVAR) approach. Instead of using a-theoretical restrictions only, some restrictions are substituted by a priori restrictions on the covariance matrix of the orthogonal errors and/or restrictions on the impulse response based on economic theory (Cooley and Dwyer, 1998).

Another development that aims at linking vector autoregressive and structural models is based on vector error correction models (VECM). The vector error correction model can be obtained by rewriting equation (11.1) as:

$$\Delta Y_t = \sum_{j=1}^{n}\Gamma_j\Delta_j Y_{t-j} + \Pi Y_{t-1} + \mu_t \tag{11.4}$$

where $\Delta Y_t = Y_t - Y_{t-1}$ and Π is the matrix representing the long-term relationship (Hamilton, 1994, p. 580). The matrix Π can be decomposed into two matrices (say) α and β, where $\beta' Y_t$ is the matrix of cointegrating vectors and α the matrix that shows how deviations impact other variables. As has been shown by Johansen and Juselius (1994) equation (11.4), which is an unrestricted reduced-form model, can, through identifying restrictions on long- and short-term reduced-form parameters, be interpreted as a representation of a structural model of the form:

$$A_0 \Delta Y_t = \sum_{j=1}^{n} A_j \Delta_j Y_{t-j} + a\beta' Y_{t-1} + \eta_t \tag{11.5}$$

with $A_j = A_0 \Gamma_j$, $a = A_0 \alpha$, $\eta_t = A_0 \mu_t$ and $\Omega_\eta = A_0 \Sigma_\mu A_0'$. The long-term parameters β are unchanged. The main problem remains the choice of A_0, for which there are several options (Thomas, 1997).

As this discussion shows, the vector autoregressive approach can be translated into terms of structural models. In energy modelling, vector autoregressive and vector error correction models are not (yet) widely used. Examples are Hondroyiannis (2004), and Cropton and Wu (2005). The main drawback remains the fact that due to data and estimation problems the models remain small (less than 10 variables), whereas energy policy analysis is concerned with many (interacting) sectors and/or many technologies.

2.3 Structural or Simultaneous Equation Models

Structural or simultaneous equation models have been popular for a long time and were widely used in the Keynesian tradition. Despite many criticisms they are still widely used in applied work. This type of model is called structural economic model or structural equation system, because they are based on models of consumer, producer and other economic actors' behaviour. Others, with the statistical aspects of the model in mind, prefer to refer to simultaneous equation models (Jacobs and Wallis, 2002). This latter view does not require the model to be based on explicit economic theory. We prefer this latter interpretation, since it better covers the approach, which not necessarily requires rigorous economic modelling. Many of the simultaneous equation models used for policy support were (and are) a mixture of partial theoretical and ad hoc explanations, that is, partly based on economic theory and partly more descriptive with some economic common sense as a foundation for the equations in the model.

With the growth of computer capacity macro- and meso-econometric models for policy analysis became very popular, but due to data problems the statistical support was often flawed. Systems with thousands of equations were developed to support policy analysis. An example is the HERMES-model developed by the European Commission (CEC, 1993). The economic theoretical foundation of this type of model is, compared with general equilibrium models, incomplete. Many of the models used are, despite their size, partial models and can be qualified as ad hoc or descriptive models that, from an economic theoretical and often also from a statistical point of view, are incomplete.

The general form of a linear simultaneous equation model, which is the most widely used model, is:

$$A(L)\,Y_t = B(L)X_t + \varepsilon_t, \tag{11.6}$$

with $Y_t \in R^n$ the vector of endogenous variables, $X_t \in R^m$ a vector of exogenous variables, and $\varepsilon_t \in R^n$ a vector of random disturbances with $E(\varepsilon_t) = 0$ and $V(\varepsilon_t) = \Sigma_\varepsilon$. $A(L)$ and $B(L)$ are matrix polynomials in the lag operator L (for example, $LX_t \equiv X_{t-1}$), and A_0 is normalized such that the diagonal elements are all equal to 1 and the matrix has full rank. It is easy to see that equation (11.5) is a restricted form of equation (11.6).

With the Lucas critique in mind, methods to improve structural models have been developed as well. The vector error correction model mentioned above is one. Note that the error term in equation (11.6) is interpreted as being the result of neglected variables (Theil, 1971, p. 104). If the model is formulated in line with economic theory these effects will be small and random, hence they can be neglected. This is in contrast to (structural) vector autoregressive models that have a different view on the role of the disturbance term in economic modelling. In terms of equation (11.6) the Lucas critique means that part of the parameters in the simultaneous equation model, for example those of the private sector, are not independent of those of the other sectors, say the government. As a result these parameters might change in case of new government policies.

2.4 Simultaneous Equation Energy Model Specifications

A large number of models for energy demand have been developed and used for either partial analysis or as part of a more comprehensive simultaneous equation model. We discuss the most widely used specifications below.

2.4.1 Expenditure systems

To model final demand by the household sector a large number of demand models based on utility maximization subject to a budget constraint have been developed. This line of reasoning started with the Stone-Geary model based on a Cobb-Douglas utility function and a linear budget constraint. Since then many models have been developed, of which Theil's Rotterdam model (Theil, 1965), and Deaton and Muelbauer's Almost Ideal Demand System (AIDS; Deaton and Muelbauer, 1980) are perhaps the most well known. A wide variety of related models have been developed, see for example Neves (1994), and various forms have been tested (Tridimas, 2000).

For the analysis of energy policy this approach has some disadvantages. First, energy is only a small fraction of consumer expenditure and small fractions tend to be poorly estimated. This problem is amplified when, instead of using the general category energy, different forms of energy are introduced. This disadvantage can be partially remedied by using an approach based on nested demand functions, where total energy demand is determined first, and subsequently the issue of how total demand is met by different forms of energy using relative prices. Expenditure systems require, however, detailed data over a longer period. Furthermore, this approach does not identify energy use in the productions sector, whereas many policy measures aim at changes in this sector. Arguably, this approach is less useful for sectoral analysis.

2.4.2 Production and cost functions

For the manufacturing (sub-)sector a large number of models has been developed as well. Two important and related approaches are the production function approach and the cost/profit function approach.

Production functions are a powerful tool to model factor demand. Normally the production function is not used directly to derive factor demand, but factor demand is based on the assumption that producers (or in a macroeconomic context, a representative producer) minimize cost or (less widely used) maximize profit. Using Shephard's lemma, share equations can be derived which can be estimated. For a comprehensive discussion of this approach we refer to Fuss and McFadden (1978). Since then many contributions and more flexible forms have been introduced. In applied work, the translog approach is the most popular. Note that the mathematical formulation of production functions and their possibilities for a nested approach to model demand is also used in applied general equilibrium models; for an example see Naqvi (1998).

2.4.3 Energy intensity

Another approach that is popular in energy demand modelling is the use of energy intensity, that is, energy use divided by real gross domestic product. Energy intensity can be analysed on the macro, sector or sub-sector level, and be related to economic growth through the development in real gross value added. For the development of future demand the adjustment of the intensities can be modelled as a function of energy prices and technology (van Groenendaal, 1998b). From an economic theoretical point of view these types of models can be regarded as ad hoc models, since they are not based on integrated optimizing behaviour as the previous two approaches are. The decision to invest in an energy technique, for new investments as well as retrofit, is however, based on a comparison of the net present values (NPVs), or any other financial criterion, of the technologies available. The technology chosen is the one with the largest NPV. This approach requires detailed information on the costs of various technologies, which is often not available. Van Groenendaal (1998b) uses information obtained through energy audits of over 240 production facilities in different sectors of the economy. The audits were used to formulate an average production process and opportunities for improvement based on various fuels. With respect to the use of detailed technology descriptions and options for improvement, this type of model could also be classified as a bottom-up model.

2.4.4 Fuel demand models

In a fuel demand model, the demand for a fuel is modelled as a function of the economic activity, its own real price, the real prices of its main competitors, and other explanatory variables considered relevant. This approach can be applied for a single fuel as well as for several fuels (Belhaj, 2002; Chow, 2001). The main difference between this approach and those based on explicit economic theory is that the explanatory variables are introduced on an ad hoc basis, although they are similar to those in derived demand equations.

2.5 Bottom-Up Models

Bottom-up or engineering models are normally large, technology-oriented models based on sector or even sub-sector activities. They use detailed descriptions of energy consumption in production and examine technological options for energy saving and fuel switching. Fuel switching or incremental investment in energy saving is based on direct cost and benefit information of the available technology options. The evolution of technologies is made explicit together with the likelihood of retrofit and replacement.

Bottom-up models often use this information to optimize investments in energy technology for a long time horizon, using exogenous information on prices of energy and economic growth. As such they result in an optimal investment schedule for all sectors, taking into account all technology and cost information. The most well known example is the MARKAL model (Fishbone and Abilock, 1981), which is a process-oriented dynamic programming model of the energy system, and is applied in many countries and has been augmented in many directions. The MARKAL family of models is currently used, among others, to construct emission policies; that is, an energy system that will keep emissions within targets.[2]

2.6 Pros and Cons of Different Models

Each of the models described above has its pros and cons, and which one is most appropriate depends on what one aims to analyse. Ideally, developing and evaluating an energy policy should be based on analyses with several of these model types and the results would hopefully converge. This is, however, not common practice.

Static general equilibrium models have economic theoretical depth in the sense that they are consistent from an economic theoretical perspective, but they are based on a very liberal view with respect to statistical methodology. The functional forms of the larger models are normally simple, such as Cobb-Douglas or constant elasticity of substitution. The policy analysis that can be performed with static general equilibrium models is of a comparative static nature. As such these analyses do not result in information about the path that will lead to the result.

The empirical foundation of static general equilibrium models is often weak and mostly based on a fit of cross-section data or picking parameter values such that the simulated values for one or a few years resemble the actual data. The models are not statistically confronted with the data as (structural) vector autoregressive and simultaneous equation models are.

The dynamic general equilibrium modelling approach seems promising, but it is not suitable for the purpose of economy-wide policy analysis, since it is still in its development phase and has only been applied to artificial and small economies. From a theoretical point of view a properly formulated structural model, such as a dynamic general equilibrium model that accounts for all the effects of policies on behavioural parameters, is difficult to estimate and even the simple models are often rejected by the data (Heckman, 2000).

A clear advantage of causal models, as well as simultaneous equation models despite their theoretical inadequacies, over bottom-up models is the fact that behaviours and prices are normally endogenous to the model. In the case of dynamic structural models this implies feedback within the system, which is absent in most bottom-up models. Bottom-up models tend to concentrate on sector-specific analyses looking at detailed technology trends, but ignoring economy-wide effects. To overcome these shortcomings separate macroeconomic models have been added, for example MARKAL-MACRO. As a result of the differences, top-down models are better suited to predict actual outcomes and bottom-up models are better to simulate the effects of changes in historical patterns of technology use.

The empirical evidence for the direct cost estimates used in bottom-up models tends to be weak and the cost development is completely or to a large extent exogenous to the model when simulating future developments. Top-down models rely on more aggregate information that is captured by substitution through relative prices and autonomous efficiency improvements. They do not allow for detailed analysis of technology use in (sub-)sectors.

A (structural) vector autoregressive model, especially when the number of endogenous variables is large, requires a large number of covariance and/or economic restrictions to identify the model. In the case of n variables, $n(n - 1)/2$ identifying covariance restrictions are required for a complete identification (Garratt et al., 2000). These restrictions are difficult to obtain and it is unclear how they are related to economic theory.

Cooley and Dwyer (1998) analysed the exact nature of the identifying restrictions in a structural vector autoregressive approach and compared these with those of a completely specified economic model. They argue that the structural vector autoregressive approach uses a mixture of theoretical and a-theoretical restrictions. From an economic point of view, the a-theoretical assumptions are concerned with the type of (non-)stationarity of the data, which can be tested, although these tests lack power. The assumptions are, however, crucial in a structural vector autoregressive approach (Cooley and Dwyer, 1998). The a-theoretical assumption that the structural shocks are orthogonal is of crucial importance as well, but this restriction cannot be tested, although it strongly affects the dynamics of the system. The economic theoretical restrictions refer to economic theory, without formulating this theory explicitly, as dynamic general equilibrium and to some extent structural equation models do. Cooley and Dwyer (1998) conclude that the structural vector autoregressive approach can easily lead to misinterpretations of what the important shocks are.

Although the structural equation modelling approach has its disadvantages, which resulted in general equilibrium and vector autoregressive models, it is still widely applied. Furthermore, the vector error correction model and other approaches are improvements in model specifications that, under particular assumptions, make simultaneous equation models useful for policy analysis.

Finally, in the literature there seems to be some convergence between (structural) vector autoregressive and structural equation modelling. Clements and Mizon (1991) suggest an approach that starts with vector autoregressive components to search for the adequate dynamic structure, which is then used as a basis to search for an adequate structural equation model. However, as the work by Heckman (2000) has shown we have still a long way to go before we have models that satisfy our needs for adequate policy analysis.

3 ENERGY MODELLING IN THE NETHERLANDS

This section discusses the main differences and similarities between the three models developed to analyse the Dutch energy policy. The goal of each of the three models is to analyse the effects of energy policies for subsectors of the Dutch manufacturing sector. The effects of economic growth and energy prices are introduced as exogenous variables. However, each of the three models has a different model structure.

NEMO aims at analysing the effects of economic growth and energy prices on a sector's energy use as part of a set of models used by the CPB to analyse the Dutch economy. Because it uses a partial equilibrium setting, each policy option has to be translated into either an effect on the prices of energy carriers (CO_2-policy), or an adjustment of the trend in the development of energy technologies (investment subsidies and IRR requirement).

In each of the three models the development of technology is exogenous. However, each model uses the same information on energy technology, that is, the information contained in the ICARUS database containing detailed information on available and expected energy-saving technologies and their direct costs. Furthermore, the database contains information on, for instance, how the technologies can be applied in various (sub-) sectors, and which technologies cannot be applied simultaneously. For an example see Michels (2000). Note that this information has also been used for other modelling efforts, such as MARKAL.

SAVE and MEI-Energy use this detailed information to simulate when a particular technology will be used. In both models the variables that determine the utilization of a technique are modelled explicitly, with the exception of energy market prices. What this means will be discussed in more detail later. Each of the models uses a vintage approach, in which each vintage has its specific energy use.

Note that none of the three models has a sound empirical basis. Each model has to some extent been calibrated, that is, (many) parameters are set such that the model mimics the data for a limited number of years.

3.1 Diffusion Mechanisms

The main differences between the models result from the ways energy-saving can be realized. Each of the models distinguishes three possibilities: new investments, improvement of the energy efficiency of existing investments or retrofit, and good housekeeping. There are, however, significant differences in the ways these effects are modelled.

3.1.1 NEMO

NEMO is a top-down approach combining microeconomic theory and information on technology contained in the ICARUS database. NEMO assumes a putty-semiputty production structure, that is, factors are substitutes *ex post* as well as *ex ante*, but *ex post* these possibilities are smaller than *ex ante* (Koopmans et al., 1999). NEMO uses a nested production function:

$$Y = F(I^o, L^o, (L^F(I^F, F)), (L^E, (I^E, E)), M)$$

where Y stands for output of the new vintage that is characterized by investment (I^o, I^F, I^E), labour (L^o, L^F, L^E), fuel use F, electricity use E and intermediate input M. The inputs capital and labour are split into parts that affect fuel use (I^F, L^F), electricity use (I^E, L^E), and inputs that do not affect energy use (I^o, L^o, M). First, investments in fuel and electricity aiming at fuel and electricity savings respectively are linked, which are then combined with fuel and electricity saving labour respectively, resulting in energy services (Koopmans and te Velde, 2001). This structure enables NEMO to simulate substitution between capital, labour and fuel respectively electricity in energy services. At the highest level, these energy services are combined with intermediate inputs M, labour L^o and capital I^o not related to energy.

The demand functions for fuel and electricity, and the related capital and labour are based on cost minimization (Koopmans et al., 1999). For new investments these are actually modelled as changes in fuel efficiency of the vintage added, compared to the base year say t_0. The changes are based on the development in energy prices, say $P^e_t/P^e_{t_0}$, and an exogenous technology trend. How price elasticities and retrofit trends for NEMO were computed using ICARUS data is explained in Koopmans et al. (1999).

For existing vintages a similar approach is used, but with a different trend and price elasticity, based on the development since the year of investment. These result in a potential for retrofit. For the actual *ex post* investments a partial adaptation model is used, which is active till the total potential for retrofit has been used.

The volumes of production, non-energy related capital, labour, and materials as well as the level of energy services required are exogenous to NEMO and derived in another model. As such NEMO only calculates the actual amount of fuel and electricity needed for the energy services required by adapting the energy intensities.

The vintage structure of NEMO implies that new technologies are introduced through new investments affecting the energy intensity of capital. This also implies that technological progress is directly linked to investment and depreciation decisions.

Apart from the *ex ante* and *ex post* diffusion mechanism, good housekeeping has been implemented as a third diffusion mechanism. This effect, which is different for each vintage, is modelled as a correction of the fuel and electricity intensities due to changes of the relative fuel/wage prices in year t, compared to the start year t_0. As a result, this effect is affected by differences in the developments of energy prices and wages only.

This brief description of NEMO shows that the diffusion of energy-saving technologies is affected mainly by substitution elasticities based on relative prices and time trends. The values of all parameters have been computed for each sector of the economy. The computations are based on the savings potential of all technologies available in ICARUS using a 15 per cent discount rate. This is the potential that can be reached. Based on this the two trend parameters were determined. Price elasticities per sector have been estimated using observations constructed from the ICARUS database.[3]

3.1.2 SAVE

SAVE distinguishes segments and a segment is (a part of) an economic sector. The distinguishing feature of a segment is normally the energy technology used. Except for very energy-intensive sectors, a segment equals a sector according to the international standard industrial classification. The basic relationship in SAVE is the final demand for energy in a particular sector or segment of the economy (say) v, which is $E^i_{v,t} = \varepsilon^i_{v,t} F_{v,t}$, where $E^i_{v,t}$ is the final demand for energy carrier i, with $i \in \{\text{electricity}, \text{heat}\}$, $\varepsilon^i_{v,t}$ is the energy intensity of segment v in year t, and $F_{v,t}$ is an indicator for the physical production of the segment and is exogenous to SAVE. The physical indicator is obtained by linking the growth in real value added of sector j, say $Y_{j,t}$, to which segment v belongs, to the trend in the ratio $F_{v,t}/Y_{i,t}$. The final demand for a complete sector is obtained by summing over the technologies applied in a sector.

An energy scenario is obtained by analysing how the energy intensity $\varepsilon^i_{v,t}$ is affected by investment decisions, taking into account restrictions on technology development due to investments in the past.

The choice for a particular energy-savings option is based on financial considerations. If a technology is profitable according to the ICARUS data, it will be applied. Furthermore, SAVE allows for the application of competing energy-saving technologies, which means technologies that exclude each other can both be applied in an economic sector or segment.

The opportunities to apply a technology are limited by organizational and technical constraints. This aspect is modelled rather crudely as a

Fisher-Pry model, used to avoid bang-bang behaviour, that is, whenever a technique is profitable all companies will apply it immediately. Bang-bang behaviour is a problem especially bottom-up models encounter without adequate delay in investment behaviour.

The vintage approach is basically the calculation of how much of the existing stock of capital will be replaced due to old age, for which part energy-saving technologies will be applied, and how much new capital is needed to accommodate economic growth. In SAVE the useful lifetime of a vintage is exogenous.

The financial evaluation of an energy-saving technology is based on a comparison of its internal rate of return (IRR) and the cost of capital (COC) of the sector, if $IRR^{i}_{v,t} > COC^{i}_{v,t}$ the technology will be applied. Furthermore, the required COC is increased if the technology is risky in the sense of being unknown. The COC used is 12 per cent, and for many technologies the risk is set at 0.25, so an IRR of 15 per cent is required. Note that this assumption is the same as the 15 per cent used to calculate the NPV of the technologies that add to the energy savings potential in NEMO. In case the government requires the application of a particular technique, for example insulation, the financial evaluation can be discarded.

The analysis of energy-saving options in SAVE is very detailed and on the level of particular technologies. This richness of detail, however, comes at a cost. It requires detailed inputs, which have to be formulated per technology option. Furthermore, it is assumed that all current and future technology options are known. These detailed descriptions allow the detailed analysis of particular government policy measures, such as building regulation or subsidizing particular technologies. The effects will, however, depend strongly on the assumptions made with respect to the exogenous variables.

3.1.3 MEI-Energy

Where SAVE tries to find a good mix between financial considerations, technical and organizational limitations, and the composition of the capital stock, MEI-Energy tries to include a wide variety of policy considerations that could affect the diffusion of energy-saving technology. The starting point is the IRR of an investment. Similar to the SAVE model, if the IRR is larger than the required discount rate, the investment will be considered. However, to avoid bang-bang solutions, a spread on the payback period companies in a sector require is used. This spread is based on empirical research among Dutch companies. The shorter the payback period of a technology is, the larger the fraction of companies that will apply the technology. Based on these financial considerations a maximum penetration (P^{emax}) can be calculated.

Table 11.1 The relationships between driving forces and policy characteristics

Driving force Characteristic	Policy pressure	Complexity of the technology	Financial pressure	Market pressure	Knowledge level	Non-governmental pressure	Attitude of the sector
Multi- vs unilateral							++
Compelling	++						+
Just			+				++
Financial support							+
Knowledge support					++		
Enforceable	++						
Scope	+			+			
Legally binding	+						
Level of enforcement	++				+		
Discrepancy between goal and current level							+

Source: Elzenga and Ros (2004).

A second limitation of realizing the potential for energy savings in a sector is based on the share of energy costs in the total. It is assumed that in the sector with the highest share, the petrochemical sector, maximum penetration (100 per cent) can be achieved. For all other sectors the maximum penetration is adjusted according to the ratio of a sector's energy cost share and the cost share in the petrochemical sector. An example: the share of energy costs in total costs in the petrochemical sector is 11.7 per cent and that of the building materials sector 4.8 per cent, so for the latter the potential is adjusted to $P^{max} = 0.41P^{emax}$.

However, P^{max} is affected by other, non-financial factors, called driving forces, as well. These driving forces are thought to be affected by policy measures. For example, policy pressure is a driving force (see Table 11.1, row 1), which is affected by several policy measures, such as long-term voluntary agreements, production permits and emission permits. These policies affect the driving forces through policy characteristics.

MEI-Energy distinguishes 10 possible characteristics of policy instruments that could affect one or more of the driving forces. These characteristics are labelled in the first column of Table 11.1. 'Multi- or unilateral' indicates to what extent the sector was involved in the formulation of a policy; high involvement will lead to better acceptance. 'Compelling' indicates how much a policy instrument limits a company's freedom of choice, whereas 'just' indicates if a rule applies to all companies equally. Furthermore, a rule can be 'legally binding', but is not necessarily strongly enforced (level of enforcement).

MEI-Energy distinguishes seven driving forces, ranging from policy pressure and complexity of the technology, to the strength of non-governmental pressure groups and the attitude of the sector (see row 1 of Table 11.1). Before using MEI-Energy for policy analysis a value between 0 and 10 has to be assigned to each of the driving forces, and the larger these values are the faster an energy-saving technology will be introduced and the larger the actual penetration will be.

Table 11.1 indicates how the characteristics of policy instruments affect the forces that drive the introduction of energy-saving technologies according to MEI-Energy. An empty cell implies that there is no effect, + that there is an effect, and ++ that there is a large effect. Table 11.1 shows that the driving force 'policy pressure' is assumed to be affected by five out of the 10 possible characteristics of a policy, whereas 'complexity of the technology' and 'non-governmental pressure' are not affected by characteristics of policy instruments.

The parameters in MEI-Energy are set by the user, who has to indicate on a scale of 0–5 (sometimes 1–5), to what extent a driving force in a particular sector is affected by the characteristic(s) that contain a + or ++ in Table 11.1. For each driving force, MEI-Energy contains a function that will calculate the value of the driving force (a number between 0 and 10), based on the valuation by the user on the 0 (or 1) to 5 scale of the effect each characteristic will have on the driving force. For example, the driving force 'market pressure' is only affected by the characteristic 'scope', where scope indicates whether a policy can be characterized as aiming only at energy saving, or includes more, such as, dematerialization. For 'market pressure' the user is asked whether the 'scope' is absent (= 1), small (= 2), average (= 3), broad (= 4) or very broad (= 5), which is then translated into the values 0, 2.5, 5, 7.5, and 10, respectively, for the driving force 'market pressure'. The more characteristics affect a driving force, the more values have to be set by the model user.

For those driving forces that have no link to policy characteristics, such as 'complexity of the technology', the user has to answer a set of questions. The questions refer to whether it affects the core of the production process, with 'yes' ($C = 1$) or 'no' ($C = 0$) as the scores, whether it is a standard technology, with 'yes' ($S = 0$) or 'no' ($S = 1$), whether production can continue during installation, with 'yes' ($I = 0$), installation can be done during maintenance ($I = 0.5$), or production has to be stopped ($I = 1$). This is then translated into a value for the driving force 'complexity of the technology' using the formula $10 \times (3 - C - S - I)/3$.

Once all driving forces have been set, they are used to calculate the maximum penetration of each energy-saving technology. As the other models do, MEI-Energy distinguishes three different kinds of energy-saving technologies: new technologies, retrofit technologies for existing investments, and good housekeeping. Each of these affects the value of p^{max}.

We will not discuss the technicalities of these calculations, but weighted sums of the values of the driving forces that are thought to affect (say) good housekeeping are divided by the maximum possible score, which results in a number between 0 and 1, indicating the level of penetration. For retrofit and new investments a penetration level is calculated in a similar way. However, the values of the driving forces are also used to calculate a penetration speed, which is between 1 and 8 years for large companies and between 2 and 10 years for small companies. If all driving forces are at their maximum level, the penetration is completed after 1 or 2 years, respectively, and if all driving forces are at their minimum level it takes 8 or 10 years, respectively, for a technology to realize its penetration potential. Note that once the driving forces have been set, they can be used to analyse several scenarios.

Our discussion of the MEI-Energy model indicates that the user has to set many parameters (the exercise has to be completed for all sectors) before MEI-energy can be used. It requires detailed knowledge of the views people in each sector have on the effects of policy measures on the behaviour of companies in the sector.

Furthermore, one can criticize the assignment of the values + and ++ in Table 11.1. For example, there is no link between the characteristics and 'non-governmental pressure'. With all the money the Dutch government spends on educating people with respect to environmental issues, in combination with other national and regional initiatives, this may be incorrect. It is unclear what information was used to fill the cells in Table 11.1.

One should note that the variety of technologies distinguished in SAVE is also present in MEI-Energy, that MEI-Energy also has a vintage structure, and finally that economic growth and energy prices are also treated as exogenous variables.

3.1.4 Comparison of diffusion mechanisms

The diffusion mechanism within NEMO is typical for a top-down model, simple and easy to understand, because it is based on substitution elasticities and trend parameters, in combination with a vintage approach. The model does not distinguish technologies explicitly, nor does it account for the conditions under which a technology can be applied.

On a high level of aggregation this is a reasonable approach, because combinations of several actual effects are covered by the parameters, as is the case with all aggregated economic models. The crucial question is, however, whether this aggregated approach can also be applied for sector analyses when the number of sectors distinguished is large. The answer is no, if, as is the case in the Netherlands, the economy is relatively small. In this case the particular technologies used will dominate the analysis. For example, there is only one genuine iron and steel factory in the Netherlands. A general analysis of this sector at the five-digit level would be dominated by this factory. Furthermore, given the particular technology used by this factory, it is clear from an engineering point of view what improvements are available. SAVE solves this by identifying segments. A description of this sector in terms of trends and price elasticities does not seem appropriate.

It is also questionable if an analysis on a higher level of aggregation, where data of the genuine iron and steel producer are combined with those of companies using scrap and electric arc technology would improve the analysis. These are two different energy-intensive production processes. On the other hand for an energy-extensive and diverse sector such as the service sector, a variety of energy-saving techniques is available, which are difficult to analyse separately. An aggregated analysis, based on elasticities and trends from the past, seems appropriate in that case.

The richness in detail on energy-saving technologies and potential developments present in ICARUS, and used in SAVE and MEI-Energy, is lost in NEMO. SAVE and MEI-Energy link the technologies per sector in ICARUS to the direct cost information of the technologies already used in a sector, in order to obtain detailed and hopefully realistic scenarios of future energy technology use and excluding unrealistic scenarios that do not take into account the current state of affairs. However, ICARUS does not contain complete descriptions of the technologies used in a sector, but only describes the usefulness of additional energy-saving options. The technologies applied in the base year are not made explicit, but only described in terms of energy intensity.

The diffusion mechanism in MEI-Energy is very different from both NEMO and SAVE. MEI-Energy aims at modelling the outcome of the decision processes within companies, using a kind of multi-criteria approach. It asks the user to evaluate per sector how policies affect the forces that drive the decision process. The latter means that, for example, the effect of NGOs on public opinion could reduce the profitability requirements in a particular sector. MEI-Energy distinguishes seven different driving forces that affect the diffusion process. Some of these can be affected by government policy and some cannot. Although this is an interesting and novel approach, it has some drawbacks.

First, none of the mechanisms has a sound empirical foundation. The MEI-Energy model has been calibrated by comparing model results with the realization over the period 1995–99, but this is insufficient to know whether the parameter setting chosen is realistic or not. The approach used is not based on any structural analysis.

Second, different parameter settings can result in the same model output in terms of energy savings. This means that for the same result (infinite) many different policy advices exist. This may seem realistic, but it makes sense only if all possible policies are made explicit. Only then can the policy-makers do what they are paid for, that is decide on a policy. However, in the current MEI-Energy setting as a simulation model, it is impossible to make explicit all policy combinations that lead to the same result.

3.2 Identifying Model Type

Each of the models is clearly a partial model; they result in energy demand but there is no feedback to other sectors of the economy. For all three models, economic growth and energy prices are exogenously determined. What they do is updating the energy (heat and electricity) intensity, which in combination with sector growth will result in energy demand.

Each model distinguishes three types of energy saving: new technology is introduced in new vintages, the efficiency of existing vintages can be improved through investments, and good housekeeping will result in efficiency improvements. In NEMO and SAVE these improvements result in principle from relative prices, whereas in MEI-Energy other forces will strengthen or diminish the potential savings based on direct cost considerations.

Each model uses the information on energy savings available in the ICARUS database. NEMO uses this to calculate a trend in energy efficiency as well as a substitution elasticity based on relative prices per sector. SAVE and MEI-Energy allow the use of all technology options per sector contained in ICARUS separately.

The formulation of NEMO is that of a partial equilibrium model, with the exception of retrofit investments where a partial adjustment model is introduced to avoid bang-bang behaviour. As such NEMO qualifies as structural equation model.

SAVE is an energy-intensity model using bottom-up information. Some will call it a bottom-up model, but the starting point is not based on descriptions of the technologies used in the base year, only the energy intensity. Furthermore, energy-saving technologies are introduced when their IRR is larger than the cost of capital.

MEI-Energy does not seem to fit any of the models discussed above. The analysis is based on a quantification of views the user has on the effects policies might have on forces that are thought to alter the purely financial decision criterion for technology use. The values of the driving forces vary from 0, which stands for no effect, to 10 for the maximum effect. Some driving forces are thought to contribute more than others. This is realized through different weights in the functions used to quantify the overall effect on the three types of energy saving. None of the functions used is based on a structural or on an empirically tested model. MEI-Energy may best be described as a multi-criteria model that elicits expert views in order to calculate energy use per sector.

Note that in none of the models is behaviour properly based on a structural model, nor do they use econometric or time series methods to analyse dynamics in the data. Furthermore, properly formulated feedback between the energy sector and the rest of the economy is missing. NEMO and SAVE are, however, part of a larger set of models that can be used for a more complete policy analysis.

4 COMPARING POLICY EVALUATIONS

This section discusses the results of the policy experiments. For all scenarios we used the same real value added growth rate and a price increase of 2 per cent annually. All scenarios are compared to this business as usual scenario. First an extra increase of energy prices by 3 per cent over business as usual (the increase of 2 per cent) is analysed. Next come three popular policy measures: a CO_2-tax of €50 per ton, a subsidy of 10 per cent on the cost of investment in energy-saving technology, and, finally, companies are obliged to invest in all energy-saving technologies available in ICARUS that have an IRR of 10 per cent or more. Before the policy results are compared, the business as usual scenario for the three models is discussed.

4.1 Business as Usual

Each of the institutes reported final energy demand in terms of heat and electricity; see Figure 11.1. Because the models analyse the future in different ways, one would expect differences in results. The differences in energy use in the start year (2000) are however too large to be explained by differences in model definition. They must be the result of differences in the definition of final use of heat and electricity. This is odd since all three models claim to analyse final energy use.

MEI-Energy is based on energy use as defined in ICARUS 4 and for 1995 energy use is equal to the sector energy use as defined in the National Energy Statistics of Statistics Netherlands. NEMO analyses final energy use as well, but including non-energy uses such as the production of ammonia. For the energy-extensive sectors analysed here, however, this is not relevant. In SAVE the uses of energy for heat purposes are lower, because final energy use is defined as use in central heating, radiators and boilers only. For the industrial sectors SAVE uses natural gas usage as an indicator, since natural gas is the main source of energy for heat. This could explain some of the differences. For electricity, however, the results of SAVE should be similar to those of the other models since all models use the same definition.

Figure 11.1a shows that the differences in energy use in the year 2000 are largest for the service sector; the final use in NEMO is 92.0 PJ, in SAVE 53.2 PJ, and in MEI-Energy 62.3 PJ. However, services are the sector for which the Statistics Netherlands data are most unreliable. Adjustments to accommodate the different model definitions are most likely the largest for this sector. The model descriptions do not, however, contain any information on the definition of final energy use.

Note that the differences between the models differ per sector. For example, for the service sector final energy use is the largest in NEMO, whereas the use of heat in SAVE and MEI-Energy for the food and beverages sector exceeds that of NEMO by 75 per cent. There are also large differences between SAVE and MEI-Energy, despite the fact that they are both bottom-up models; the use of heat in the sector other metal is according to MEI-Energy almost 50 per cent larger than in SAVE and 5 per cent smaller in food and beverages.

When we analyse the growth in final energy use (see also Table 11.2), we again observe large differences between the three models (see graphs a, c and d of Figure 11.1). For the service sector the differences are the largest, but where NEMO and SAVE show positive growth figures for heat, MEI shows a reduction in heat use. For the sector other metal, Figure 11.1b indicates similar growth but from different starting points, but this similarity is somewhat misleading. As Table 11.2 shows, the cumulative growth figures differ considerably.

Owing to the different starting points of the three models, differences in nominal development exist. However, as Table 11.2 shows there are also large differences between the growth percentages for the different sectors

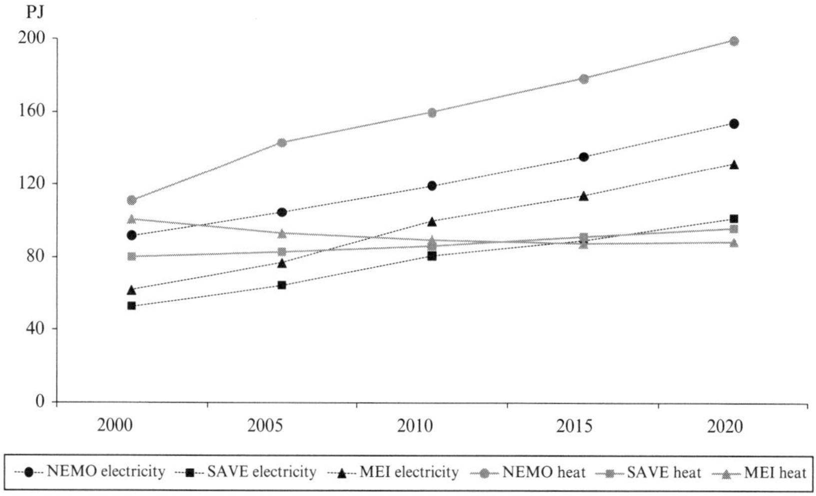

Figure 11.1a Energy use in the service sector

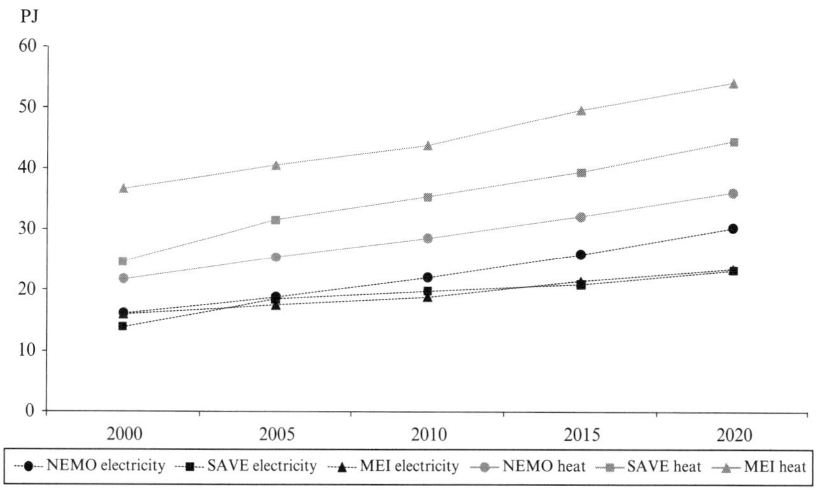

Figure 11.1b Energy use in the other metal sector

despite the fact that all models use the same economic growth and price scenario. So, it is apparent that policy analyses will result in different views.

Since RIVM was still fine-tuning MEI-Energy when the experiment was conducted, the results for this model for the service sector have to be interpreted with care. This energy-extensive sector seems by nature, given

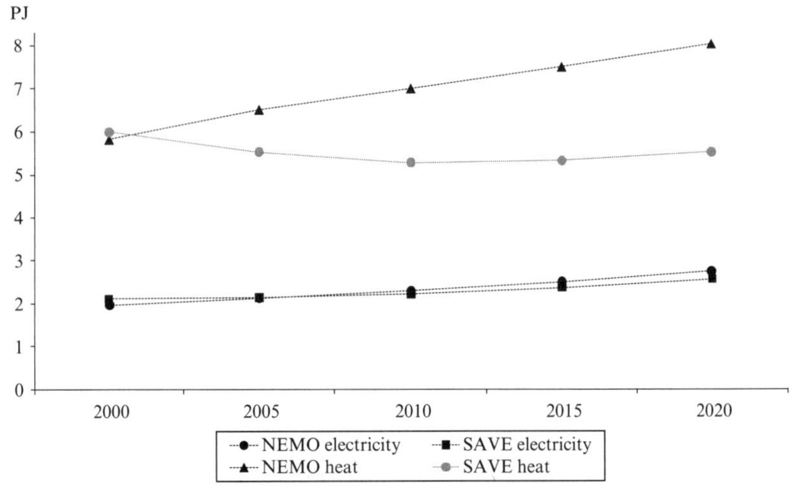

Figure 11.1c *Energy use in the textiles sector*

Table 11.2 *Growth in electricity and heat demand in PJ and percentage of 2000 under the business as usual scenario*

Sector	NEMO		SAVE		MEI	
	Electricity	Heat	Electricity	Heat	Electricity	Heat
Services, PJ	62.3	88.7	48.4	15.7	69.8	−12.3
% of 2000	67.7	79.9	91.0	19.5	112.0	−12.2
Other metal, PJ	14.1	14.3	9.3	19.7	7.3	17.5
% of 2000	86.7	65.7	48.6	66.5	45.7	47.7
Textiles, PJ	0.8	2.2	0.4	−0.5	–	–
% of 2000	38.9	37.8	40.0	11.0	–	–
Food and	6.4	10.5	5.2	0.5	4.6	15.0
beverages, PJ	27.8	23.7	23.9	11.5	21.1	20.2
% of 2000						

many different technologies and low energy intensity, more sensitive to the factors MEI-Energy wants to incorporate than the other sectors are.

4.2 The Effect of a 3 Per Cent Extra Increase in Energy Prices

Table 11.3 contains the comparison of the business as usual scenario and an extra increase in energy prices of 3 per cent (on top of the 2 per cent increase). Because changes in demand on top of the autonomous trends

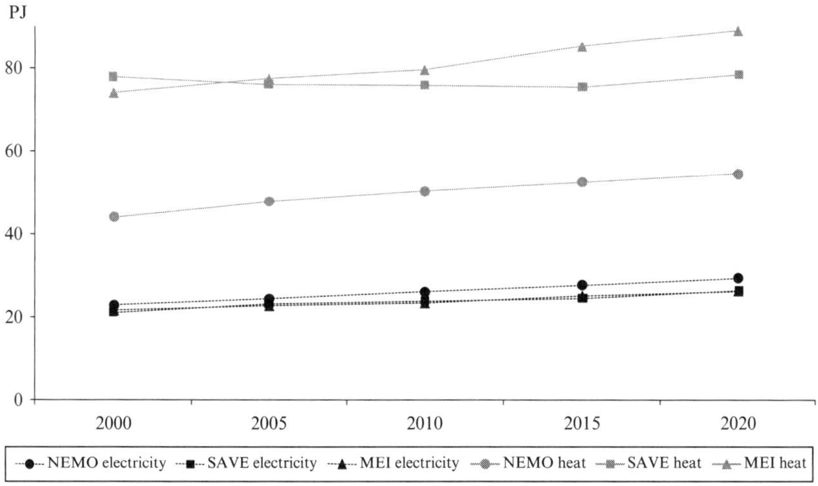

Figure 11.1d Energy use in the sector food and beverages

Table 11.3 The effect of a 3 per cent extra increase in energy prices, percentage

Sector	NEMO		SAVE		MEI	
	Electricity	Heat	Electricity	Heat	Electricity	Heat
Services	−10.2	−7.0	–	–	−0.6	−4.1
Other metal	−4.8	−7.6	−11.0	−0.6	0.0	−0.3
Textiles	−4.0	−12.6	−1.4	0.0	–	–
Food and beverages	−6.9	−11.5	−2.5	−3.6	0.0	0.0

are only possible due to changes in energy prices, a price increase has a clear effect on energy use in NEMO. In comparison to the business as usual scenario the largest decrease in electricity demand is in the service sector (−10.2 percentage points by 2020), and the smallest (−4 percentage points) in the textiles sector. For heat this is −12.6 percentage points for the other metal sector and −7 percentage points for the service sector. So, all effects are as one would expect from an economic point of view.

The results of the SAVE model show a different picture. There is a large price effect (−11 percentage points) for the sector other metal for electricity, as one would expect given the technologies used in this sector. There is some effect for electricity and heat demand in the food and beverages sector. Unfortunately there are no data available for the service sector.

MEI-Energy hardly reacts to the change in energy prices. This is the result of the fact that according to the information in the ICARUS database most technologies available are profitable. An increase in profitability due to an increase in energy prices will therefore have a negligible effect within MEI-Energy, since this only increases the profitability. As was shown by DeCanio (1998) organizational and institutional factors are also important factors in a company's decision to invest. Although MEI is an effort to include these effects, the starting point seems to be that in our business as usual scenario the financial incentive in ICARUS is sufficient. Only for heat in the service sector do prices affect usage (a 4.1 percentage point decrease).

This analysis shows important differences between the models. In NEMO, changes in energy prices clearly affect demand. In SAVE and MEI-Energy the effects of price increases are limited. This is mainly due to limitations in the modelling of economic decision-making and the assumption that technologies with a positive NPV will be applied. As argued above, this reasoning has some flaws. It does not take into account the option value of not investing now. Furthermore, it does not account for the costs of gathering information on energy-saving technologies and their application in the production process (DeCanio, 1998). For energy-intensive sectors, where energy costs are substantial, one can expect the company to be knowledgeable with respect to technologies. For energy-extensive companies the collection of information on energy technologies and the preparation of decisions will be costly, compared to the information used in ICARUS's standard NPV calculation. Not taking these costs into account in combination with the fact that the savings potential in money terms is relatively small since energy costs are small, makes the NPV evaluation an inappropriate starting point for analysis. These differences in modelling decision making will result in differences in policy advice when analysing the same question.

4.3 Comparing Policies

Because the definition of final energy use of the three models differs, the comparison is based on index numbers. The results for each of the models in 2000 is set equal to unity, and the development until 2020 due to the policy measures are analysed as changes compared to this index. With three models, four sectors, three policy measures, and two forms of energy, we have 72 model outputs. The most interesting results will be discussed. Table 11.4 shows the effect the three policy measures have on

Table 11.4 The effect of the policy measures in 2020 compared to the business as usual scenario, percentage

Sector	Policy	NEMO		SAVE		MEI	
		Electricity	Heat	Electricity	Heat	Electricity	Heat
Services	CO_2-tax	−3.74	−3.07	−0.44	−2.07	−2.47	−14.99
	Subsidy	−3.12	−1.68	−0.16	−0.31	−2.78	−14.57
	IRR			−1.71	−11.52	NR	NR
Other metal	CO_2-tax	−2.07	−3.67	−0.68	−0.52	0.00	−0.80
	Subsidy	−1.34	−1.02	−0.19	−0.18	−0.18	−0.24
	IRR			−0.19	−0.18	−3.79	−6.61
Textiles	CO_2-tax	−1.55	−4.30	−0.65	0.00	–	–
	Subsidy	−0.25	−0.14	−0.16	0.00	–	–
	IRR			0.00	0.00	–	–
Food and beverages	CO_2-tax	−3.67	−13.26	−0.87	−4.52	0.00	0.05
	Subsidy	−3.10	−1.05	0.10	−1.39	0.01	0.05
	IRR			−0.16	−1.24	−0.23	−0.80

the result for 2020, compared to the business as usual scenario, or BAU for short. The content of a cell is defined as:

$$\left(\frac{X_{2020}^{\text{Policy}}}{X_{2020}^{\text{BAU}}} - 1\right) \times 100\%$$

4.3.1 CO_2-tax

The CO_2-tax will increase the price of energy. For SAVE and MEI-Energy this measure results in €2.81 per GJ_{th} and €5.50 per GJ_e for all sectors. For NEMO the effect on energy prices differs per sector, ranging from €2.80 per GJ_e in the services sector to €3.17 in the other metal sector, and €1.13 per GJ_{th} for food and beverages to €4.43 for the service sector. So, price increases are steep, ranging from 35 per cent to 96 per cent for electricity and from 59 per cent to 133 per cent for heat.

As expected, there is a clear effect when analysing the price increases with NEMO for all sectors and both forms of energy (see Table 11.4, Column NEMO). For SAVE and MEI-Energy this is much less visible. In MEI-Energy because it assumes that the cost data in the ICARUS database are sufficient to evaluate profitability and most saving options are already profitable (that is, they have a short payback period). Since SAVE uses the same data, to some extent this argument holds for SAVE as well, but the diffusion mechanism in SAVE still reacts to price policies, although the effects are small. The use of heat in the service sector shows some price sensitivity, heat use reduces by 2.07 per cent, as does the food and beverages sector, where heat use reduces by 4.5 per cent (see Table 11.4, Column SAVE). The effect of price increases on electricity use is negligible. For MEI-Energy the effect is negligible, with the exception of the service sector, but as mentioned above these results need to be interpreted cautiously. The effect could be due to the fact that this sector is more sensitive to the driving forces identified in MEI-Energy, but for this sector the model may need more fine-tuning.

Despite the large increases in prices of 35 per cent or more, the effects of these increases are limited in all three models. This is partly due to the vintage structure that is incorporated in all three models, because it reduces the response speed. Apparently the autonomous trends used in each of the models are the main drivers of the development in energy use. In addition, the fact that there is no feedback mechanism between energy price increases and the growth of each sector (which is exogenous) might hamper the analysis. In sum, however, taking all this into consideration, it is still remarkable that large increases in prices do not affect the utilization of energy-saving technologies.

4.3.2 Subsidizing investment

The policy to subsidize investments in energy-saving technology has a strong effect for the service sector in MEI-Energy, especially for heat (−14.57 per cent). For the other sectors there is no effect. For the textiles sector Table 11.4 lists 'NA'. This can be interpreted as 0 per cent, and is due to the fact that the ICARUS database contains no information on energy saving potentials for this sector. With SAVE using the same data, it is obvious that the effect there is small too.

4.3.3 The IRR policy

In the MEI-Energy model the 'command and control' policy that all technologies with an IRR of 10 per cent or more have to be implemented leads to results that make no sense (indicated by 'NR'). As stated above, the model is not well tuned yet for this sector, leading to unacceptable results for this policy.

The analysis with SAVE results in a clear effect for heat consumption in the service sector (−11.52 per cent), and a much smaller effect for electricity (−1.71 per cent). In the food and beverages sector there is a small effect for thermal use (−1.24 per cent) and no effect for electricity use. There is no effect for the sector other metal either, which might be due to the fact that this sector is more energy-minded given the technologies used in this sector. As expected, given the fact that there are no energy-saving technology data in the ICARUS database, there is no effect for the textiles sector.

When we compare the results of SAVE to those of MEI-Energy for the services and the other metal sectors this policy has a clear effect in MEI-Energy, but not in SAVE. Translating the IRR policy into terms of the NEMO model is difficult. The only translation possible is the same as the translation of the subsidy policy, so NEMO cannot distinguish between these two policies. This lack of detail in top-down models compared to bottom-up models is normal and it is due to the fact that top-down models cannot distinguish individual technologies.

5 CONCLUSIONS

This contribution started off by reviewing modelling options and their pros and cons for energy modelling. Subsequently, three specific models (NEMO, SAVE and MEI-Energy) developed to analyse Dutch energy policy proposals were discussed, and the models were used in an experiment to compare the effects of alternative energy policies. The experiment, conducted by the institutes that have built the models, comprised three policies for four different energy-extensive sectors and the results have been reported on two forms of energy, electricity and heat. The three models differ in structure: NEMO is a top-down model where technology is modelled indirectly through trends and substitution elasticities, and SAVE and MEI-Energy are models using bottom-up detailed and explicit descriptions of energy-saving techniques.

However, the models also have some features in common. All three are partial models using a vintage structure. Furthermore, the three models have at their core an update of a sector's energy intensity based on three investment options: new vintages, retrofit of existing vintages and good housekeeping. They also use the technology descriptions and related direct cost and benefit data from the ICARUS database, either explicitly (SAVE and MEI-Energy) or implicitly (NEMO), as a starting point to analyse energy savings. Furthermore, they use the same economic growth and energy price data as exogenous inputs.

The most important difference between the models is the way diffusion of energy-saving technologies has been modelled and what factors influence this process. In NEMO this process is basically an autonomous trend, complemented with a vintage approach and price elasticities computed using the cost–benefit data in the ICARUS database (Koopmans et al., 1999). In SAVE three diffusion mechanisms have been modelled. The financial evaluation based on the IRR is limited by non-economic factors, basically a Fisher-Prey type diffusion model that limits the opportunities for energy-saving technologies through a growth path, in combination with a limitation in the part of the capital stock that can be replaced in a particular year. MEI-Energy uses a similar approach. The financial data of ICARUS are again used to evaluate the investment opportunities in a vintage setting, but these are limited or enhanced through an evaluation of the forces that (might) affect the financial evaluation of companies in a particular sector.

In SAVE and MEI-Energy exogenous economic growth is translated into physical units, such as tons produced or number of square meters heated. This translation is based on developments in the past and actually accounts for dematerialization trends (less product per unit real gross value added) in the past.

This translation into physical units instead of using real gross value added is typical for the difference between economic and engineering models. Economic models internalize assumptions as much as possible, whereas an engineering model often resembles a toolbox with assumptions explicitly translated into model input (van Beeck, 1999).

Although SAVE and MEI-Energy have a similar basic starting point for diffusion, MEI-Energy tries to model all forces that affect an investment decision, ranging from the complexity of the technology to the pressure of NGOs on the private and the government sector. They also explicitly include whether a government policy affects a driving force. The way in which this is included in the model is, however, not based on tested hypotheses about behaviour, but implemented in a rather ad hoc fashion. In this respect MEI-Energy distinguishes itself very clearly from traditional top-down and bottom-up energy models.

The parameter values used in each of the models are not statistically determined. All three models can be looked upon as simulation models with parameters set in such a way that (at least) for some years in the recent past the model output 'fits' the data.

The analyses with the three models for the four energy-extensive sectors (services, other metal, food, textiles and beverages) and three policies (CO_2-tax, investment subsidy, and investing in all technologies with an IRR of 10 per cent or more) show remarkable differences. Remarkable, given the similar approach (updating energy efficiency based on profitability) and limited scope (energy use per sector) of the three models, the use of the same technology data from the ICARUS database, and the use of the same exogenous economic growth and energy price scenario.

The starting point for the simulation, energy use in 2000, differs considerably as well, despite the fact that all three models claim to model final energy use. Especially for heat applications the differences are considerable. Only SAVE makes some remarks about correcting the content of the concepts used, but for energy-extensive sectors this does not explain the differences. In addition, the relative growth in energy use in the business as usual scenario shows remarkable differences.

Similarly, the differences in effect of policies that affect energy prices are remarkable. In MEI-Energy there is no effect and in SAVE only a very limited effect occurs. This can be explained by the fact that many energy saving options in the ICARUS database are already considered profitable in the business as usual scenario according to direct cost and benefit data available in ICARUS. Higher energy prices only increase the profitability and within the vintage structure modelled there is only limited room for new investments. Given the structure of NEMO, where relative prices guide substitution, it is obvious that prices have a much larger effect on energy savings.

It is questionable whether the NPV, the IRR or the payback-period are adequate criteria to model investment decisions, especially for energy-extensive sectors. As argued by van Groenendaal (1998a), only when energy is a substantial cost component (more than 7 per cent of total cost) and strongly related to the production technique used, will companies be aware of energy-saving technologies. The smaller the share of energy in total costs the smaller the knowledge on energy-saving options will be in a sector. Additional costs, such as gathering information, are not taken into account in the NPV evaluations used, resulting in what is known as 'the energy efficiency gap'. However, these inefficiencies can be relatively large given the wide variety of savings possibilities in energy-extensive sectors. This reduces the profitability considerably. However, Aalbers et al. (2004) report that companies (almost 50 per cent in their sample) in the Netherlands do not always use the NPV or payback as criterion to evaluate investments in energy technology.

There are also more fundamental objections against this type of criterion. As was shown by Dixit and Pindyck (1994) the NPV does not account for the option value of postponing the investment in energy saving, which is often an irreversible process. The value of energy price uncertainty can be considerable (van Soest and Bulte, 2001).

The criticism on financial criteria, such as the NPV and IRR, however, is less serious than some researchers want us to believe. The reason is that companies choose their discount factor for the NPV, the acceptance level for the IRR or the payback period such that these account for risk. In this way uncertainty is accounted for, albeit in a way that many academics consider insufficient.

MEI-Energy tries to explicitly model the non-financial internal and external forces that affect the financial criterion to invest in energy savings.

In most models these different ways are covered by one or more parameters, but they are hardly ever made explicit. The user of the model assigns values to the various parameters that affect the speed of technology diffusion. This approach resembles a multi-criteria decision model. However, we do not know whether this is a realistic representation of this process, owing to lack of empirical data.

There are other objections to the approach used in MEI-Energy as well. For MEI-Energy a large number of parameter settings (if not infinitely many) are feasible that result in the same energy savings outcome. However, only the setting used by the analyst will be known. Other feasible policies leading to the same result are not. Policy-makers might however prefer one of the other feasible policies. MEI-Energy is an interesting modelling exercise, but still needs more work to make it easily applicable. Furthermore, the exercise for the service sector has shown that the model is not yet well tuned.

In sum, the approaches of the three models differ considerably, as do their results. The fact that the three research institutes work together on energy matters, makes the differences even more astounding. It is obvious, and this is not a new conclusion (Grubb et al., 1993; Hourcade et al., 1996), that the top-down model reacts primarily to price changes and the bottom-up models react better to command-and-control type policies. For policy advice this result means that almost any combination of a more general economic policy based on prices and command-and-control type policies can be supported, depending on the right mix of results. In energy policy in the Netherlands, model choice does create degrees of freedom for policy-makers.

ACKNOWLEDGEMENTS

This research was executed with a grant from the Netherlands Organization for Scientific Research (NWO).

NOTES

1. The 1993 SBI classification system used by Statistics Netherlands is based on the International Standard Industrial Classification of All Economic Activities of the UN (ISIC Rev. 3.1).
2. For more information we refer to http://www.etsap.org/markal/main.html.
3. For a precise description of the calibration process, see Koopmans and te Velde (2001).

REFERENCES

Aalbers, R., H.L.F. de Groot and H.R.J. Vollebergh (2004), 'Effectiviteit van Energie-Investeringssubsidies op Technologiekeuze in Nederland', in H.R.J. Vollebergh (ed.), *Milieubeleid, Technologische Ontwikkeling en de Nederlandse Economie*, The Hague: SDU, 2004.

Adams, P.D., P.B. Dixon, D. McDonald, G.A. Meagher and B.R. Parmenter (1994), 'Forecasts for the Australian economy using the MONASH model', *International Journal of Forecasting*, **10**, 557–71.

Belhaj, M (2002), 'Vehicle and fuel demand in Morocco', *Energy Policy*, **30**, 1163–71.

Bunn, D.W. and E.R. Larsen (1997), *Systems Modelling for Energy Policy*, Chichester: John Wiley and Sons.

Chow, L.C. (2001), 'A study of sectoral energy consumption in Hong Kong (1984–97) with special emphasis on the household sector', *Energy Policy*, **29**, 1099–110.

Clements, M.P. and G.E. Mizon (1991), 'Empirical analysis of macroeconomic time series: VAR and structural models', *European Economic Review*, **35**, 887–932.

Commission of the European Communities (CEC) (1993), *Harmonized Econometric Research for Modelling Economic Systems*, Amsterdam: North-Holland.

Cooley, T.F. and M. Dwyer (1998), 'Business cycle analysis without much theory: a look at structural VARs', *Journal of Econometrics*, **83**, 57–88.

Cropton, P. and Y. Wu (2005), 'Energy consumption in China: past trends and future directions', *Energy Economics*, **27**, 195–208.

Daniëls, B.W. and A.W.N. van Dril (2007), 'Save production: a bottom-up energy model for Dutch industry and agriculture', *Energy Economics*, **29**, 847–67.

Deaton, A. and J. Muelbauer (1980), 'An almost ideal system', *American Economic Review*, **70**, 312–26.

DeCanio, S.J. (1998), 'The efficiency paradox: bureaucratic and organizational barriers to profitable energy-saving investments', *Energy Policy*, **26**, 441–54.

Diebold, F.X. (1997), 'The past, the present, and future of macroeconomic forecasting', http://www.ssc.upenn.edu/~diebold/papers/paper19/ jep2.pdf.

Dixit, A.K. and R.S. Pindyck (1994), *Investment under Uncertainty*, Princeton, NJ: Princeton University Press.

Elzenga, H. and J. Ros (2004), 'MEI-Energie: RIVM's Energiebesparingsmodel', *Kwartaalschrift Economie*, **1**, 168–89.

Fishbone, L.G. and H. Abilock (1981), 'MARKAL a linear programming model for energy systems analysis: technical description of the BNL version', *International Journal of Energy Research*, **5**, 353–75.

Fox, T. and D. Fullerton (1991), 'The irrelevance of detail in a computable general equilibrium model', *Economics Letters*, **36**, 67–71.

Fuss, M.A. and D.L. McFadden (1978), *Production Economics: A Dual Approach to Theory and Applications*, Amsterdam: North-Holland.

Garratt, A., K. Lee, M.H. Pesaran and Y. Shin (2000), 'The structural cointegrating VAR approach to macroeconometric modelling', in S. Holly and M. Weale (eds), *Econometric Modelling: Techniques and Applications*, Cambridge: Cambridge University Press.

Gottschalk, J. (2001), 'An introduction to the SVAR methodology: identification, interpretation and limitations of SVAR models', Working Paper No. 1072, Kiel: Kiel Institute of World Economics.

Grubb, M., J. Edmonds, P. ten Brink and M. Morrison (1993), 'The cost of limiting fossil-fuel emissions: a survey and analysis', *Annual Review of Energy and the Environment*, **18**, 397–478.

Hamilton, J.D. (1994), *Time Series Analysis*, Princeton, NJ: Princeton University Press.

Heckman, J.J. (2000), 'Causal parameters and policy analysis in economics: a twentieth century retrospective', *The Quarterly Journal of Economics*, **115**, 45–97.

Hondroyiannis, G. (2004), 'Estimating residential demand for electricity in Greece', *Energy Economics*, **26**, 319–34.

Hourcade, J.C. (1996), 'Estimating the cost of mitigating greenhouse gasses', in J.P. Bruce, H.Lee and E.F. Haites (eds), *Climate Change 1995: Economic and Social Dimensions of Climate Change, Contribution of Working Group III to the Second Assessment Report of the IPCC*, Cambridge: Cambridge University Press.

Jacobs, J. and K.F. Wallis (2002), 'Comparing SVARs and SEMs: more shocking stories', paper presented at the 57th European Meeting of the Econometric Society, Venice, 25–28 August.

Johansen, S. and K. Juselius (1994), 'Identification of the long-run and short-run structure: an application to the ISLM model', *Journal of Econometrics*, **63**, 7–36.

Kleinpeter, M. (1995), *Energy Planning and Policy*, Chichester: John Wiley and Sons.

Koopmans, C.C. and D.W. te Velde (2001), 'Bridging the energy efficiency gap: using bottom-up information in a top-down energy demand model', *Energy Economics*, **23**, 57–75.

Koopmans, C.C., D.W. te Velde, W. Groot and J.H.A. Hendriks (1999), 'NEMO: Netherlands energy demand model: a top-down model based on bottom-up information', Research Memorandum 155, The Hague: Netherlands Bureau for Economic Policy Analysis (CPB).

Lucas, R.J. (1976), 'Econometric policy evaluations: a critique', *Carnegie-Rochester Conference Series on Public Policy*, **1**, 19–46.

McKintrick, R.R. (1998), 'The econometric critique of computable general equilibrium modelling: the role of functional forms', *Economic Modelling*, **15**, 543–73.

Michels, K. (2000), 'ICARUS-4 Sector Study for the Iron and Steel Industry, Report no. NWS-E-2000-10, Utrecht: Department of Science, Technology and Society, Utrecht University.

Naqvi, F. (1998), 'A computable general equilibrium model of energy, economy and equity interactions in Pakistan', *Energy Economics*, **20**, 347–73.

Neves, P.D. (1994), 'A class of differential demand systems', *Economics Letters*, **44**, 83–6.

Shoven, J.B. and J. Whalley (1992), *Applying General Equilibrium*, Cambridge: Cambridge University Press.

Sims, C.A. (1980), 'Macroeconomics and reality', *Econometrica*, **48**, 1–48.

Sims, C.A. (1996), 'Macroeconomics and methodology', *Journal of Economic Perspectives*, **10**, 105–20.

Theil, H. (1965), 'The information approach to demand analysis', *Econometrica*, **33**, 67–87.

Theil, H. (1971), *Principles of Econometrics*, Amsterdam: North-Holland.

Thomas, R. (1997), *The Demand for M4: A Sectoral Analysis*, London: Bank of England.
Tridimas, G. (2000), 'The analysis of consumer demand in Greece: model selection and dynamic specification', *Economic Modelling*, **17**, 455–71.
van Beeck, N. (1999), *Classification of Energy Models, Research Memorandum 777*, Tilburg: Tilburg University.
van Bergeijk, P.A.G. and J.M. Berk (2001), 'The Lucas critique: an empirical investigation of the impact of European monetary integration on the term structure', Working Paper No. ISSN 1424–0459, Zurich: University of Zurich, Institute for Empirical Research in Economics.
van Dril, A.W.N., F.M.J.A. Diepstraten and M. Beeldman (1995), 'SAVE-Module Produktiebedrijven: De modellering van Energieverbruikontwikkelingen', ECN-1-94-043, Petten: ECN-Beleidsstudies, Energieonderzoek Centrum Nederland.
van Groenendaal, W.J.H. (1998a), *The Economic Appraisal of Natural Gas Projects*, Oxford: Oxford University Press.
van Groenendaal, W.J.H. (1998b), 'Estimating NPV variability for deterministic models', *European Journal of Operations Research*, **107**, 202–13.
van Soest, D.P. and E.H. Bulte (2001), 'Does the energy-efficiency paradox exist? Technological progress and uncertainty', *Environmental and Resource Economics*, **18**, 101–12.
van Wijk, J.J., R.F.J.M. Engelen and J.P.M. Ros (2001), 'Model for effectiveness of policy instruments for the energy-saving in industry', RIVM Report 778011004, Bilthoven: National Institute for Public Health and the Environment (RIVM).

Epilogue

12. Conclusions and future research

Raymond J.G.M. Florax, Henri L.F. de Groot and Peter Mulder

1 POSITIONING

Maintaining energy security and adaptation and mitigation to the global process of climate change constitute serious challenges for the twenty-first century. The failure to reach an agreement on fighting climate change in Copenhagen in December 2009 reveals the political complexity and the important economic stakes involved. Despite the complexity of the energy security and climate change issues, there is general consensus that innovation and adoption of new technologies resulting in substantial improvements in energy efficiency is critical in finding long-term solutions for these problems. There is also an increasing recognition that the development of new technologies alone is not sufficient. New technologies need to be adopted and integrated into already existing production processes. This implementation is costly and far from automatic. We are therefore in need of policies conducive to fostering dissemination and adoption of both new and already existing technologies.

This book contains theoretical and empirical contributions offering insights into energy efficiency trends across countries and sectors, and into aspects of the complex interaction between technological change, energy policy, investment behaviour and the resulting trends in energy efficiency. It deviates from many other studies by its focus on small and medium-sized enterprises in energy-extensive sectors (for instance, the light industry and the service sector). Although total energy use in these sectors is substantial (around 20 per cent of economy-wide energy use, and about one-third of corporate energy use) and potential savings are therefore considerable, little is known about the penetration and adoption of energy-saving technologies in these sectors.

This book aims to combine empirical evidence, theory and policy analyses, both in terms of impact evaluation and energy models, centred around the complexities involved in adoption of energy efficient technologies. The collection of chapters provides a coherent view on the various issues and complexities associated with the development of effective energy policies. This is achieved by studying in the subsequent parts of this book:

- long-term comparative developments of energy efficiency,
- the factors that prevent firms from adopting those technologies that will contribute to improved energy efficiency and sustainable energy use, and
- the appropriateness of specific policies aimed at technology diffusion.

The book fits a rich tradition of studies on energy efficiency developments. Previous studies have already convincingly shown that with the current state of technological knowledge, substantial improvements in energy efficiency are feasible from a technological point of view. In addition, these improvements are cost-effective from an economic point of view (Schipper and Meyers, 1992). The existence of a broad range of technologies that can be employed cost-effectively according to standard net present value computations but are nevertheless not adopted by firms is by now a well recognised phenomenon known as the energy efficiency paradox (Blok et al., 2004; Gillissen et al., 1995; Velthuijsen, 1995; Worrell et al., 1997). Potential explanations for this paradox that have been put forward to explain this seemingly irrational behaviour are the presence of uncertainty with respect to future energy prices, environmental policies, technology developments or the quality of the technology to be applied, lack of information on the available technologies, hidden costs of information and implementation, limited attention of firms for energy efficiency, difficulties of incorporating the technology within the existing production process, organizational restraints, the presence of current machinery which needs to be replaced and depreciated before new technologies will be applied, financial constraints, and so on. Although there is likely potential truth and relevance in all these factors, there is little systematic empirical information about the relative (quantitative) importance of these factors. Little is also known about whether the barriers that prevent firms from adopting technologies differ between, for instance, regions, sectors, technological characteristics and firm characteristics. Are there, for example, differences in adoption behaviour between large and small firms, or between energy-intensive and energy-extensive firms, or between competitive versus non-competitive sectors, or between internationally competing versus nationally competing firms? Or is adoption behaviour different with respect to technologies varying in the extent of required integration in the production process? Without detailed knowledge of why firms do not adequately invest in energy-efficient technology, designing tailor-made policies to overcome the adoption barriers is difficult.

The goals of this book have been to find out in more detail what factors are responsible for the fact that promising technologies are not always adopted, and to study the desirability and effectiveness of policies aimed at improving energy efficiency, taking into account the bottlenecks that prevent firms from adopting, as well as the extent to which these factors are dependent on space, sector structure, type of firm, technology, or variable through time.

2 MAIN CONCLUSIONS

The analyses in this book have been naturally organized in three different parts. The first part has identified some of the major stylized facts regarding energy efficiency and its development both over space as well as time.

A first general message to be taken from Part I of this book is that improvements in energy efficiency have played an important role in keeping energy consumption within bounds. A striking result is that energy consumption would have been 60 per cent higher in 2001 than in 1971 if energy productivity had stayed constant at its 1971 level. For China, this increase would have been even more dramatic with approximately 260 per cent. Energy productivity improvements have thus played an important role in combining economic growth, energy security and the adaptation and mitigation of climate change. Despite these impressive achievements, convergence of energy productivity has been limited, as evidenced by large and persistent cross-country differences in energy productivity. Technological options to enhance energy efficiency seem to be available but are apparently not adopted.

A second message centres on the complexity of properly defining and measuring energy efficiency. These complexities become evident in comparing Chapters 2 and 3, which are based on economic indicators at a detailed sectoral level, with Chapters 4 and 5, which concentrate on the options to measure energy efficiency at an even more detailed level using physical indicators for energy extensive industries. The physical indicators have the advantage that they come as close as possible to characterizing actual production processes in physical terms, but at the expense of huge efforts that are required to gather comparable data for long time series and a sufficiently broad set of countries.

Part II centres around the notion that many energy-efficient technologies are available, but at the same time not adopted. This part of the book aims to provide explanations for this so-called energy efficiency paradox.

The contributions in this part of the book add to the existing literature in different ways.

The first chapter develops a theoretical model to show that the introduction of new technology is a *casus origine* of industry structure, and may easily be implicated by extension to differences in adoption timing between firms and a generally perceived adoption lag. The main conclusion of this chapter is that firm size is driven by technological opportunities, a phenomenon that is sometimes referred to as the Schumpeterian hypothesis. This conclusion is supported by the weight of empirical evidence on adoption timing and diffusion.

The two subsequent chapters look deeper into one particular barrier to adoption of new technologies, namely, uncertainty. The effect of uncertainty on investment spending has been a heavily debated issue since the early 1970s. Many of the theoretical insights developed over the years provide an ambiguous picture on the direction of the effect. The chapter analyses the heterogeneity in the outcomes of empirical studies by means of a meta-analysis. The results suggest that, in general, the relationship between uncertainty and investment spending is negative. Although there are several empirically relevant sources of systematic variation in the sample of elasticities, none of them appear to influence the outcome of a study in such a way that a positive and statistically significant elasticity results. Although only limited evidence is available so far (de Groot et al., 2001; Koetse et al., 2006), the results seem to corroborate the inference that increased uncertainty has a negative impact on energy-saving technology adoption.

The final chapter in Part II combines a simple theoretical model with empirical evidence on policy-induced uncertainty. More specifically, it looks at the environmental consequences of energy market liberalization resulting in decreasing energy prices and increased uncertainty on prices and hence the returns to investment. Using US data, it shows that a simple calibrated model of investment under uncertainty predicts that liberalization of the natural gas market has a substantial detrimental impact on the willingness to invest in more energy-efficient technologies. This is not only caused by a fall in the price level, but also by substantially increased variability resulting in a large increase in the implementation lag. This means that from a social point of view, there is a trade-off between energy market liberalization and environmental protection for two reasons. The first reason is that, obviously, lower energy prices depress the return on investment in energy-saving technologies. However, our analysis shows that there is a second reason for the trade-off to occur: the volatility of the energy price has also increased over the liberalization period. The increased variance results in a higher mark-up on top of the standard expected net present value criterion. As a result, liberalization results in a higher critical energy price to trigger investments in energy-saving technologies. In assessing the desirability of energy market liberalization, not only the environmental consequences of a lower energy price should be taken into account but also its higher volatility.

Building on the empirical and theoretical insights in Parts I and II, Part III of this book turns to policy implications with a special focus on implications for reducing energy use in small and medium-sized enterprises in energy-extensive industries. This is done by elaborating on evidence on the effectiveness of specific instruments as well as discussing the possibilities for *ex ante* evaluation of the impact of policies using applied general equilibrium models. The Netherlands provide a suitable case given the many policy instruments that have been developed and suggested, and also given the rich tradition in developing a variety of models, both top-down as well as bottom-up to assess the impact of policies.

Chapter 9 focuses on subsidies, which are typically met with scepticism by economists given their conviction that subsidies are dominated in terms of effectiveness as well as in terms of efficiency by measures such as environmental taxes and tradable permits. In this chapter, it is shown that this view only holds to some extent if the objective is to induce firms to adopt new, environmentally friendly technologies. Announced levels of stringency, in terms of tax rates or the total amounts of permits distributed, may not always be credible: as soon as a new technology is adopted on a fairly large scale, industry-wide emissions of polluting substances fall substantially, and politicians may be induced to set taxes below (or distribute permits up and above) the announced level. Investment subsidies do not suffer from such time-inconsistency problems, and hence may be preferable to other instruments. In addition, subsidies may be the appropriate instrument in the case where positive externalities with respect to adoption exist. For instance, the efficiency of technologies might increase if agents obtain more experience with respect to their use and/or the externalities involved in this learning process justify subsidies. In addition, the authors argue that some of the traditional arguments against subsidies may be less damaging than they appear at first sight. Perhaps, the fact that subsidies rank high on the list of most frequently used policy instruments is, in practice, less cumbersome than economists usually tend to think.

The subsequent chapter applies theory-based evaluation, using a substantive collection of data from primary sources. An important lesson from this chapter is that complete and coherent information to carry out such evaluations is often lacking, resulting in considerable uncertainty margins. In order to reduce these uncertainties in future policy evaluations, it is necessary to have the necessary monitoring already in place while the policy instrument is in effect. This requires upfront thinking about the appropriate evaluation procedure. In addition, target setting should be improved, particularly in view of making policy evaluations feasible and meaningful. Notwithstanding these limitations, the evaluation of Dutch policies to reduce carbon dioxide emissions in the service sector shows that the combination of these policies has been effective in reducing emissions. The most substantive contribution to meeting the overall policy objectives comes from economic instruments such as subsidies and taxes, but the use of normative instruments, specifically standards and environmental permits, contributed as well.

Chapter 11 builds on a rich Dutch tradition of developing energy models to provide *ex ante* evaluations of the effectiveness and efficiency of energy policies. It discusses the different modelling approaches and their pros and cons. It concludes with a structured comparison of the impact of various policy measures as predicted by the different policy simulation models (NEMO, SAVE and MEI-Energy). The three models differ in structure: NEMO is a top-down model where technology is modelled indirectly through trends and substitution elasticities, and SAVE and MEI-Energy are models using bottom-up detailed and explicit descriptions of energy-saving techniques. However, the models also have some features in common. All three are partial equilibrium models using a vintage structure. Furthermore, the three models have at their core an update of a sector's energy intensity based on three investment options: new vintages, retrofit of existing vintages, and good housekeeping. The most important difference between the models is the way in which diffusion of energy-saving technologies has been modelled and what factors influence this diffusion process. In NEMO this process is basically an autonomous trend, complemented with a vintage approach and price elasticities computed using the cost–benefit data from the ICARUS database. In SAVE, three diffusion mechanisms have been modelled. The financial evaluation based on the IRR is limited by non-economic factors, in combination with a limitation in the part of the capital stock that can be replaced in a particular year. MEI-Energy uses a similar approach. Although SAVE and MEI-Energy have a similar basic starting point for diffusion, MEI-Energy tries to model all forces that affect an investment decision, ranging from the complexity of the technology to the pressure of non-governmental organization (NGOs) on the private and the government sector. They also explicitly include whether a government policy affects a driving force. The analyses with the three models for the four energy-extensive sectors (services, other metal, food, textiles and beverages) and three policies (CO_2-tax, investment subsidy, and investing in all technologies with an internal rate of return of 10 per cent or more) show a number of differences. The differences are remarkable, given the similar approach (updating energy efficiency based on profitability), the limited scope (energy use per sector), the use of the same technology data from the ICARUS database and the use of the same exogenous economic growth and energy price scenario, in all three models. The fact that the three research institutes work together on energy matters, makes the differences even more astounding. It is obvious that the top-down model reacts primarily to price changes and the bottom-up models react better to command and control type policies. For policy advice this result means that almost any combination of a more general economic policy based on prices and command-and-control type policies can be supported, depending on the right mix of results. In energy policy in the Netherlands, model choice does apparently create degrees of freedom for policy-makers.

3 THE ROAD AHEAD

In view of the challenges the world faces in coping with the increased scarcity of energy and the changes in climatic conditions, enhancing our knowledge about the complex interplay between technological development, behaviour of economic agents and policy measures is essential. This book has aimed to contribute to our knowledge, but many open issues still maintain. We see several challenges ahead.

First, there is clearly more need for careful monitoring of energy efficiency at a refined and consistent sectoral and spatial detail. Good information is relevant for describing and understanding the dynamic evolution of energy efficiency, assessing the effectiveness of policies and also for carefully calibrating models. This requires integration of data from different sources, ranging from economic data to information on, for instance, energy use and emissions. Despite a sizeable amount of research time spent on obtaining and constructing reliable time series with a sufficient level of detail to allow the drivers behind energy efficiency changes to be analysed, data impediments remain to constitute a major deterrent to performing this type of analysis for different sectors and/or other geographical areas. This information is also crucial for clarifying key issues related to the proper measurement of the physical dimensions of technology that were central in Part I of this book and form the basis for any policy evaluation.

Second, although we have shed further light on several barriers and their empirical relevance, more research is needed, based on detailed firm-level data. In our view, too much attention is still devoted to innovation without acknowledging that adoptions are innovations in themselves at the level of an individual firm. Further understanding these barriers and their relevance and importance can provide policy-makers with policy instruments to help reducing the energy-efficiency paradox.

Third, although this book has made a useful step towards assessing the relevance of small and medium-sized enterprises in energy-extensive sectors, clearly more effort is needed in those directions, not the least because of the increasing dominance of these sectors in modern societies. Apart from collecting data, this also requires a change of mind, away from the natural tendency to focus on the big polluters for whom problems of coping with energy efficiency may well be least severe, given the economic incentives abound.

Fourth, there is huge scope for more extensive utilization of micro data, which increasingly become available for various research purposes. Such information can make important contributions to *ex post* policy evaluations and to monitoring responses to policy measures, capturing real changes in technologies used at the firm level.

Fifth, and largely due to a lack of good and reliable information, our understanding of the processes through which technologies diffuse and knowledge spillovers occur is still weak. Moreover, insights into the extent to which differences in energy intensity are caused by inefficiencies on the one hand and spatial differences in for example temperature on the other hand are still not very well understood. This complicates policy initiatives using benchmarking as a tool to get information on the possibilities to increase energy efficiency at acceptable costs.

Finally, the comparison of different modelling approaches and the associated differences in model outcomes illustrates that we are still far away from a properly calibrated model capturing the essential behavioural and technological characteristics driving energy efficiency. Continued efforts of developing such models, integrating the different approaches and understanding the sources of differences between the various approaches is of critical importance given the need of proper *ex ante* policy evaluations to inform successful future energy policies.

REFERENCES

Blok, K., H.L.F. de Groot, E.E.M. Luiten and M.G. Rietbergen (2004), *The Effectiveness of Policy Instruments for Energy-Efficiency Improvement in Firms*, Dordrecht, The Netherlands: Kluwer Academic Publishers.

Gillissen, M., H. Opschoor, J. Farla and K. Blok (1995), *Energy Conservation and-Investment Behaviour of Firms*, Amsterdam: Economic and Social Institute (ESI), VU University Amsterdam and Utrecht: Department of Science, Technology and Society, Utrecht University.

Groot, H.L.F. de, E.T. Verhoef and P. Nijkamp (2001), 'Energy saving by firms: decision-making, barriers and policies', *Energy Economics*, **23**, 717–40.

Koetse, M.J., A.J. van der Vlist and H.L.F. de Groot (2006), 'The impact of perceived expectations and uncertainty on firm investment', *Small Business Economics*, **26**, 365–76.

Schipper, L. and S. Meyers (1992), *Energy Efficiency and Human Activity: Past Trends and Future Prospects*, Cambridge: Cambridge University Press.

Velthuijsen, J.W. (1995), 'Determinants of investments in energy conservation', Amsterdam: Foundation for Economic Research (SEO), University of Amsterdam, The Netherlands.

Worrell, E., L. Price, N. Martin, J. Farla and R. Schaeffer (1997), 'Energy intensity in the iron and steel industry: a comparison of physical and economic indices', *Energy Policy*, **25**, 727–44.

Index